SAVING ASIA'S THREATENED BIRDS

A GUIDE FOR GOVERNMENT AND CIVIL SOCIETY

A project of the BirdLife Asia Partnership

Funding for the project has been provided by the Critical Ecosystem Partnership Fund (CEPF), a joint initiative of Conservation International, the Global Environment Facility, the Government of Japan, the John D. and Catherine T. MacArthur Foundation and the World Bank

with additional support from the
Asia Bird Fund of BirdLife International

This book is based on *Threatened birds of Asia: the BirdLife International Red Data Book*, which received major sponsorship from the Ministry of the Environment, Government of Japan

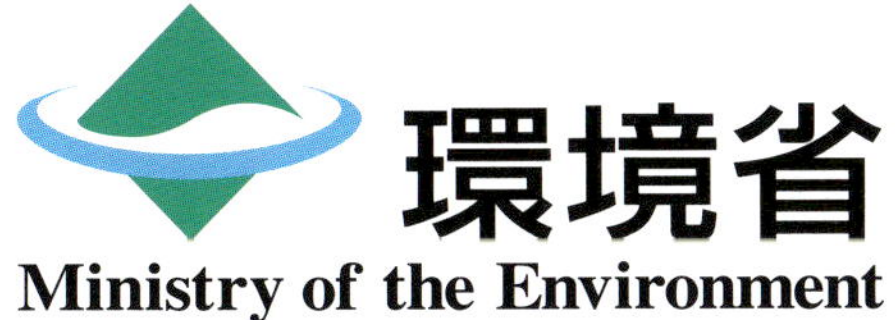

through the
Wild Bird Society of Japan
(BirdLife Partner in Japan)

SAVING ASIA'S THREATENED BIRDS

A GUIDE FOR GOVERNMENT AND CIVIL SOCIETY

Compiled by M. J. Crosby

This review has benefited considerably from the guidance and contributions of N. Ichida, R. F. A. Grimmett, N. J. Collar, S. Chan and J. A. Tobias, and the input of the following: A. Ahmed, D. Allen, A. V. Andreev, H. S. Baral, M. Barter, A. Bräunlich, T. M. Brooks, A. Chaudhury, P. Davidson, G. W. H. Davison, Ding Changqing, Ding Ping, J. W. Duckworth, J. C. Eames, Fang Woei-horng, J. H. Fanshawe, J. R. Fellowes, J. Harkness, P. Herkenrath, Htin Hla, C. Inskipp, T. P. Inskipp, Y. Kanai, A. A. Khan, Y. Kominami, A. Kumar, M. W. N. Lau, W. A. Laurie, Lu Zhi, N. A. D. Mallari, P. J. K. McGowan, N. Moores, D. Mudappa, T. Mundkur, R. Nawaz, D. Nel, J. O'Sullivan, D. J. Pain, O. Pfister, A. Plantilla, D. E. Pritchard, A. R. Rahmani, T. R. S. Raman, Rudyanto, H. S. Sangha, A. C. Sebastian, L. L. Severinghaus, S. R. Severinghaus, T. Shivanand, S. Subramanya, G. Sundar, B. R. Tabaranza, Jr., P. Thompson, A. W. Tordoff, U. Treesucon, M. Ueta, Wang Qishan, Yeap Chin Aik, Zafar-ul Islam, Zhang Zhengwang and Zheng Guangmei.

Picture research by R. D. Thomas and P. Benstead

Maps by M. Balman, H. Dobie and S. Green

Design and layout by P. C. Creed of **Nature**Bureau

Recommended citation
BirdLife International (2003) *Saving Asia's threatened birds: a guide for government and civil society.* Cambridge, U.K.: BirdLife International.

BirdLife International, Wellbrook Court, Girton Road, Cambridge CB3 0NA, United Kingdom
Tel: +44 1223 277318 Fax: +44 1223 277200 Email: birdlife@birdlife.org.uk
Internet: www.birdlife.org

BirdLife International is a UK-registered charity, no. 1042125

ISBN 0-946888-47-7

British Library-in-Publication Data
A catalogue record for this book is available from the British Library

First published 2003 by BirdLife International

Designed and produced by the **Nature**Bureau, 36 Kingfisher Court, Hambridge Road, Newbury, Berkshire RG14 5SJ, United Kingdom

Available from the Natural History Book Service Ltd, 2–3 Wills Road, Totnes, Devon TQ9 5XN, UK. Tel: +44 1803 865913
Fax: +44 1803 865280 Email: nhbs@nhbs.co.uk
Internet: www.nhbs.com/services/birdlife.html

CONTENTS

FOREWORD BY HIH PRINCESS TAKAMADO

Birds are so special. Throughout the ages, they have continued to touch our hearts. They have always been close to us, although sometimes the proximity is not a physical one. They appear in the mythology and legends of all the countries in the world and they have been the inspiration behind many a creative work, be it in the form of poetry, music, painting or even fashion.

Asia is no exception and we are proud of our heritage and the place that birds occupy in our tradition. Colourful pittas and majestic raptors, amazing hornbills, elegant migratory cranes and the various offshore seabirds make our lives richer by their sheer existence. With our undulating terrain, our luscious forests and our long coastlines, Asia is blessed with a uniquely varied and abundant number of bird species. It is this treasure trove that we are in danger of losing.

In June 2001, we launched a very detailed Red Data Book called *Threatened Birds of Asia*. The book is a monumental achievement, representing years of co-operation and dedicated hard work at an international and regional level, with more than 1,000 Asian ornithologists, conservationists and bird-lovers participating. It is the single most detailed publication ever produced on the subject of birds or conservation and it uncovers, analyses and assesses all the evidence, presenting it together with all the sources. Such scrupulously exact information is a crucial part of the red-listing process but, as a guide to the highest priorities and to the most important actions, *Threatened Birds of Asia* is not the easiest of documents to use – if only because of its sheer weight!

It is therefore with great pride and satisfaction that we now present to you our new book, *Saving Asia's Threatened Birds*. This highly focused, elegant reworking of the Red Data Book consists of 250 pages of key data and critical information. I am certain that it will serve as a 'field guide' for decision-makers as they seek to target their energies and resources towards safeguarding the most threatened bird species and protecting Important Bird Areas throughout Asia.

If one could have a bird's-eye view of the world, what would one see? Despite the ravages inflicted upon it, the earth is a breathtakingly beautiful planet. A myriad of different life forms weave an intricate web of existence, creating an unbelievable ecosystem in which we, the human species, play a potentially important part. If we have taken upon ourselves the leadership role, then we have a duty to all other existing life forms to maintain, or at least not to disturb, the harmony and balance on earth. It often strikes me that in this otherwise harmonious world, we are the only ones who are off key. We are the ones playing the discordant notes.

Now, as environmental issues grow into global concerns, it is imperative that we act with intelligent integrity and I am pleased to be a part of the BirdLife Partnership in its efforts to guide the world in this direction. I do hope and pray that many of you will join me.

HIH Princess Takamado
Honorary Patron
BirdLife Asia Rare Bird Club

FOREWORD BY MICHAEL RANDS

In BirdLife's commitment to birds and to people, science has always played a crucial role. Our Globally Threatened Species Programme is the core of our scientific endeavour. For 40 years BirdLife has been producing the international Bird Red Data Book, and the information generated by this work has been the fuel that has driven much of our field action, advocacy and strategic thinking.

Threatened Birds of Asia was a monumental study of the plight of bird species in the world's most heavily populated and fastest-developing region. This new strategy draws on that work and brings together every crucial element in it for the conservation of Asia's threatened birds. It identifies all the important actions that need to be taken, and it does this in a way which is clear and accessible to its users. It is also a vital tool for the monitoring of progress towards the many and various targets and goals it identifies.

Three primary issues highlighted by this strategy fire my determination. First, the lowland dipterocarp rainforests of Sundaland—Peninsular Malaysia, Sumatra and Borneo—are among the very richest in biodiversity on earth, but they are disappearing so fast that we scarcely have time to draw breath. This strategy demands *true* sustainability and *real* corporate responsibility in the region in order to prevent any further destruction of habitats that are vital not just to birds but to thousands of species of animal and plant, and indeed to the long-term welfare of the indigenous and local peoples of the region.

Second, it is imperative that the governments and NGOs of Asia take new steps towards improving the gaps in their protected area coverage. Everyone acknowledges that parks and reserves are crucial guarantors of biodiversity on this planet but, as this strategy makes very plain, there are still many critically important areas which remain outside the formal systems of protection that Asia currently possesses. This situation must change.

Third, the dimensions of the bird trade in Asia can no longer be countenanced. This problem affects sites and habitats as much as it does species, and much of it is beyond the control of international regulation. New vigour, and new rigour, are needed to enforce the many existing national and international laws on wildlife trafficking, and this must be accompanied by a sustained and well-targeted series of advocacy and educational campaigns to reduce demand. Details are in this strategy.

The work that has gone into producing this document has been intense, and I congratulate all who have participated in its realisation, especially the BirdLife Network throughout Asia and the secretariat staff that worked with them and helped create this tremendous strategy. I also extend warmest thanks to our colleagues in Conservation International and particularly the Critical Ecosystem Partnership Fund for their magnificent support. Altogether, this document is a major step forward for bird conservation in Asia. But the key thing now is to implement it. I hope everyone who uses it will be inspired to play their part in rising to this challenge.

Michael Rands
Director and Chief Executive
BirdLife International

FOREWORD BY JORGEN THOMSEN

BirdLife International and Conservation International have for years been allies in global wildlife conservation, but as our individual strengths have grown and our activities expanded, our paths have not diverged; rather, our closeness has only become stronger. The reason for this is very simple: both organisations share the same fundamental philosophy—that species must be saved, that extinction is not an option. On this single premise we do not compromise. It brings us together in ever greater solidarity.

CI's mission is to focus attention—ours, and the rest of the world's—on what we term the "Hotspots", areas of the planet which are supremely rich in biological diversity and seriously in danger from human development. To this end, CI created the Critical Ecosystem Partnership Fund (CEPF)—an alliance involving ourselves, the World Bank, the Global Environment Facility, the MacArthur Foundation and the Japanese Government—in order to support key conservation initiatives in the many Hotspots around the world.

The opportunity for CEPF to support BirdLife's *Saving Asia's Threatened Birds* was too good to miss. What CI has always admired and indeed envied about BirdLife is the extraordinary quality of its science. In our own work we repeatedly make use of seminal BirdLife documents such as *Endemic Bird Areas of the World*, *Important Bird Areas in Africa*, *Threatened Birds of the World* and, of course, the detailed Red Data Books on which these other documents extensively draw. So when we learnt that BirdLife was to embark on the production of a strategy which would so clearly articulate a major suite of key actions required to conserve the rarest bird species and most threatened avian habitats in Asia, we were eager to demonstrate our commitment to such a necessary and predictably rock-solid scientific initiative.

And of course these actions will benefit many other life-forms, not only the birds: most of the sites where threatened birds occur are shared with dozens of less familiar species of animal and plant, often no less at risk. If the birds act as flagships for these other species, then only good can come of it. Moreover, a very high proportion of the actions will fall within Hotspots, and it is most helpful to CI to have them set out in a clear, systematic manner. Not only does this allow still further convergence between our two organisations, but it also means that we can speak with the greater authority of our collective voices when we advocate these actions to the appropriate figures and institutions in the rest of the world.

So CI is proud to have supported BirdLife in the production of this strategy. It is a new route map for both our organisations. The way forward is clearly not easy and it is certainly not for the faint-hearted; but BirdLife and CI relish the challenge, and we expect to continue the journey together, in increasingly close mutual support, for many years to come.

Jorgen B. Thomsen
CI Senior Vice President
CEPF Executive Director

ACKNOWLEDGEMENTS

The major sponsor of this project was the Critical Ecosystem Partnership Fund, a joint initiative of Conservation International, the Global Environment Facility, the Government of Japan, the MacArthur Foundation and the World Bank. BirdLife International expresses its warm thanks to all the people connected to the Fund who have supported the project over its development and implementation, notably Judy A. Mills, Jorgen Thomsen and Leanne Miller, and also Thomas M. Brooks of Conservation International. A fundamental goal of the Fund is to ensure civil society is engaged in biodiversity conservation, and we hope that this book will make a significant contribution towards that goal in Asia.

The current project is a follow-up to *Threatened birds of Asia: the BirdLife International Red Data Book*, which was a project of the BirdLife Asia Partnership, spearheaded by the Wild Bird Society of Japan under the direction of Noritaka Ichida. The major sponsor of that project was the Ministry of the Environment, Government of Japan, and BirdLife International would once again like to thank the Ministry, and the other funders of the project, namely – Anthony Collerton, ASEAN Region Center for Biodiversity Conservation, British Airways Assisting Conservation (BAAC), British Embassy, Manila, Bromley Trust, Club 300, Conservation International (Center for Applied Biodiversity Science), Dansk Ornitologisk Forening (DOF), Derek A. Holmes memorial collection, Finnish Department of International Cooperation, Garfield Foundation, Karen Hsu, Kleinwort Charitable Trust, Loke Wan Tho Memorial Foundation, Henry Luce Foundation, Inc., Louise M. Parent Fund, Mary Gordon Roberts Fund, Nippon Telegraph and Telephone Corporation (NTT-ME), Royal Netherlands Embassy, Jakarta, Royal Society for the Protection of Birds (RSPB), Sarnia Charitable Trust, Swedish International Development Agency (Sida), Toshiba, Tung Foundation, Wild Bird Federation Taiwan, Hans Wilsdorf Foundation, Vogelbescherming Nederland and an anonymous donor.

The information and ideas synthesised in *Threatened birds of Asia: the BirdLife International Red Data Book* came from many sources in and outside the Asian region. The initial, and in many countries the main, source of data were the c.160 national compilers and principal data contributors who are listed on the opening pages of that book, many of whom also contributed to the current project and are acknowledged below. It is not possible here to list all of these people, or the 1,000 plus other people who contributed data to the Red Data Book. However, they are all to be congratulated for their monumental efforts in collating that huge body of work, and for providing information and perceptions on the conservation of the region's birds that provided the basis for the current analysis and synthesis.

The current project was initiated at a planning workshop held in Tokyo in June 2001, immediately after the launch of *Threatened birds of Asia: the BirdLife International Red Data Book*. The workshop participants were: Aldrin Mallari (Haribon Foundation), Asad Rahmani (Bombay Natural History Society), Cristy Juan (BirdLife Secretariat), Fang Woei-horng (Wild Bird Federation Taiwan), Joe Tobias (BirdLife Secretariat), Karen Hsu (Sponsor to Red Data Book), Kazuo Koyama (Wild Bird Society of Japan), Mike Crosby (BirdLife Secretariat), Makoto Kawanabe (Wild Bird Society of Japan), Nigel Collar (BirdLife Secretariat), Noritaka Ichida (BirdLife Asia Council), Richard Grimmett (BirdLife Secretariat), Rudyanto (BirdLife Secretariat), Simba Chan (Wild Bird Society of Japan), Yutaka Kanai (Wild Bird Society of Japan). A second planning workshop was held in Cambridge, U.K. later in the same month, and included Thomas Brooks (Center for Applied Biodiversity Science, Conservation International), Philip McGowan (World Pheasant Association), Steve Parr (Royal Society for the Protection of Birds), and many of the BirdLife Cambridge Secretariat staff acknowledged below. We would like to thank all participants in these important planning workshops.

The editing of this book was a difficult challenge, to balance the requirement for succinct, user-friendly text with the need to provide sufficient detail for readers to understand the conservation issues that are covered. Joe Tobias prepared the first draft accounts for many of the habitat regions in South and South-East Asia, and his efforts are gratefully acknowledged. The following reviewers played a vital role in the editing process, and provided many new ideas for conservation actions to address the environmental problems faced in certain habitat regions. They are therefore listed with the codes for the habitat accounts which they commented on: A. Ahmed (F05), D. Allen (F09, W19), A. V. Andreev (F01, G01, W01, W02, S01), H. S. Baral (F04, G02), M. Barter (W06, W08), A. Bräunlich (G01, W05), S. Chan (all habitat accounts), A. Chaudhury (F04, F06, G02), N. J. Collar (all habitat accounts), P. Davidson (F03, F04, F06, F07, G02, G03, W10, W15, W16, W18), G. W. H. Davison (F07), Ding Changqing (W07), Ding Ping (F03, W10), J. W. Duckworth (F04, F06, W18), J. C. Eames (F03, F04, F06, F07, G02, G03, W10, W15, W16, W18), Fang Woei-horng (F03, W10, S01), J. R. Fellowes (F03, F04, F06, W10), R. F. A. Grimmett (all habitat accounts), J. Harkness (W08), Htin Hla (F04, W15), C. Inskipp (F04, G02), T. P. Inskipp (introductory sections), Y. Kanai (F01), A. A. Khan (F04, G02, G03, W11), Y. Kominami (F02, W02, W04), A. Kumar (F05), M. W. N. Lau (F03, F04, F06, W10), W. A. Laurie (G01, W05), Lu Zhi (W09), N. A. D. Mallari (F09, W19), P. J. K. McGowan (F03, F04, F06, F07, F09), N. Moores (W06), D. Mudappa (F05), T. Mundkur (introductory sections, W06, W11, W12), R. Nawaz (F04), D. Nel (S01), D. J. Pain (G03), O. Pfister (W09), A. Plantilla (F09, W19), A. R. Rahmani (F05, G02, G03), T. R. S. Raman (F05), Rudyanto (F07), H. S. Sangha (W09), A. C. Sebastian (F07), L. L. Severinghaus (F03), S. R. Severinghaus (G01, W05), T. Shivanand (F05), S. Subramanya (F05, W12, W13), G. Sundar (W12), B. R. Tabaranza, Jr. (F09, W19), P. Thompson (F06, W14, W15), A. W. Tordoff (F03, F04, F06, F07, G02, G03, W10, W15, W16, W18), U. Treesucon (F04), M. Ueta (W02), Wang Qishan (W03, W05, W06, W08, W10), Yeap Chin Aik (F07, W20), Zafar-ul Islam (F04, F05, G02, G03), Zhang Zhengwang (F01, F03, F04, F06, G01, W05, W06, W07, W08, W09, W10) and Zheng Guangmei (F03, F04, F06, W06, W07, W08, W09).

The introductory sections of the book covering *Policy approaches to biodiversity conservation* and *Conventions and related mechanisms* were drafted by John Fanshawe and Peter Herkenrath of the BirdLife International Secretariat, and were further developed by Dave Pritchard and John O'Sullivan of the RSPB. We believe that these sections will prove to be of great interest and value to many readers, and are especially grateful to the team that painstakingly put them together.

As always, colleagues in the BirdLife International Secretariat have provided vital support to the project, particularly during the stressful few weeks while the book was being finalised. We would like to thank the following: Adrian Long, Alison Stattersfield, April Faulkner, Beverly Childs, Chris Mills, Chris Spreadbury, David Thomas, Elizabeth Ansell, Gary Allport, Janet Chow, Joan Clements, John Fanshawe, Jonathan Ekstrom, Leon Bennun, Lincoln Fishpool, Marco Lambertini, Mark Balman, Martin Sneary, Mich Boyt, Mike Evans, Mike Rands, Mwangi Githiru, Nicholas Wilkinson, Peter Herkenrath, Richard Thomas, Rob Pople, Roger Safford, Rosina Abudulai, Sarah Kendall, Stuart Butchart and Sue Shutes.

BirdLife gratefully acknowledges the photographers who either sent examples of their work for possible inclusion in this publication, or helped in other ways to source images. The following are all particularly thanked (numbers after photographer's names relate to the page or pages on which their work appears): A. Compost (102), Adrian Long (11), Alan Lewis, Alexander Andreev, Ali Hassan Habib, Allan Michaud (223), Amano Samarpan, Ameen Ahmed (69,71), Amjad Aslam, Anish Andheria, Anwaruddin Choudhury, Arun P Singh, Asad Rahmani (9,73,125,129,130,132,133,134), Axel Bräunlich (137), B. P. Rajesh, Bas van Balen (87), Boonsiri-UNEP/Still Pictures (16), Carol Inskipp, Chang Shou-Hua (237), Chris Schenk (138), Christian Artuso (92), Christoph Zöckler (15), Christopher Gow, Claudia Mettke-Hofmann (136), Colin Poole, Colin Trainor (95), Dai Bo (60), Dave Showler, Derek Scott, Des Allen (112), Dev Ghimire, Di Yun/China Features (174), Dipankar Ghose, Eberhard Curio (107), Eleanor Briggs (1,7,213,221,224,225), Eric Lott, Eugene Potapov (139,146), F. A. Clements (6), Fang Woei-horng, Farah Ishtiaq (74), Frank Todd (48,160), Gehan de Silva Wijeyeratne, Gerhard Hofmann (136), Gerry Gomez (80), Graham Robertson (240), Guy Duke (14), Guy Dutson (105), Guy Petherbridge (6), H. E. McClure (220), Haixiang Zhou (7,10,161,165,166), He Fenqi (159), Hem Sagar Baral, Igor Karyakin, Ingar Josten Ølen (140), J. C. Eames (17,79,124), Jacob Wijpkema (12,86,117,154, 156,193,203), James Eaton, James Zeng-huang, Jan Willem den Besten, Jean Howman (67), Jeremy Holden (88), Jim Walseth, Jim Wardill (100), Joe Blossom (229), Joe Tobias (11), John Corder, John Holmes (68,163,179,184), Jon Hornbuckle (144,153,200,205,231), Jon Riley (97), Jugal Tiwari, Kanit Khanikul (86), Khalid Rafeek, Koji Ono (235, 239), Koustubh Sharma, Kushal Mookherjee, Lin Jianyang, M. R. Chaitra, Marco Lambertini (1,8,29,32,88,89,90,91), Mark Edwards (9,39,82), Martin Hale (49,185,230), Michael Poulsen (14,34,103,110), Mike Ball, Mike Crosby (13,55,58,65,66,75,217,232), Monirul Khan, Muhammad Farooq, Ng Soon Chye, Nial Moores, Nigel Redman, O. Morvan/Pygargue Productions (108), Ong Kiem Sian, Otto Pfister (43,46,116,119,122,128,131,177,180,191,194, 199,201), P. Round (83), P. Jepson (98), P. B. Taylor (143), Palitha, Paritosh Khanvilkar, Parvish Pandya, Paul Goriup (8), Paul Thompson (207,209), Perwez Iqubal (121), Pete Davidson, Pete Morris (53,226), Pete Wood, Peter Los (173), Phil Benstead (93,97,219), Ph. Garguil/Pygargue Productions (108), Priya Raja, Ray Tipper (115,181,183, 187,197,210), Richard Porter, Ron Saldino (171), Rosemary Low (17), Sarah Fowler (12), Seb Buckton, Shehzad Noorani/Still Pictures (5), Shimpei Watanabe (210), Simba Chan (13,35,147,149,151,165,215,216,227), Simon Cook (135), Slim Sreedharan, Smith Sutibut (77), Steve Pryor, Stuart Butchart (101), Taej Mundkur, Takao Baba (155,238), Takashi Kurosaki (211), Takuki Hanashiro (2,10,36,51), Tee Lian Huat (15), Terry Cooper, Thripthy Thomas, Tim Inskipp, Tim Laman (107, 109), Tim Loseby (16,71,72,196, 204,206,208,215), Toby Sinclair, Tom Brooks, Tony Martin (189), Tung-Huei Kuo (186), Ute Bradter (113,114,157), Vijay Cavale, Viktor Gritsyuk, Wen-Hsin Huang (57,150,175), Xi Zhinong (167,168,169,170), Yasuyuki Makino (45,54), Yu Yat-tung, Zafar-ul Islam, Zainal Abidin Jaafar and Zulfikar Ali (189).

We would like to thank Peter Creed, Helen Dobie and Simon Green of the **Nature**Bureau, not only for the design, layout, maps and graphic production, but also for their great dedication, flexibility and patience.

SUMMARY

ASIA: BIRDS, HABITATS AND PEOPLE (pp.3–17)

Asia has a great diversity of habitats, ranging from Arctic tundra to tropical forests, and including the highest mountains in the world. This is reflected in the region's immense richness in birds and other wildlife. Asia also has a very large and rapidly growing human population, and many of the world's most dynamic economies. As a consequence, the region is experiencing rapid environmental change and many of its habitats and the biodiversity that they support are under great pressure. Every country in Asia has populations of threatened birds, and needs to take action for the conservation of these species and their habitats.

THREATENED BIRDS OF ASIA (pp.18–19)

BirdLife International has been documenting the conservation status of the world's birds since the 1970s, in partnership with IUCN–The World Conservation Union, in a series of regional Red Data Books and global checklists of threatened birds. The most recent of these, *Threatened birds of Asia: the BirdLife International Red Data Book*, was completed by the BirdLife Asia Partnership in 2001. It covers more then 300 threatened bird species, and contains many thousands of recommendations for the conservation of these birds, and their habitats and key sites. These proposals for conservation action are synthesised and further developed in the current review.

KEY HABITATS FOR ASIA'S THREATENED BIRDS (pp.19–24)

Analysis of the distributions and habitat requirements of Asia's threatened birds has identified nine major forest regions, three grassland regions and 20 wetland regions as priority areas for conservation. There is also a group of threatened seabirds. Many of these habitat regions correspond closely to one or more of Conservation International's 25 global Biodiversity Hotspots, BirdLife's 218 Endemic Bird Areas and WWF's Global 200 Ecoregions. The recommendations made for threatened birds, sites and habitats in the forest, grassland and wetland regions are relevant to all people working in these priority areas for the conservation of biodiversity and the sustainability in the use of natural resources.

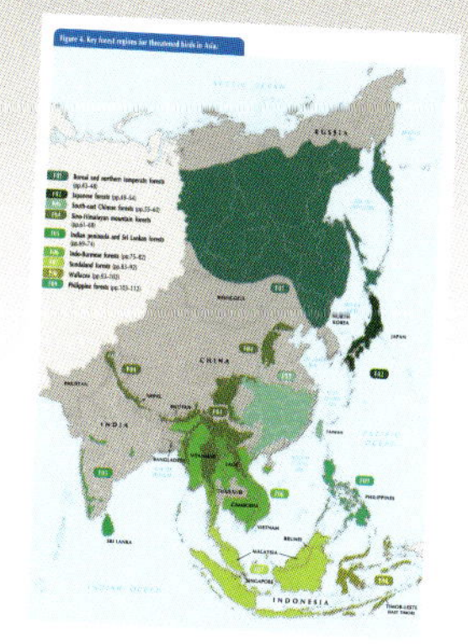

POLICY APPROACHES TO BIODIVERSITY CONSERVATION (pp.25–30)

The current review focuses primarily on the direct pressures on Asia's threatened birds and their habitats, and on how these should be addressed. However, in the longer term the underlying and indirect causes of biodiversity loss, for example rising consumption, undervaluation and perverse subsidies, will need to be tackled. Ultimately, fundamental changes in landuse and resource utilisation are needed, through policy and planning, and social, political and economic reform. Specifically, the value of area-based approaches to conservation policy is explored, covering Important Bird Areas and other networks of key sites for biodiversity, and how these can be used to achieve conservation on the ground.

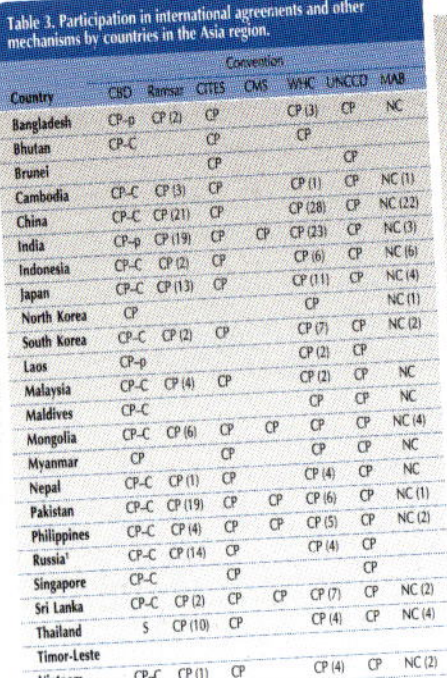

Table 3. Participation in international agreements and other mechanisms by countries in the Asia region.

Country	CBD	Ramsar	CITES	CMS	WHC	UNCCD	MAB
Bangladesh	CP-p	CP (2)	CP		CP (3)	CP	NC
Bhutan	CP-C		CP		CP		
Brunei			CP			CP	
Cambodia	CP-C	CP (3)	CP		CP (1)	CP	NC (1)
China	CP-C	CP (21)	CP		CP (28)	CP	NC (22)
India	CP-p	CP (19)	CP	CP	CP (23)	CP	NC (3)
Indonesia	CP-C	CP (2)	CP		CP (6)	CP	NC (6)
Japan	CP-C	CP (13)	CP		CP (11)	CP	NC (4)
North Korea	CP				CP		NC (1)
South Korea	CP-C	CP (2)	CP		CP (7)	CP	NC (2)
Laos	CP-p				CP (2)	CP	
Malaysia	CP-C	CP (4)	CP		CP (2)	CP	NC
Maldives	CP-C				CP	CP	NC
Mongolia	CP-C	CP (6)	CP	CP	CP	CP	NC (4)
Myanmar	CP		CP		CP	CP	NC
Nepal	CP-C	CP (1)	CP		CP (4)	CP	NC
Pakistan	CP-C	CP (19)	CP	CP	CP (6)	CP	NC (1)
Philippines	CP-C	CP (4)	CP	CP	CP (5)	CP	NC (2)
Russia[1]	CP-C	CP (14)	CP		CP (4)	CP	
Singapore	CP-C		CP			CP	
Sri Lanka	CP-C	CP (2)	CP	CP	CP (7)	CP	NC (2)
Thailand	S	CP (10)	CP		CP (4)	CP	NC (4)
Timor-Leste							
Vietnam	CP-C	CP (1)	CP		CP (4)	CP	NC (2)

CONVENTIONS AND RELATED MECHANISMS (pp.30–36)

Various international conventions and other mechanisms are relevant to the conservation of threatened species (e.g. the Convention on International Trade in Endangered Species), sites (e.g. the Ramsar Convention on Wetlands) and habitats (e.g. the United Nations Convention to Combat Desertification). Implementation by contracting parties is currently patchy, and guidance is provided to governments and civil society on what immediate action might be taken to advance these mechanisms for the benefit of threatened birds in the Asia region.

PRIORITIES TO PREVENT THE EXTINCTION OF ASIAN BIRDS (pp.36–40)

Remarkably few birds are proven to have become extinct in the Asia region in historical times, but more than one hundred species are now Critical or Endangered. Several of these have not been recorded in recent decades, and may already have disappeared. The issues that are driving these species towards extinction include forestry mismanagement and illegal logging, and conversion for agriculture and plantations, particularly in the tropical forests of Indonesia, the Philippines and mainland South-East Asia. Urban, industrial and infrastructural development is affecting natural habitats in many parts of Asia, including major dam and irrigation projects and the construction of roads into previously inaccessible areas. Planned large-scale reclamation of coastal wetlands could have a huge impact on the migratory waterbirds of the East Asia-Australasia flyway. This section identifies the issues that could cause birds extinctions in Asia, and highlights the priority conservation actions that must be taken to prevent this.

ACTION FOR ASIA'S KEY BIRD HABITATS (forests: pp.43–112, grasslands: pp.113–136, wetlands: pp.137–234, seabirds: pp.235–240)

Each regional account documents and maps the groups of threatened bird species that it supports, and the habitats and sites that are critical for their survival. Most crucially, the accounts focus on the major land-use issues affecting the habitats of threatened birds, as well as any direct threats to the birds themselves. Proposals are made on how to reduce or eliminate the negative impacts of activities causing habitat loss and degradation. Important gaps in coverage of threatened bird species by national protected areas networks are identified, with proposals on where new parks and reserves might be established. Unsustainable exploitation is identified as a major problem for certain threatened birds, and recommendations are made for its control. The conservation of many threatened Asian birds is hindered by incomplete data, and surveys or ecological studies are identified to address the most important gaps in knowledge.

ASIA: BIRDS, HABITATS AND PEOPLE

BIRDS AND HABITATS

The Asia region[1] extends from the tropics to the Arctic, and includes the highest mountain ranges in the world. The great diversity of climates and habitats is reflected in a great diversity of birds (and other animals and plants), with about 2,700 species occurring in the region. A high proportion of these birds are confined to forests, particularly the tropical rain forests and dry forests in the south, although the subtropical, temperate and boreal forests further north each support many characteristic species. Asia's grasslands, wetlands and seas also support groups of birds that are specialised to these habitats.

Some Asian birds are very widespread, for example many species nest throughout the vast boreal forests of eastern Russia and migrate to the south for the winter. Many others are much more restricted in distribution, being confined, for example, to a single island group or mountain range. These restricted-range species are most numerous on the tropical archipelagos of Indonesia and the Philippines, and these island nations are consequently immensely rich in species in relation to their land areas. The tropical and subtropical mountains of the Himalayas, south-west China and northern South-East Asia are also very diverse, because different forest and grassland types (each inhabited by distinct groups of birds and other wildlife) are found close together in the different altitudinal zones on their slopes.

Natural habitats in Asia have been greatly affected by human activities. Some parts of the region are particularly suitable for agriculture, for example the temperate and subtropical lowlands of eastern China and the rich volcanic soils of Java, and have consequently been intensively cultivated for hundreds or even thousands of years. Other habitats were relatively untouched until recent decades, for example the tropical forests in many parts of Indonesia and the Philippines, but the growth of international trade in timber and other commodities has led to their rapid exploitation. As a result, some of the birds which inhabit the most intensely exploited habitats are now threatened with extinction, particularly those which are most specialised and least able to adapt to changes in their environment, or the species with particularly restricted ranges.

PEOPLE

Asia possesses a remarkable diversity of people and cultures, with more than 500 ethno-linguistic groups in the region. This diversity is no less important and worthy of celebration than the diversity of the wildlife, and no less challenging to preserve. Fortunately, birds and other wildlife are widely valued in Asia for economic, cultural, ethical and spiritual reasons. Over the centuries birds have inspired artists, for example in China and Japan where cranes have traditionally been regarded as holy birds, and Mandarin Duck as a symbol of marriage. In Indonesia and China, there is a long tradition of keeping cagebirds for the beauty of their appearance and their songs. Visitors to the northern Indian subcontinent are often amazed at the ability of wildlife, including large waterbirds and birds-of-prey, to co-exist with man in intensively utilised landscapes, as a result of enlightened attitudes to nature enshrined within the Hindu philosophical tradition.

Asia is currently experiencing a period of rapid change, linked to dynamic economic growth. Half of the world's people live in the region, with China and India both having human populations in excess of one billion. The combination of economic development and increasing human population is putting unprecedented pressure on the environment. Traditional land-use practices and associated beliefs are gradually breaking down in response to the need to increase production. The demand for raw materials has led to massive expansion in logging and conversion for cash crops such as oil palm and coffee. Hunting of birds for food and trapping for the wild bird trade are increasingly becoming unsustainable, through a combination of declining habitat availability and bird numbers, and improved technology to catch and transport the birds.

Despite the high estimation of birds in many Asian cultures, the natural environments in which birds live tend to be accorded far less value. Thus, although there is increasing awareness amongst Asian governments and civil society that the region faces extensive environmental problems, there is little official acknowledgement that these problems are the result of the economically unsustainable utilisation of species and habitats for short-term growth. In many countries, the reduction of natural habitats and the consequent declines in wildlife are all too apparent, and, as the environment deteriorates, it is steadily losing its capacity to provide resources and ecological services (such as reliable water supplies). Nevertheless, the value of environmental services is simply not acknowledged in most economies, and serious intervention to ensure the purity and permanence of major ecosystems does not figure as a priority in most Asian economies, which are all geared up for rapid growth and expansion. There is therefore a massive challenge to conservation to establish the environment where it should be, at the heart of policy for all governments aiming for prudent and sustainable development.

Of course, many people in the Asia region understand that the remaining natural habitats and the wildlife that they support need to be carefully protected and managed, and the most damaging economic activities and policies modified or even stopped. As economies develop, large middle classes are emerging in the region's cities, people with leisure time to devote to birdwatching and other pastimes which re-connect them to the natural environment. Through environmental NGOs, they are growing into a strong lobby for sustainable development, to balance economic growth with the need to protect habitats and wildlife.

[1] The Asia region as defined by BirdLife International extends from Pakistan to Indonesia (excluding Irian Jaya), and northwards to China, Mongolia and Russia east of the Yenisey River.

Figure 1. The Asia region.

WHY CONSERVE BIRDS AND NATURAL HABITATS?

Birds, like all elements of biodiversity, should be conserved for the richness and diversity they contribute to human experience (Collar 2003). Moreover, the natural habitats which birds share with other animals and plants are of immense economic value. A recent study (Costanza *et al.* 1997) estimated the combined value of 17 different ecosystem services – such as climate regulation, water supply and food production – to lie between US$16 and 54 trillion per year, around twice the entire world's Gross National Product. These services are not traded in markets, and carry no price tags to alert society to changes in their supply or to the deterioration of the ecosystems which generate them. In the future, we will need the genetic diversity of the natural world to be able to respond to climate change, diseases and crop pests.

Birds themselves are of great economic, aesthetic and cultural value. The world's commonest bird is the domestic chicken whose wild ancestor, the Red Junglefowl of tropical Asia, was domesticated around 5,000 years ago; chicken meat and eggs are an important source of protein throughout the world. The beauty of birds has inspired artists over the centuries, and bird images are frequently used to adorn everyday objects like money and postage stamps. Ever increasing numbers of people belong to bird societies: the Wild Bird Society of Japan (BirdLife in Japan) has around 50,000 members, and similar organisations elsewhere in Asia are growing rapidly. The birdwatching industry is becoming a major economic force, both globally and in Asia.

Birds are good indicators, and can be used to identify the most biologically rich areas, as well as environmental changes and problems. They are found in almost all natural habitats, they are high in the food chain and thus reflect changes lower down, a wealth of data have been collected by ornithologists, and their conservation status is well known relative to other taxa. In general, places that are rich in bird species are also rich for other forms of biodiversity, so birds can be used as indicators to locate these important areas. Studying birds can tell us about the habitats on which we all depend, and the loss of Asia's threatened birds from many parts of the region is a measure of a more general deterioration in other biodiversity and the natural environment.

KEY SPECIES AND HABITATS BY TERRITORY

This section covers all of the countries and territories in the Asia region (except the Maldives[1]), with information on the numbers of threatened bird species that they support[2], and the forest, grassland and wetland regions[3] that lie within their boundaries. It is designed to help find the sections of this book that cover the territory or territories that are of interest to the reader.

[1] Two Vulnerable species occur in the Maldives, both as rare passage migrants, and the country is therefore not covered in this section.
[2] Complete lists of threatened and Lower Risk bird species are given for Asian countries and territories in BirdLife International (2001): note that the species totals for some territories include birds which do not occur in the Asia region (e.g. the Indonesia totals include Irian Jayan species);
[3] Codes F01–F09 refer to forests, G01–G03 to grasslands, W01–W20 to wetlands and S01 to seabirds; see pp.19–21 for details of how these habitat regions were defined.

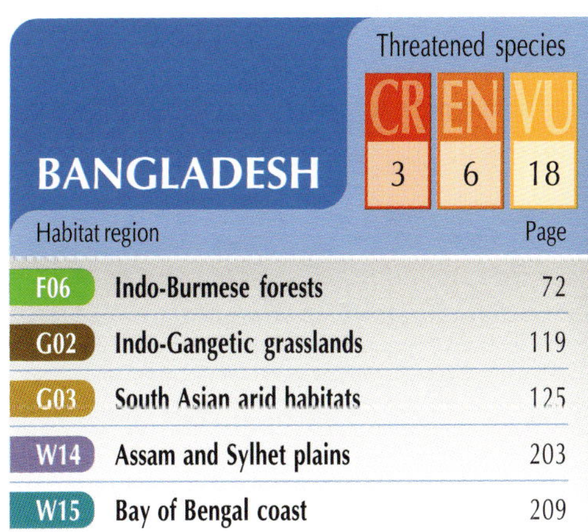

BANGLADESH

Threatened species		
CR	EN	VU
3	6	18

Habitat region		Page
F06	Indo-Burmese forests	72
G02	Indo-Gangetic grasslands	119
G03	South Asian arid habitats	125
W14	Assam and Sylhet plains	203
W15	Bay of Bengal coast	209

PHOTO: SHEHZAD NOORANI/STILL PICTURES

Bangladesh has already lost much of its natural forests, grasslands and other habitats, but it remains vitally important for several threatened species. The country's coastal wetlands support the largest known concentrations of two shorebirds, Spotted Greenshank and Spoon-billed Sandpiper, and the *haor* wetlands in the north-east support important populations of several other waterbirds, notably Pallas's Fish-eagle.

River traffic on the Buriganga river near Dhaka.

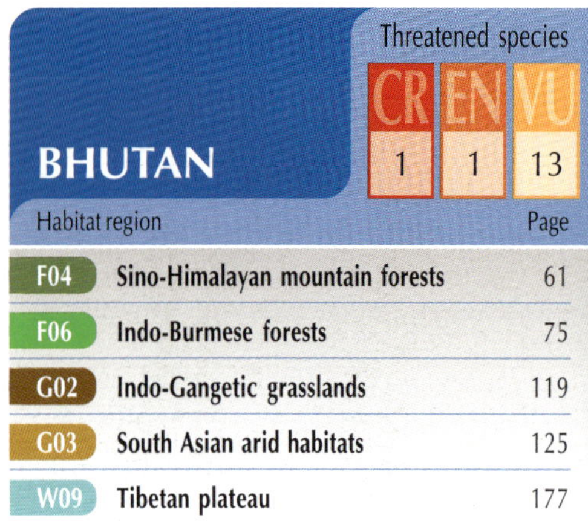

BHUTAN

Threatened species	CR	EN	VU
	1	1	13

Habitat region		Page
F04	Sino-Himalayan mountain forests	61
F06	Indo-Burmese forests	75
G02	Indo-Gangetic grasslands	119
G03	South Asian arid habitats	125
W09	Tibetan plateau	177

PHOTO: F. A. CLEMENTS/BIRDLIFE

Bhutan has an admirable national policy to maintain forests in over 60% of the country, and has an extensive protected areas system. The country is therefore extremely important for the conservation of several threatened montane forest species, and is a stronghold for birds such as Rufous-necked Hornbill. There are also important wintering flocks of Black-necked Crane in several high-altitude valleys.

Several mountain valleys in Bhutan support wintering flocks of Black-necked Cranes.

BRUNEI

Threatened species	CR	EN	VU
	—	3	19

Habitat region		Page
F07	Sundaland forests	83
W20	Sundaland wetlands	231

PHOTO: GUY PETHERBRIDGE

The tropical rain forests of the Sundaland (or Sundaic) region are being lost at an alarming rate, particularly in the lowlands, and many of the birds which rely on this habitat are threatened. However, the forests are relatively secure in Brunei, and are becoming an increasingly important stronghold for threatened birds such as Storm's Stork.

The Omar Ali Saifuddin mosque dominates the skyline in Bandar Seri Begawan, the capital of Brunei.

CAMBODIA

Threatened species	CR	EN	VU
	4	5	13

	Habitat region	Page
F06	Indo-Burmese forests	75
W18	Lower Mekong basin	221

PHOTO: ELEANOR BRIGGS

Cambodia is the only country in South-East Asia that still has extensive undeveloped wetlands and dry dipterocarp forests, and healthy populations of large waterbirds. The most notable of these are Giant Ibis and White-shouldered Ibis, the former of which is now virtually confined to Cambodia, and the country is also a stronghold for Greater Adjutant, Spot-billed Pelican and Bengal Florican. Cambodia has the some of the healthiest populations of vultures in South-East Asia, which are set to become increasingly important because they are isolated from the factors which appear to be causing South Asian vulture populations to crash. The extensive undisturbed moist forests in the Cardamom mountains in the south-west are the global stronghold of Chestnut-headed Partridge.

A carving of Sarus Cranes at the Bayon temple, Angkor.

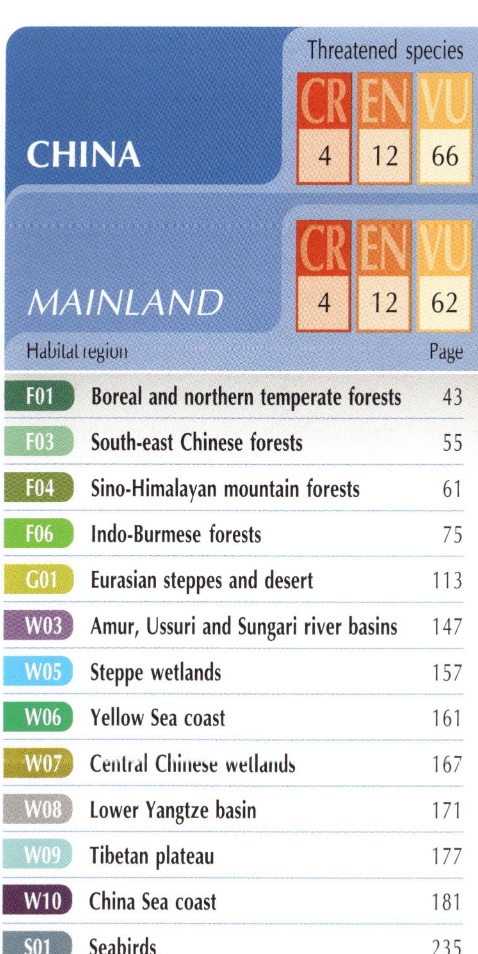

CHINA

Threatened species	CR	EN	VU
CHINA	4	12	66
MAINLAND	4	12	62

	Habitat region	Page
F01	Boreal and northern temperate forests	43
F03	South-east Chinese forests	55
F04	Sino-Himalayan mountain forests	61
F06	Indo-Burmese forests	75
G01	Eurasian steppes and desert	113
W03	Amur, Ussuri and Sungari river basins	147
W05	Steppe wetlands	157
W06	Yellow Sea coast	161
W07	Central Chinese wetlands	167
W08	Lower Yangtze basin	171
W09	Tibetan plateau	177
W10	China Sea coast	181
S01	Seabirds	235

PHOTO: HAIXIANG ZHOU

Natural habitats in China have suffered widespread clearance and degradation, and heavy hunting pressure, but many areas are still immensely rich in wildlife. The mountains support many threatened forest birds, including south-west Chinese endemics such as Chinese Monal and Grey-hooded Parrotbill, and south-east Chinese specialities such as White-eared Night-heron and Reeves's Pheasant. There have been many recent studies of China's pheasants, but the conservation requirements of the threatened passerines are generally poorly understood. The coastal and riverine wetlands hold large concentrations of cranes and other threatened waterbirds, including the only wild population of Crested Ibis, and almost the entire wintering populations of Oriental Stork, Swan Goose and Siberian Crane. In recent decades, the Chinese government has declared many hundreds of new protected areas, and a logging ban, which provide a major opportunity to ensure the long-term survival of the country's unique biodiversity.

Cranes have traditionally been regarded as holy birds in China and Japan.

HONG KONG AND MACAU

Threatened species	CR	EN	VU
	—	4	10

Habitat region		Page
W10	China Sea coast	181

PHOTO: PAUL GORIUP/NATUREBUREAU

The cities of Hong Kong and Macau lie either side of the Pearl river delta, the largest intertidal area in southern China. Despite the high human population density, both territories have important wetlands. Inner Deep Bay in Hong Kong is well protected, and supports important non-breeding populations of several threatened waterbirds, notably Black-faced Spoonbill. The wetlands in Macau are relatively small, and are under pressure from development, but also support a wintering population of Black-faced Spoonbills.

Inner Deep Bay is one of the richest wetlands on the China Sea coast, despite the proximity of Hong Kong and other large cities.

TAIWAN

Threatened species	CR	EN	VU
	1	7	15

Habitat region		Page
F03	South-east Chinese forests	55
W10	China Sea coast	181
S01	Seabirds	235

PHOTO: MARCO LAMBERTINI/BIRDLIFE

The wetlands of Taiwan, particularly on the west coast, support non-breeding populations of several threatened waterbirds, most notably about half of the world population of Black-faced Spoonbill. The only known breeding site of Chinese Crested-tern is on the Mazu Dao islands. The forests of Taiwan support 15 or more endemic species, but there is an extensive network of protected areas and high public awareness of wildlife conservation; only one of these birds is globally threatened, Taiwan Bulbul, and this is because of hybridisation with the closely related Chinese Bulbul, rather than habitat loss.

The Taiwan Birdwatching Fair attracts tens of thousands of visitors each year.

INDIA

Threatened species		
CR	EN	VU
8	10	55

Habitat region		Page
F04	Sino-Himalayan mountain forests	61
F05	Indian peninsula and Sri Lankan forests	69
F06	Indo-Burmese forests	75
G02	Indo-Gangetic grasslands	119
G03	South Asian arid habitats	125
W09	Tibetan plateau	177
W12	North Indian wetlands	191
W13	South Indian and Sri Lankan wetlands	197
W14	Assam and Sylhet plains	203
W15	Bay of Bengal coast	209

PHOTO: ASAD RAHMANI

Threatened birds are found virtually throughout India. Forests in the Western and Eastern Himalayas support groups of species with small and declining ranges, including several partridges and pheasants. The recently rediscovered Forest Owlet inhabits a few forest fragments in Central India, and a group of birds endemic to the Western Ghats are also globally threatened. The grasslands and wetlands of the northern plains are strongholds for birds such as Bengal Florican and Greater Adjutant. Grasslands and semi-deserts in eastern India are vital for Great Indian Bustard and Lesser Florican, and arid thorn forests in the south support the poorly known Jerdon's Courser and several other threatened birds. The populations of three *Gyps* vulture species have recently crashed in the subcontinent, for reasons that have yet to be determined. Although India has a well developed protected areas system, there is immense pressure from population growth and economic development, and the country faces a huge challenge to balance human needs with the protection and management of natural habitats.

Grasslands in India are vital for Great Indian Bustard and several other threatened species.

INDONESIA

Threatened species		
CR	EN	VU
14	30	73

Habitat region		Page
F07	Sundaland forests	83
F08	Wallacea	93
W20	Sundaland wetlands	231

PHOTO: MARK EDWARDS/BIRDLIFE

Indonesia is a vast archipelago of more than 17,000 islands. It is notable for the very high levels of endemism, with many species confined to certain islands or island groups. It supports 117 globally threatened bird species, similar to Brazil but far higher than any other country in the world. Many species unique to the lowland forests of the Sundaland (or Sundaic) region in the west of the country are declining rapidly because of large-scale forest clearance, as well as the effects of forest fires. Some of the birds unique to smaller islands in eastern Indonesia are also being badly affected by deforestation, and several of these islands support remarkable concentrations of highly threatened species, notably Talaud and Sangihe. The keeping of cagebirds is traditional in Indonesia, but some species are currently being captured at unsustainable levels, particularly parrots in eastern Indonesia. There in an urgent need to further develop the national protected areas system, particularly in the east of the country, and to improve controls on the exploitation of forests and other natural resources.

Rice terraces near Bali Barat National Park.

JAPAN

Threatened species

CR	EN	VU
3	9	29

	Habitat region	Page
F01	Boreal and northern temperate forests	43
F02	Japanese forests	49
W02	Sea of Okhotsk and Sea of Japan coasts	141
W04	Japanese wetlands	153
W10	China Sea coast	181
S01	Seabirds	235

PHOTO: TAKUKI HANASHIRO

Only limited areas of natural forest and wetlands remain in the lowlands of Japan, and several large waterbirds became extinct there as a result of habitat loss and hunting, including Oriental Stork (as a breeder) and Crested Ibis. The population of Short-tailed Albatross crashed to near extinction because of massive exploitation for its feathers, but careful protection in recent decades has allowed its numbers to slowly recover. Deforestation has now almost ceased in most parts of the country, which has a well developed protected areas system. However, several endemic birds of the Nansei Shoto and Izu islands are highly threatened by localised forest clearance, as well as by introduced predators. Japan still supports some important waterbird populations, notably the remarkable concentrations of wintering cranes at Izumi on Kyushu, and of resident Red-crowned Cranes on Hokkaido, which are conservation success stories but (at least in the case of Izumi) partially reflect a lack of natural wetlands elsewhere in the country.

Okinawa Rail is confined to the subtropical forests of northern Okinawa, in the Nansei Shoto islands.

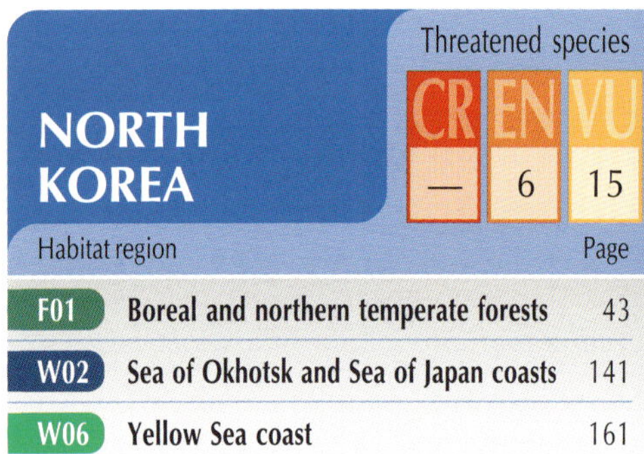

NORTH KOREA

Threatened species

CR	EN	VU
—	6	15

	Habitat region	Page
F01	Boreal and northern temperate forests	43
W02	Sea of Okhotsk and Sea of Japan coasts	141
W06	Yellow Sea coast	161

PHOTO: HAIXIANG ZHOU

Most of world's Black-faced Spoonbills and Chinese Egrets nest on islets off the west coast of Korea, and several of the most important colonies are in North Korea. The spoonbills and egrets fly to intertidal flats on the mainland to feed, and these coastal wetlands are also visited by large numbers of cranes and waterfowl on migration. The Demilitarised Zone (DMZ) is important for many waterbirds, including wintering flocks of cranes and breeding Black-faced Spoonbills; the key sites within this zone need to be protected from development should there be a change in the political situation in Korea in the future. The forests in the north of the Korean peninsula support a breeding population of Scaly-sided Merganser, and Rufous-backed Bunting formerly occurred in the far north-east, but there is little recent information on its status there.

Black-faced Spoonbills nest on islets off the west coast of Korea, and fly to the mainland to feed on intertidal flats.

PHOTO: ADRIAN LONG/BIRDLIFE

The Yellow Sea wetlands on the western and southern coasts of the Korean peninsula are vital for the migratory waterbirds of the east Asia flyway, including threatened species such as Spotted Greenshank and Spoon-billed Sandpiper, but are steadily being reclaimed for development. The coastal wetlands of South Korea are also important for both Saunders's and Relict Gulls, and the large concentrations of waterbirds that winter there include most of the global population of Baikal Teal. There are colonies of Black-faced Spoonbills and Chinese Egrets off the west coast, mainly near to the Demilitarised Zone (DMZ), and Styan's Grasshopper Warblers nest on offshore islets in the south-west. Flocks of White-naped and Red-crowned Cranes winter near to the DMZ, and Hooded Cranes near the south coast.

The intertidal wetlands on the Yellow Sea coast of Korea support vast numbers of migratory waterbirds.

PHOTO: JOE TOBIAS

Laos retains relatively extensive semi-evergreen and dry dipterocarp forests, with important populations of threatened forest birds such as White-winged Duck and Rufous-necked Hornbill. The country is currently developing an extensive protected areas system, but needs to develop the infrastructure to effectively manage these reserves, in particular to control the heavy hunting pressure that has already greatly reduced the populations of most large birds. Laos supports some important waterbird populations, particularly on the floodplains close to the Cambodian border.

A village in the hill forests on the Nakai Plateau, central Laos.

MALAYSIA

Threatened species

CR	EN	VU
3	4	33

Habitat region		Page
F07	Sundaland forests	83
W20	Sundaland wetlands	231

PHOTO: SARAH FOWLER/NATUREBUREAU

Malaysia has important populations of many threatened rain forest birds. These include three species which are endemic to the Malaysian peninsula, the montane Mountain Peacock-pheasant and Malayan Whistling-thrush, and the lowland Malaysian Peacock-pheasant, and several which are confined to the island of Borneo. Natural forests have been reduced and degraded in many parts of the country and, although the rate of deforestation has slowed in peninsular Malaysia, logging and forest conversion to plantations are still affecting large areas of East Malaysia. The coastal wetlands in both Peninsular and East Malaysia are important for several threatened waterbirds.

The lower Kinabatangan river in Sabah, which supports Storm's Stork and other threatened lowland rain forest birds.

MONGOLIA

Threatened species

CR	EN	VU
1	4	13

Habitat region		Page
F01	Boreal and northern temperate forests	43
G01	Eurasian steppes and desert	113
W05	Steppe wetlands	157

PHOTO: JACOB WIJPKEMA

Mongolia retains extensive relatively unspoiled areas of steppe and boreal forest. It is the Asian stronghold for several threatened grassland and wetland species, including White-headed Duck, Lesser Kestrel, Great Bustard, Relict Gull and White-throated Bushchat, and it has the only surviving breeding population of Dalmatian Pelican in eastern Asia. Overgrazing and the use of rodenticides to control vole outbreaks are affecting steppe habitats and fauna in some areas. In the longer term, the Mongolian government has plans to develop the country economically, a process that needs to be carefully planned and managed to prevent unnecessary damage to the country's natural habitats.

The steppes of Mongolia have important populations of several threatened grassland and wetland birds.

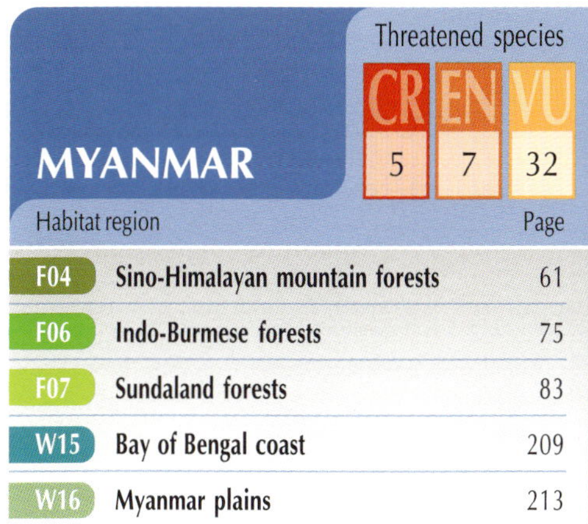

MYANMAR

Threatened species

CR	EN	VU
5	7	32

Habitat region		Page
F04	Sino-Himalayan mountain forests	61
F06	Indo-Burmese forests	75
F07	Sundaland forests	83
W15	Bay of Bengal coast	209
W16	Myanmar plains	213

PHOTO: SIMBA CHAN

Myanmar is a remarkably diverse country for it size, with threatened species in its lowland and montane forests and in coastal and riverine wetlands. It could prove to be the stronghold of several relatively widespread threatened birds, such as White-bellied Heron, Hume's Pheasant, Green Peafowl and Pale-capped Pigeon, its coastal wetlands are likely to be important for Spotted Greenshank and Spoon-billed Sandpiper, and there is even a possibility that Pink-headed Duck may survive in the remote wetlands in the north. However, there is little recent information on the national status of most of the threatened birds, and surveys are urgently required to locate important populations and key sites. Gurney's Pitta was rediscovered in southern Myanmar in 2003, after a gap of almost 90 years, and further work is required to identify the actions required for its protection.

Protected area staff at Moyingyi Wildlife Reserve on a water-monitoring training course.

NEPAL

Threatened species

CR	EN	VU
2	3	21

Habitat region		Page
F04	Sino-Himalayan mountain forests	61
G02	Indo-Gangetic grasslands	119
G03	South Asian arid habitats	125
W12	North Indian wetlands	191

PHOTO: MIKE CROSBY/BIRDLIFE

The montane forests of Nepal support important populations of several threatened birds, but are being cleared and degraded in some areas as a result of conversion for agriculture, livestock grazing and cutting for timber and fuel. Hunting is also a problem. The southern lowlands of the country (the *terai*) are densely populated, and virtually all of the remaining natural grasslands are inside a few large protected areas. These are very important for several threatened grassland specialists and waterbirds, including Bengal Florican, but their protection and management are a major challenge because of the intense pressure from human utilisation.

Machapuchare (the fishtail mountain) in Annapurna Conservation Area is one of many holy mountains in Nepal.

PAKISTAN

Threatened species

CR	EN	VU
3	3	20

	Habitat region	Page
F04	Sino-Himalayan mountain forests	61
G03	South Asian arid habitats	125
W11	Indus basin	187

PHOTO: GUY DUKE

Pakistan is an arid country, and its forests are mainly confined to the mountains in the north. These Himalayan forests are now much reduced in extent, but the remaining fragments support important populations of several threatened species, notably Western Tragopan. A number of threatened waterbirds inhabit the wetlands in the Indus valley and on the coast, including the most important Asian populations of White-headed Duck and Marbled Teal, but many of them have declined because of wetland drainage and degradation, and hunting.

Villagers in the Palas valley, which supports the largest population of Western Tragopan in the world.

PHILIPPINES

Threatened species

CR	EN	VU
12	14	44

	Habitat region	Page
F09	Philippine forests	103
W19	Philippine wetlands	227

PHOTO: MICHAEL POULSEN/BIRDLIFE

The Philippines is remarkable for its very high levels of endemism, including many species which are confined to certain islands or island groups within the archipelago. It has a very high total of threatened species for a country of its size, mainly because of the widespread clearance and degradation of its tropical forests. Deforestation is particularly severe in the Western Visayas, Mindoro and the Sulu archipelago, and the effective protection of the small remaining forests on these islands is vital for endemics such as Negros Bleeding-heart, Black-hooded Coucal, Sulu Hornbill and Cebu Flowerpecker. On Luzon, Samar, Leyte and Mindanao, a network of forests needs to be protected to prevent the extinction of Philippine Eagle and many other threatened birds. Unsustainable capture for the wild bird trade is causing a dramatic decline in the range and numbers of Philippine Cockatoo, and many threatened birds are hunted for food and sport.

Mossy forest in Mount Pulog National Park, in the Cordillera Central mountains on Luzon.

RUSSIA

Threatened species	CR	EN	VU
	3	10	29

	Habitat region	Page
F01	Boreal and northern temperate forests	43
G01	Eurasian steppes and desert	113
W01	Arctic tundra	137
W02	Sea of Okhotsk and Sea of Japan coasts	141
W03	Amur, Ussuri and Sungari river basins	147
W05	Steppe wetlands	157
S01	Seabirds	235

PHOTO: CHRISTOPH ZÖCKLER

The tundra, boreal forest, wetland and steppe habitats in Eastern Siberia support huge numbers of breeding birds, many of which migrate to East and South-East Asia in winter. Large areas of these habitats remain pristine, but forestry and industrial and agricultural development near to the Chinese border and along the east coast are reducing and degrading natural forests, grasslands and wetlands. Several threatened species are concentrated in this part of the country, although others (including Siberian Crane and Spoon-billed Sandpiper, which nest in the northern tundra) are declining mostly because of threats on their passage and wintering grounds.

The tundra and boreal forests in eastern Russia are the breeding grounds of huge numbers of migratory birds, including several threatened species.

SINGAPORE

Threatened species	CR	EN	VU
	—	1	7

	Habitat region	Page
F07	Sundaland forests	83
W20	Sundaland wetlands	231

PHOTO: TEE LIAN HUAT

Although the remaining forests in Singapore are small and fragmented, they support an important population of Straw-headed Bulbul, which is declining rapidly elsewhere because of trapping for the wild bird trade. The coastal wetlands regularly support small numbers of the threatened Chinese Egret, with occasional records of Spoon-billed Sandpiper.

The Straw-headed Bulbul population in Singapore has become increasingly important as the species has declined elsewhere.

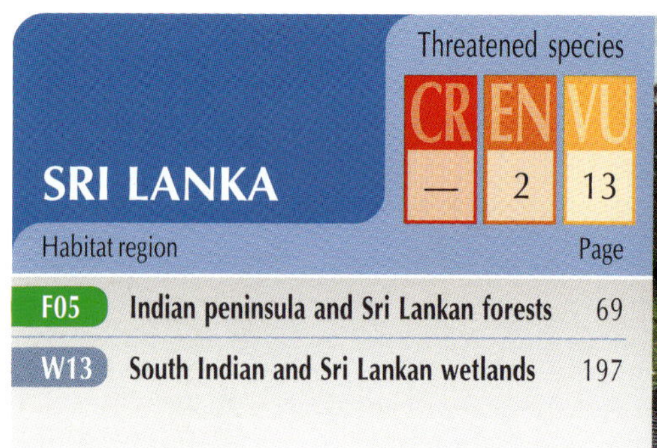

SRI LANKA

Threatened species	CR	EN	VU
	—	2	13

Habitat region		Page
F05	Indian peninsula and Sri Lankan forests	69
W13	South Indian and Sri Lankan wetlands	197

PHOTO: TIM LOSEBY

Sri Lanka has over twenty endemic bird species, many of which are confined to the rain forests of the wet zone, in the south and west of the island. These forests are highly fragmented, with few large blocks remaining and only a small number of protected areas, and consequently seven of the endemic birds are globally threatened. The government has imposed a moratorium on logging, which needs to be strictly enforced to prevent further habitat loss. Sri Lanka is also one of the strongholds of Spot-billed Pelican, and the major colonies and feeding areas of this waterbird need to be protected.

The wetlands of Sri Lanka, including in Yala National Park, are a global stronghold for Spot-billed Pelican.

THAILAND

Threatened species	CR	EN	VU
	5	8	31

Habitat region		Page
F04	Sino-Himalayan mountain forests	61
F06	Indo-Burmese forests	75
F07	Sundaland forests	83
W17	Thailand wetlands	217
W20	Sundaland wetlands	231

PHOTO: BOONSIRI-UNEP/STILL PICTURES

Many threatened forest birds occur in Thailand, but most have declined there because of rapid deforestation in recent decades. However, several forest species still have important populations, mainly inside protected areas, most notably the population of Gurney's Pitta at Khao Nor Chuchi in peninsular Thailand. The freshwater and coastal wetlands have also been seriously affected by development, as well as heavy hunting pressure, and several threatened waterbirds no longer breed in the country. The only records of the enigmatic White-eyed River-martin were from central Thailand in the 1960s and 1970s and, if it survives, it is assumed to nest along large rivers in Thailand or another South-East Asian country.

Buddhist monks encircle a threatened forest in Thailand to protest against its destruction.

TIMOR-LESTE (EAST TIMOR)

Threatened species	CR	EN	VU
	1	3	2

Habitat region		Page
F08	Wallacea	93

PHOTO: ROSEMARY LOW

This newly independent country retains more extensive forests than western Timor (in Indonesia), and it is likely to support the most important populations of some of the threatened species which are confined to the islands of Timor and Wetar, including Timor Green-pigeon and Timor Imperial-pigeon. Timor-Leste still holds a healthy population of Yellow-crested Cockatoo, which has declined rapidly elsewhere because of capture for the wild bird trade. There is little recent information on the distribution and status of the threatened birds, so surveys are required to identify the richest areas of forest and the measures needed for their protection.

Timor-Leste is one of the few parts of Wallacea that retains a healthy population of Yellow-crested Cockatoo.

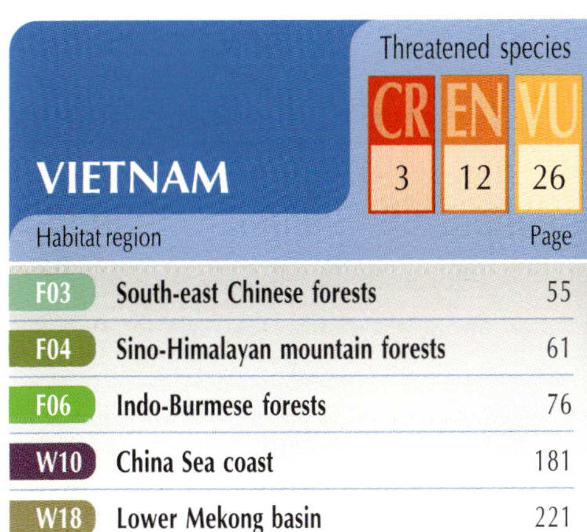

VIETNAM

Threatened species	CR	EN	VU
	3	12	26

Habitat region		Page
F03	South-east Chinese forests	55
F04	Sino-Himalayan mountain forests	61
F06	Indo-Burmese forests	76
W10	China Sea coast	181
W18	Lower Mekong basin	221

PHOTO: J. C. EAMES

The evergreen and semi-evergreen forests of Vietnam support many threatened forest birds, including groups of nationally endemic (or near endemic) species in the Annamese lowlands, Kon Tum plateau, South Vietnamese lowlands and Da Lat plateau. Vietnam's forests have been greatly reduced and fragmented, particularly in the lowlands, and several of these birds are highly threatened, being confined to a handful of key sites. The coastal wetlands support concentrations of several threatened waterbirds, notably the Red River delta in northern Vietnam and the Mekong delta in the south.

Together with nearby Ke Go Nature Reserve, the forests at Khe Net are the most extensive remaining in the Annamese lowlands Endemic Bird Area.

ASIA'S THREATENED BIRDS AND THEIR HABITATS

THE THREATENED BIRDS OF ASIA

BirdLife's Threatened Species Programme

BirdLife International has been analysing and documenting the status of the world's threatened bird species since the 1970s. The results have been published in a series of global checklists and regional Red Data Books. The most recent global checklist was the *Threatened birds of the world* (BirdLife International 2000; see http://www.birdlife.org), which includes short accounts for all of the world's 1,186 threatened birds. The *Threatened birds of Asia: the BirdLife International Red Data Book* (BirdLife International 2001; see www.rdb.or.id) is a huge work (two volumes totalling over 3,000 pages), which includes detailed accounts on 323 threatened species which occur in the Asia region.

BirdLife collates information on threatened birds from a global network of experts, including members of the IUCN/SSC Specialist Groups, and from publications and unpublished sources. This information is used to assess each species's IUCN Red List Category (and hence extinction risk) using standard quantitative criteria based on population size, population trends and range size (see Box 1). Most importantly, the wealth of information generated by this programme is also used to focus global conservation efforts and to guide BirdLife's priorities for action. *Threatened birds of Asia* contains numerous proposals for projects, programmes and policies for the conservation of the region's birds, based upon the assembled evidence and the expertise of members of BirdLife's Asia network. A synthesis of these conservation

Box 1. How globally threatened birds are identified.

BirdLife International uses the IUCN Red List Categories and Criteria developed by IUCN (the World Conservation Union) to classify species at high risk of global extinction (see IUCN 2001, http://www.iucn.org/themes/ssc/redlists/RLcats2001booklet.html), and is the official Listing Authority for birds for the IUCN Red List. The four main types of criteria used to identify threatened species are:

A Rapid population reduction
B Small range and fragmented, declining or fluctuating
C Small population and declining
D Very small population or range

Within each of these, more detailed numerical criteria are used to determine whether species should be categorised as one of the following threatened or non-threatened categories:

Threatened categories

- **Critically Endangered** (from now on referred to as 'Critical'): Species faces an extremely high risk of extinction in the wild in the immediate future.
- **Endangered**: Species is not Critical, but faces a very high risk of extinction in the wild in the near future.
- **Vulnerable**: Species is not Critical or Endangered, but faces a high risk of extinction in the wild in the medium-term future.

Non-threatened categories (that are also covered in this publication)

- **Conservation Dependent**[1]: Species which are the focus of a continuing species-specific or habitat-specific conservation programme, the cessation of which would result in the species qualifying for one of the threatened categories within five years.
- **Near Threatened**: Species which do not qualify for any threatened categories, but are close to qualifying for or are likely to qualify for a threatened category in the near future.
- **Data Deficient**: Taxa which have inadequate information to make a direct, or indirect, assessment of its risk of extinction based on its distribution and/or population status. Listing of taxa in this category indicates that more information is required and acknowledges the possibility that future research will show that threatened classification is appropriate.

[1] This category was used (for one species in the Asia region) in BirdLife International (2001), following IUCN/SSC (1994), but is not included in the latest IUCN Red List Categories and Criteria (IUCN 2001).

recommendations (together with new proposals provided by reviewers during the current project) is made here, designed to make the main conclusions more accessible and immediate to a much wider audience than those with specialist interests in threatened birds and conservation.

One in eight of all bird species in the Asia region is globally threatened

A total of 324 bird species (or c.12% of the Asian avifauna) is globally threatened with extinction[1], including 41 that are Critical, 66 Endangered and 217 Vulnerable[2]. An additional 317 (Near Threatened) species are close to qualifying as globally threatened. For 23 (Data Deficient) species, there is inadequate information to make a direct, or indirect, assessment of their probability of extinction, but these too may be at risk. All told, 664 (c.25%) species in the Asian avifauna are of conservation concern at the global level.

Figure 1. Numbers of bird species in the Asia region by IUCN Red List category.

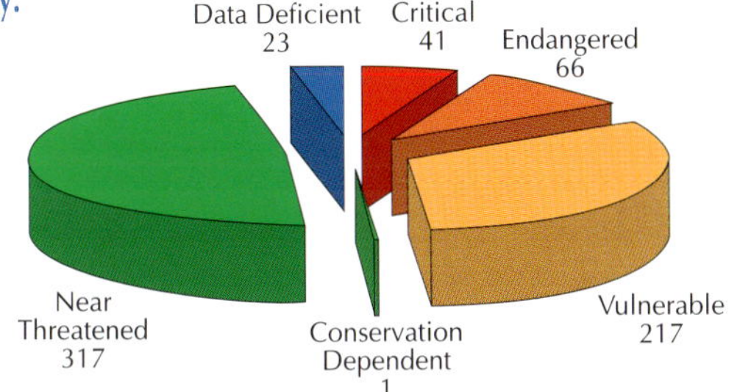

All territories are important for threatened species

Indonesia has the highest total of threatened species in the region (117 species), followed by mainland China (78), India (73) and the Philippines (70). In terms of all species of global conservation concern (i.e. those classified as Critical, Endangered, Vulnerable, Conservation Dependent, Near Threatened or Data Deficient), Indonesia again has a very high total of 322 species, followed by Malaysia (142) and Thailand (137). For Critical and Endangered species only, Indonesia has the most (44 species), followed by the Philippines (26) and India (18). For Critical and Endangered national endemics, Indonesia again has the most (32 species), followed by the Philippines (21). Although the majority of Critical and Endangered species are endemic to single territories, 35 species (>30%) occur in two or more territories and therefore the political responsibility for their survival is shared.

Figure 2. Territories ordered by their total number of globally threatened species.

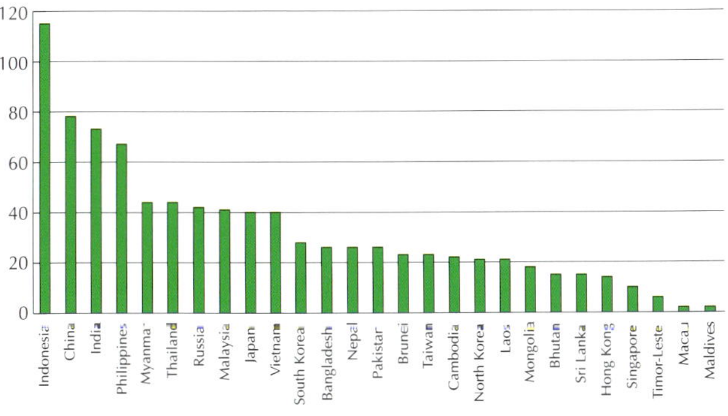

The major habitats for threatened species

Forests are by far the most important of all the habitats for threatened bird species in the Asia region, holding the greatest proportion (c.80% in total, for 75% the habitat is essential for their survival). The majority of threatened forest birds occur in the tropics (>90%) and in moist forest types (>80%), and the single most important forest type is tropical lowland moist forest with c.70% of threatened forest species (compared to c.40% globally). Tropical montane moist, tropical dry, mangrove and temperate forests also support some threatened forest birds. Grasslands, savanna and shrublands are used by nearly 30% of threatened species, but for nearly half of these birds these habitats are of only minor importance. Wetlands are under great pressure in the Asia region, with c.20% of threatened bird species found in such habitats (compared to c.10% globally), including freshwater lakes, rivers and marshes, and coastal lagoons and intertidal flats. Many large waterbirds are edging very close to extinction through the disturbance or conversion of their habitats, as well as intense hunting pressure in most areas, involving disproportionately large numbers of storks, herons, ibises, ducks, geese, cranes, gulls, waders and terns. Although artificial habitats (such as plantations, arable land, artificial wetlands etc.) apparently feature quite highly, they are also of minor importance for the great majority (88%) of threatened species occurring in them, meaning that it is unlikely that these species can survive without adjacent natural or semi-natural habitats for feeding and/or breeding.

Figure 3. The habitats upon which threatened Asian birds depend.

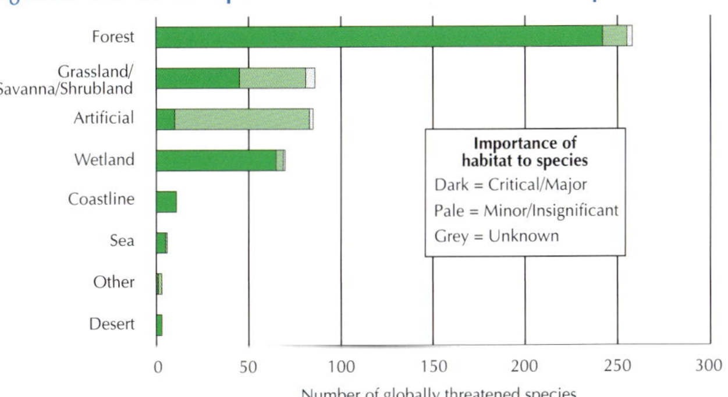

KEY FORESTS, GRASSLANDS AND WETLANDS

Most threatened bird species in Asia are specialised in their habitat requirements, and are totally dependent for their survival on a particular type of forest, grassland or wetland. The continuing loss and damage to these habitats is by far the most important threat that they face. *Threatened birds of Asia: the BirdLife International Red Data Book* includes information on every site where the threatened species have been recorded, as well as maps showing their known ranges. An analysis of these distributional data, together with information on the ecological requirements of each species, has been used to define the key forest, grassland and wetland regions for threatened birds throughout Asia. The boundaries of these habitat regions are based upon the combined distributions of all of the species that were used to define them.

[1] Note that the analysis in this book covers 303 of these species, excluding 21 (two Eurasian, one Pacific and 18 Irian Jayan) that are marginal to the Asia region;

[2] Since the publication of the *Threatened birds of Asia*, Scaly-sided Merganser has been moved from Vulnerable to Endangered, and the newly-described Chestnut-eared Laughingthrush *Garrulax konkakinhensis* (Eames and Eames 2001) has been evaluated as Vulnerable (IUCN 2002).

Table 1. The relationships between habitat regions, Endemic Bird Areas, CI Hotspots and Global 200 Ecoregions.

	Habitat region	RDB species	EBAs	CI Hotspot	Global 200 Ecoregions
F01	Boreal and northern temperate forests	6 (4)	—	—	68: Russian Far East temperate forest 82: Central and Western Siberian boreal forests and taiga
F02	Japanese forests	11 (10)	3	—	41: Nansei Shoto archipelago moist forests
F03	South-east Chinese forests	12 (10)	3	—	36.5: South-east China subtropical forests
F04	Sino-Himalayan mountain forests	28 (28)	7	South-central China mountains	36: Northern Indochina subtropical moist forests 37: North-east India and Myanmar hill forests 72: Western Himalayan temperate forests 75: Sichuan and Yunnan temperate forests 76: Eastern Himalayan broadleaf and conifer forests 91: Eastern Himalayan alpine meadows
F05	Indian peninsula and Sri Lankan forests	14 (12)	2	Western Ghats and Sri Lanka	25: Western Ghats moist forest 26: Sri Lanka moist forest
F06	Indo-Burmese forests	24 (18)	6	Indo-Burma	24: Annamite range moist forests 38: Andaman islands moist forests 40: Hainan island moist forests 55: East Indochina monsoon and dry forests
F07	Sundaland forests	47 (38)	5	Sundaland	28: Peninsular Malaysian lowland and montane forests 30: Central and south Sumatran lowland and montane forests 32: North Borneo lowland and montane forests
F08	Wallacea	51 (49)	10	Wallacea	34: Sulawesi moist forests 35: Moluccas moist forests 56: Lesser Sundas dry and monsoon forests
F09	Philippine forests	58 (54)	7	Philippines	33: Philippines moist forests
G01	Eurasian steppe and desert	5 (1)	—	—	93: Daurian steppe
G02	Indo-Gangetic grasslands	11 (9)	1	[Indo-Burma]	103: Terai-duar savannas and grasslands
G03	South Asian arid habitats	12 (10)	—	—	—
W01	Arctic tundra	4 (2)	—	—	87: Chukotsky coastal tundra 88: Taimyr coastal tundra 177: Lena river delta
W02	Sea of Okhotsk and Sea of Japan coasts	15 (2)	—	—	223: Sea of Okhotsk
W03	Amur, Ussuri and Sungari river basins	13 (1)	—	—	144: Russian Far East rivers and wetlands
W04	Japanese wetlands	16	—	—	—
W05	Steppe wetlands	12 (1)	—	—	93: Daurian steppe
W06	Yellow Sea coast	21 (2)	—	—	213: Yellow Sea and East China Sea
W07	Central Chinese wetlands	10 (1)	—	—	—
W08	Lower Yangtze basin	11	—	—	153: Central Yangtze river
W09	Tibetan plateau	2 (1)	—	—	92: Tibetan plateau steppe
W10	China Sea coast	15	—	[Indo-Burma]	—
W11	Indus basin	5	—	—	181: Indus river delta and Rann of Kutch
W12	North Indian wetlands	8	—	—	—
W13	South Indian and Sri Lankan wetlands	3	—	[Western Ghats and Sri Lanka]	—
W14	Assam and Sylhet plains	8	—	Indo-Burma	—
W15	Bay of Bengal coast	5	—	[Indo-Burma]	179: Sundarbans mangroves and Ganges and Brahmaputra deltas
W16	Myanmar plains	8	—	Indo-Burma	—
W17	Thailand wetlands	8 (1)	—	Indo-Burma	—
W18	Lower Mekong basin	9 (1)	—	Indo-Burma	—
W19	Philippine wetlands	5 (1)	—	Philippines	162: Palawan and Mindanao islands streams and lakes
W20	Sundaland wetlands	7 (1)	1	Sundaland	142: Sundaland rivers and swamps
S01	Seabirds	7 (3)	—	—	—

Key: RDB species: the total number of threatened bird species which regularly occur in each habitat region; the figures in brackets are the number of species which are unique (at least as breeding birds) to the region. EBAs: See the relevant habitat accounts for the names of EBAs. CI Hotspot: names in square brackets only overlap with part of the habitat region.

Totals of nine forest regions, three grassland regions and 20 wetland regions have been identified, as well as a small group of threatened seabirds (Table 1, Figures 4–6). All of these habitat regions support groups of threatened birds which occur in the same habitats and share broadly similar distributions. The nine forest regions and three grassland regions support between five and 58 threatened species, a high proportion of which are confined to just one of these regions. All of the forest regions have some relatively widespread threatened birds, but most of them also have groups of species with restricted ranges that are confined to just a small part of the forest region, in the Endemic Bird Areas (EBAs) and Secondary Areas (SAs) described by Stattersfield *et al.* (1998). The 20 wetland regions support between two and 21 threatened waterbirds, few of which are confined to a single wetland region, as Asia's threatened waterbirds tend to have relatively large ranges and many of them are migratory. Some of the habitat regions are entirely confined to a single country, but most are shared between several different countries.

The analysis and documentation of threatened species within habitat regions has a number of advantages:

- Many of the threatened birds in the habitat regions share broadly similar conservation needs, and it is therefore more efficient to consider the conservation of these 33 regions rather than to individually cover the 303 species that they support;
- This approach allows a direct focus on the major land-use issues affecting the habitats of threatened birds, thereby giving greater accessibility to users less immediately concerned with bird or biodiversity conservation issues;
- The habitat regions are accurately mapped, which makes it easier to relate the conservation of threatened birds with land-use planning processes (e.g. EIAs) and other projects (e.g. sustainable development and livelihood projects with a particular geographical scope);
- This analysis provides a clear geographical focus for policy and advocacy approaches to conservation (e.g. for advocacy to promote changes to national forestry policy in Cambodia);
- The habitat regions provide a framework to relate the conservation requirements of threatened birds with those of other taxonomic groups (e.g. threatened primates) and other area-based conservation priority setting analyses (see below).

Several international conservation organisations have completed analyses to identify priority areas for conservation, which are compared to the forest, grassland and wetland regions in Table 1. BirdLife International's **Endemic Bird Areas**[1] are an integral part of the habitat region analysis of the current review—eight of the nine forest regions contain EBAs, as do single grassland and wetland regions. All six of Conservation International's (CI) **Hotspots**[2] in Asia overlap with one of the forest regions, the Indo-Burma Hotspot overlaps with part of a grassland region and five wetland regions, and the Western Ghats and Sri Lanka, Sundaland and Wallacea Hotspots all overlap with a wetland region. All nine forest regions include one or more of the WWF **Global 200 Ecoregions**[3], as do two of the three grassland regions, and 11 of the 20 wetland regions. The conservation actions proposed for threatened birds and their habitats in the following sections of this book therefore provide guidance for biodiversity conservation in all CI Hotspots and a high proportion of WWF Global 200 Ecoregions in the Asia region.

KEY SITES FOR THREATENED BIRDS

BirdLife International's Important Bird Areas (IBA) Programme is a world-wide initiative aimed at identifying, documenting and protecting a network of critical sites for birds. A site is recognised as an IBA only if it meets at least one of four internationally agreed, standardised criteria, is amenable to conservation action and management, and is large enough to support populations of the key bird species for which it was identified. One of the IBA criteria focuses on sites that regularly hold significant numbers of one or more globally threatened bird species. For many threatened bird species, especially those with restricted ranges and strict habitat requirements, effective protection and management of the IBAs which have been selected for them are key measures for their survival.

BirdLife's Asia network has already made considerable progress in the identification and documentation of the region's IBAs, and a directory of *Important Bird Areas in Asia* is scheduled for publication early in 2004. The preliminary lists of IBAs for each Asian country have been used in the current review to help identify the most outstanding sites for threatened birds in each forest, grassland and wetland region, and for seabirds. These 311 IBAs were selected (through consultation with regional experts) to ensure that every threatened species is covered by at least one IBA (although it was not possible to select sites for a few of the most poorly known birds). In general, the IBAs with the most extensive and highest quality natural habitats have been chosen, but in areas where natural habitats are fragmented it was sometimes necessary to select several smaller IBAs to provide a minimum level of coverage to the threatened species. In the wetland regions, IBAs have been chosen which regularly support globally outstanding (breeding, passage or wintering) congregations of threatened waterbirds. No account has been taken of the extent to which a site is protected or threatened.

Initiatives are already underway in many Asian countries for the conservation of IBAs. In Indonesia, Vietnam and Cambodia, IBAs have recently been designated as new protected areas. In the Philippines, the Haribon Foundation is working with local governments for the designation and management of IBAs under the Local Government Code legislation (see region F09). The concept of Site Support Groups (SSGs) is being developed by BirdLife's network in the Asia region, e.g. in Vietnam, Cambodia and Indonesia (see p.30).

[1] BirdLife's Endemic Bird Areas (EBAs) are defined as areas which encompass the complete ranges of two or more restricted-range species; these are defined as species with a total global breeding range estimated at below 50,000 km^2 (see ICBP 1992, Stattersfield *et al.* 1998);

[2] Conservation International's 25 Global Hotspots are regions that harbour a great diversity of endemic species and, at the same time, have been significantly impacted and altered by human activities. Plant diversity is the biological basis for hotspot designation: to qualify as a hotspot, a region must support 1,500 endemic plant species (0.5% of the global total), and have lost more than 70% of its original habitat (see Mittermeier *et al.* 1999);

[3] WWF's Global 200 Ecoregions are a subset of the world's 1,000+ ecoregions. Ecoregions are large areas of relatively uniform climate that harbour a characteristic set of species and ecological communities, and the Global 200 includes the most biologically outstanding of these areas, selected to represent all terrestrial, freshwater and marine habitats (see Olson and Dinerstein 1998, Wikramanayake *et al.* 2002).

Figure 4. Key forest regions for threatened birds in Asia.

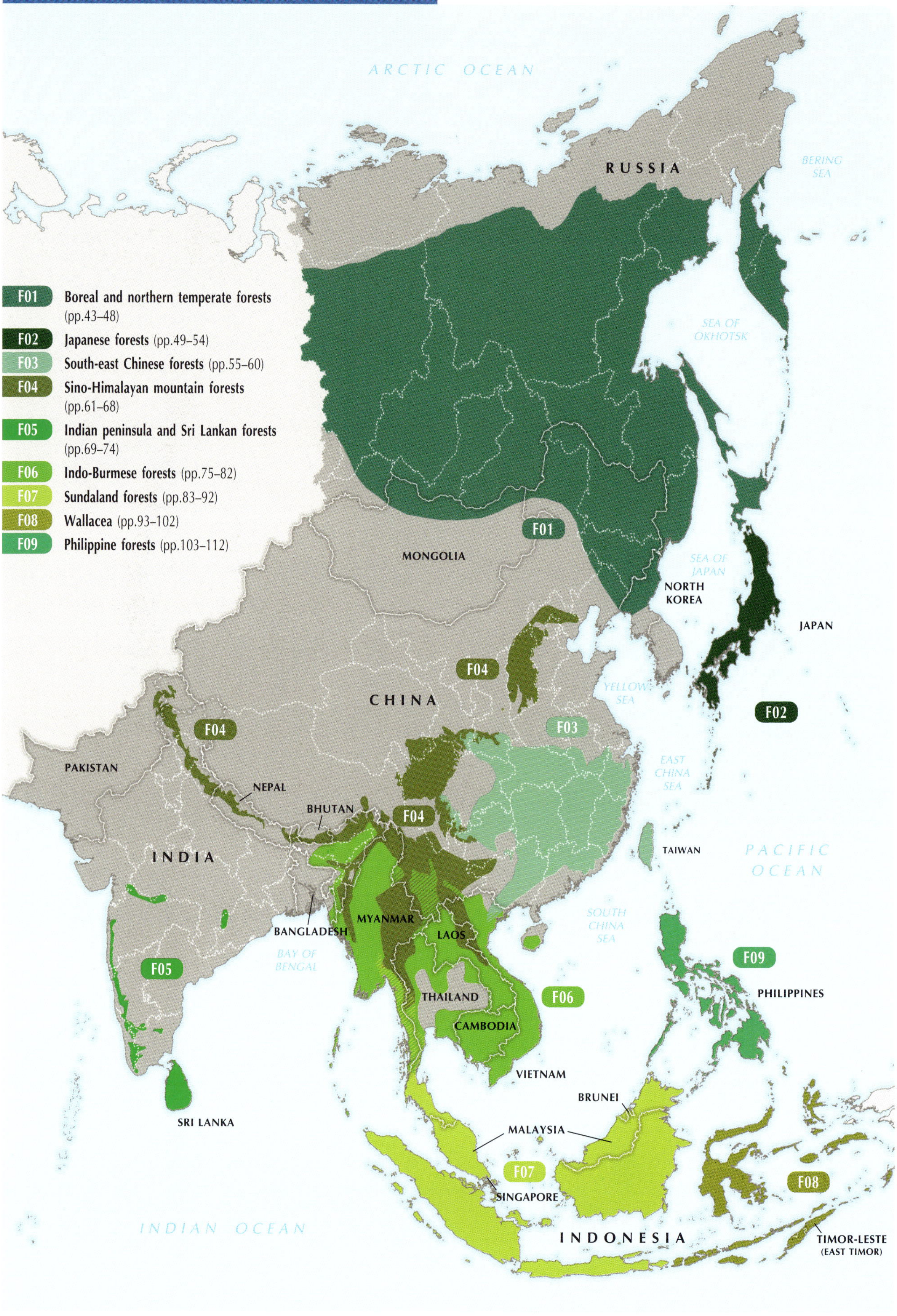

- F01 Boreal and northern temperate forests (pp.43–48)
- F02 Japanese forests (pp.49–54)
- F03 South-east Chinese forests (pp.55–60)
- F04 Sino-Himalayan mountain forests (pp.61–68)
- F05 Indian peninsula and Sri Lankan forests (pp.69–74)
- F06 Indo-Burmese forests (pp.75–82)
- F07 Sundaland forests (pp.83–92)
- F08 Wallacea (pp.93–102)
- F09 Philippine forests (pp.103–112)

ARCTIC OCEAN
RUSSIA
BERING SEA
SEA OF OKHOTSK
F01
MONGOLIA
SEA OF JAPAN
NORTH KOREA
JAPAN
F04
CHINA
YELLOW SEA
F02
F04
F03
PAKISTAN
NEPAL
EAST CHINA SEA
BHUTAN
F04
INDIA
TAIWAN
PACIFIC OCEAN
SOUTH CHINA SEA
MYANMAR
BANGLADESH
LAOS
BAY OF BENGAL
F05
F09
THAILAND
F06
PHILIPPINES
CAMBODIA
VIETNAM
BRUNEI
SRI LANKA
MALAYSIA
F07
F08
SINGAPORE
INDIAN OCEAN
INDONESIA
TIMOR-LESTE (EAST TIMOR)

Figure 5. Key grassland regions for threatened birds in Asia.

Figure 6. Key wetland regions for threatened birds in Asia.

CONSERVATION ISSUES AND STRATEGIC SOLUTIONS

THREATS TO ASIA'S BIRDS

Clearance, conversion and degradation of natural forests, grasslands and wetlands are by far the most important causes of endangerment to birds in the Asia region, affecting nearly all species classified as Critical, Endangered and Vulnerable (see Figure 1). Exploitation for human use is the second most common category of threat, affecting more than 50% of all threatened bird species; of these, c.70% are hunted for food and sport and c.30% captured for the wild bird trade. The sustainability of such exploitation is often difficult to assess, but it is feared that hunting and trade are reducing numbers of many threatened species, particularly in areas where their habitats and populations have been fragmented. Levels of exploitation of some species are clearly too high to be sustained, and are judged to be the single most serious threat that they face. Invasive species (introduced non-native animals and plants) are a potential threat to c.10% of Asia's threatened birds, but their precise impact on the populations of most of these birds is poorly understood. The most serious problem known is being caused by introduced predators on the Nansei Shoto and Izu islands in Japan. It is possible that some threatened birds are being affected by introduced predators or competitors elsewhere in Asia, e.g. on the islands of eastern Indonesia and the Philippines.

Figure 1. The main threats to globally threatened Asian birds.

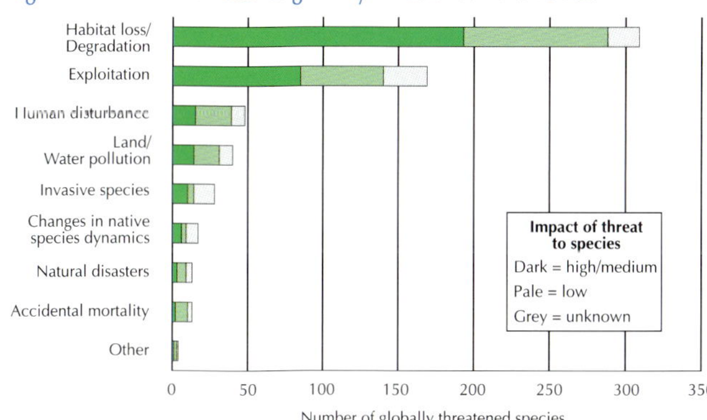

Figure 2. Main causes of habitat loss for globally threatened Asian birds.

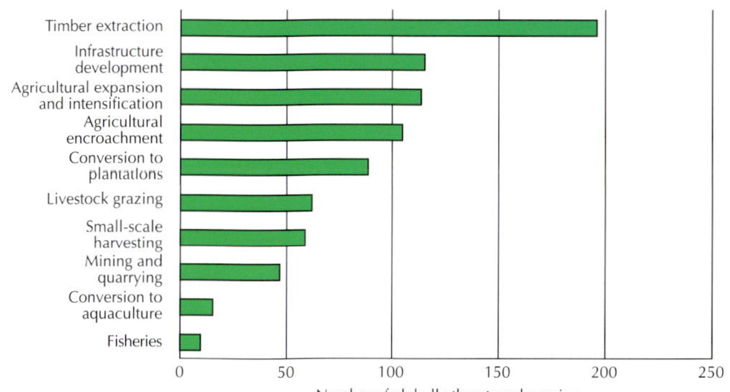

The main causes of habitat loss and degradation are shown in Figure 2. A high proportion of Asia's threatened birds inhabit forests, and cannot survive where their habitat is clear-felled for timber and/or replaced by plantations, agriculture or development (including industrial, infrastructural, urban and mining). Many forest birds are intolerant of high levels of habitat degradation, and may be badly affected (at least until the forest has had time to recover) by selective logging, shifting agriculture, forest grazing and even subsistence collection of fuelwood and timber. The main pressure on Asia's waterbirds is wetland drainage and conversion, including the infilling (or 'reclamation') of intertidal coastal wetlands, principally for agriculture and aquaculture (i.e. shrimp- and fish-ponds). Dams and irrigation projects are also negatively affecting wetlands and other habitats in parts of the region. Grasslands are being converted for agriculture and plantations, and affected by excessive grazing pressure, grass harvesting and inappropriate burning.

POLICY APPROACHES TO BIODIVERSITY CONSERVATION

The forest, grassland, wetland and seabird accounts in this book summarise the main issues affecting threatened birds throughout the Asia region. On the basis of the detailed evidence compiled in *Threatened birds of Asia: the BirdLife International Red Data Book*, together with additional material provided by reviewers during the current project, recommendations are made in each account for conservation actions to address the most important issues. The following section provides an Asia-wide overview of the direct threats identified in these accounts and outlines some of the policy-level approaches that need to be taken.

It is beyond the scope of the current work to consider in any detail the complex social and economic factors that are the ultimate causes of habitat loss and other threats. These will, however, ultimately need to be addressed if biodiversity loss in the region is to be reversed. A brief overview of some of these issues is also provided below.

Direct threats to biodiversity: habitat loss

Figure 2 identifies 10 main causes of habitat loss and degradation in the Asia region, which are analysed in more detail in Table 1. General policy-level approaches are proposed in Table 1 that should be used to address each of these issues, and help reduce their negative effects on threatened birds and their habitats. Many of these direct causes of habitat loss are discussed in more detail in the forest, grassland, wetland and seabird accounts.

Direct threats to biodiversity: species

Figure 1 identifies several threats which directly impact on wild bird populations in the Asia region, three of which are analysed in more detail in Table 2. General policy-level approaches are proposed that could be used to address each of these issues, and help reduce their negative effects on threatened birds.

Table 1. Policy-level recommendations to address the main direct causes of habitat loss and degradation in the Asia region.

Overview of issues (and habitats impacted)	Policy recommendations
Timber extraction (Forest)	
Clear-cutting and selective logging for timber have had a massive impact on Asia's forests, especially the lowland moist tropical forests in South-East Asia. Forestry is typically the first major pressure in forest areas, but often opens them up to further degradation or clearance (see below). By the end of the 20th Century, depletion of natural forests had led to a substantial decline in the forestry industry in large parts of Asia (e.g. the Philippines, Vietnam, Sumatra), with logging moratoria in north-east India, Thailand and China. This has resulted in increased pressure on forests in Cambodia, Laos, Myanmar and parts of Indonesia. Many Asian countries have regulations and programmes designed to advance sustainable forest management. However, enforcement is often poor, and illegal logging predominates in some regions (e.g. Indonesia, Cambodia), undermining efforts for sustainable management and leading to substantial loss of government revenues. The pulp and paper industry is currently expanding in tropical Asia, with several large mills operational or under construction on Sumatra, Borneo and Mindanao. Large areas of native forest are being cleared for pulp fibre, and often replaced by timber plantations to supply the long-term needs of the industry. However, over-capacity in the industry implies a long-term, but unacceptable, reliance on native forest. *Prospects for the 21st Century* Regional demand for timber is likely to remain high, and for pulp and paper to increase, in part because of the expanding Chinese and Indian economies and the logging moratoria there and elsewhere. Pressure, including illegal logging, is likely to increase on the extensive remaining forests in eastern Russia, Laos, Cambodia and eastern Indonesia, and remain high in East Malaysia and Kalimantan.	➤ Establish permanent national forest estates, protected through legislation and land-use planning (with a particular focus on remaining areas of primary forest). ➤ Strengthen and expand sustainable forest management practices, including: reduced-impact logging; sufficiently long cutting cycles to allow logged forest to recover; replanting of cleared areas using native tree species or promotion of natural regeneration; integration of biodiversity conservation into national forest programmes. ➤ Advance forest certification, based on best practices in forest management and independent monitoring (see, e.g., http://www.fscoax.org/). Increase public preference for certified timber, and encourage industry commitment (retailers, manufactures, investors, lending institutions, buyers, and producers) to rapidly phase out the use of timber from non-certified sources. ➤ Develop inter-governmental initiatives to address illegal logging. ➤ Consider logging bans or moratoria where timber stocks are severely depleted and/or high levels of illegal logging are undermining any attempts to put in place sustainable forest management practices, e.g. Sumatra. ➤ Seek an agreement, involving both the industry and national governments, to freeze expansion of the pulp and paper industry in South-East Asia until plantations are in place to supply current pulp fibre needs. ➤ Encourage the pulp and paper industry (investors, lending institutions, buyers, and producers) to commit to rapidly phasing out the cutting of natural forest for pulp fibre. ➤ Conduct wood supply assessments of major pulp and paper plants to ensure against illegal sourcing and as a basis for monitoring and compliance. ➤ Enhance efforts for paper recycling, especially in the region's demand economies (Japan, South Korea, and increasingly China).
Infrastructure development (Forest; Grassland; Wetland)	
Many Asian countries are currently experiencing rapid economic growth, and the associated urban, industrial and infrastructural developments are directly reducing natural habitats. New roads and dams often open up previously inaccessible areas to habitat degradation or clearance, and could negatively affect many key areas for threatened birds throughout Asia, e.g. Kerinci Seblat National Park, Sumatra, Indonesia, and Northern Sierra Madre Nature Park in the Philippines. Dams may have several other impacts, including flooding of habitat, changes to river flow and displacement of people, but may provide an incentive to reduce deforestation and hence sediment load upriver (see p.35: World Commission on Dams). *Prospects for the 21st Century* Many parts of Asia are likely to continue to develop rapidly, with improved infrastructure (and political change) allowing development in areas that are currently inaccessible, e.g. parts of eastern Russia, Mongolia, Kuril Islands.	➤ Make development programmes and projects subject to strategic and/or environmental impact assessments, with mitigation for any negative impacts on biodiversity. Projects that would seriously damage natural habitats and biodiversity should not be advanced. ➤ Seek to integrate biodiversity conservation with regional development, by incorporating the strict protection or sustainable management of key areas of natural habitat (including IBAs) in national and provincial development plans. ➤ Review and revise development plans that include large-scale conversion of natural habitats (e.g. coastal reclamation in East Asia), to minimise the negative impacts on biodiversity. ➤ Prepare plans to integrate biodiversity conservation with development in currently inaccessible areas that might become accessible in the future, e.g. Korean Demilitarised Zone, Kuril islands.
Agricultural expansion and intensification (large scale) (Forest; Grassland; Wetland)	
Huge areas of natural habitat in Asia have been converted to agriculture, notably for rice cultivation in the lowlands of China and South and South-East Asia. Large state-sponsored schemes continue to lead to conversion, e.g. transmigration to forest areas in Indonesia, irrigation schemes in arid habitats in western India.	➤ Make large agricultural development programmes and projects subject to strategic and/or environmental impact assessments. ➤ Incorporate habitat protection into regional land-use planning processes, with regulations and policies for habitat protection and management, e.g. the retention of traditional non-intensive forms of land management (where they are beneficial to wildlife) in biodiversity-rich areas. ➤ Develop and enforce regulations to control the use of agrochemicals, particularly in biodiversity-rich areas.
Agricultural encroachment (small scale) (Forest; Grassland; Wetland)	
Shifting agriculture has been the principal cause of forest clearance and degradation in the eastern Himalayas and the mountains of Indochina and parts of Indonesia. In the Philippines, there are many landless *kaingin* farmers, who rapidly occupy and cultivate areas opened up by new roads. The expansion of smallholdings to grow vegetables is currently having a major impact on hill forests in the Philippines and Indonesia. *Prospects for the 21st Century* The impact of shifting agriculture and smallholdings is likely to reduce as lifestyles change.	➤ Advance sustainable livelihood schemes in upland areas with high biodiversity values, e.g. agro-forestry, improved agricultural techniques. ➤ Implement land reforms to increase security of land tenure (and incentive to invest in existing land holdings), and reduce dependency on marginal forest areas.

Table 1 ... continued. Policy-level recommendations to address the main direct causes of habitat loss and degradation in the Asia region.

Overview of issues (and habitats impacted)	Policy recommendations
Conversion to plantations (Forest; Grassland)	
There has been extensive replacement of natural forests (and grasslands) by a variety of plantation crops in Asia, including tea in the hills of India, China, Sri Lanka and Java, teak in India, Myanmar and Java, and rubber in South-East Asia, especially the Thai-Malay peninsula and Sumatra. Inappropriate afforestation of natural non-forest habitats is a problem in some parts of Asia, e.g. in Vietnam, where seasonally inundated grasslands and intertidal mudflats important for threatened birds are being planted with *Melaleuca* and mangroves. In recent decades, conversion to oil palm plantations has had a huge impact in the Thai-Malay peninsula and Sumatra. Coffee plantations have been expanded in hill forest areas of, e.g., Vietnam and Sumatra, and other plantation crops have had significant localised impact (e.g. cinnamon in Sumatra, pineapple in the southern Philippines, cacao and bananas in eastern Indonesia). Fire has been used widely in South-East Asia to clear forest for plantations and agricultural activities. This had a devastating impact in the late 20th Century (especially during the 1987/88 and 1997/98 El Niño events in Indonesia). *Prospects for the 21st Century* The demand for most commodities is likely to increase, especially from China and India, with remaining extensive forests vulnerable to conversion, particularly following logging (e.g. Kalimantan, eastern Indonesia and southern Myanmar to oil palm, Cambodian dry forests to teak and pulp wood plantations). There is concern that the frequency and severity of El Niño events will increase with global warming, and that extensive burnt-over forests from past events are now more susceptible to fire.	➤ Promote industry, government and donor agency commitment to halt forest clearance for plantations (especially buyers, investors, and growers of oil palm, pulp wood, coffee, rubber, tea, but also cacao, bananas, cinnamon). This could include public pressure on the industry (possibly involving certification schemes), the removal of subsidies which encourage conversion to plantations, and government moratoria on the allocation of forest land for conversion. ➤ Introduce safeguards into national and provincial reforestation policies to prevent inappropriate afforestation of natural non-forest habitats that are important for threatened birds. ➤ Encourage plantation companies to take greater corporate responsibility for forest conservation, including by developing 'best practice' in the establishment and management of plantations, e.g. retention of natural forest patches, use of native rather than exotic shade trees. ➤ Strictly control the use of fire to clear land for plantations, including through adherence to the ASEAN policy of zero burning. Greatly enhance forest fire monitoring, prevention and control mechanisms.
Livestock grazing (Forest; Grassland; Wetland)	
Grazing by livestock affects many habitats in Asia. In some places, e.g. the grasslands of northern and western India, well-managed grazing has the potential to improve wildlife habitat. However, in many cases excessive stocking levels are degrading forests (e.g. throughout the Indian subcontinent, Nusa Tenggara in Indonesia, the Visayas in the Philippines) and grasslands (e.g. in parts of Mongolia, China and India). Burning is widely used to improve pasture for livestock, but can also degrade the habitat of grassland birds. In the steppes of Mongolia, rodenticides have recently been used to control outbreaks of voles (linked to overgrazing), which has caused mortality of wildlife.	➤ Develop national or provincial grazing policies where required, with the aim of improving livestock husbandry, particularly to minimise damage to (and improve management of) natural habitats in areas of high biodiversity importance. ➤ Promote ecological management of grazing, including rotational grazing regimes that can provide optimum habitat for many grassland birds, and reduce outbreaks of voles (and the need to use rodenticides for their control). ➤ Promote the concept and use of fewer, better quality livestock. ➤ Improve controls on, and management of, the use of fire in maintaining pasture for grazing.
Small-scale harvesting (Forest; Grassland; Wetland)	
Harvesting of timber, grass and other natural products for fuel, building materials and other products is putting considerable pressure on many forests, grasslands and wetlands in Asia. This has increased as the region's population has grown, and natural habitats have been reduced, e.g. in eastern China and many parts of Indochina. It is set to continue in line with human population growth, especially in the Indian subcontinent.	➤ Promote community forestry schemes to provide a sustainable source of forest products for local consumption, especially near to key sites for biodiversity conservation. ➤ Introduce fuel-efficient stoves, and develop sustainable alternatives to wood as a source of fuel, e.g. biogas, solar power.
Mining and quarrying (Forest; Grassland; Wetland)	
Mining and quarrying has caused localised, but significant, loss of forest and other habitats in many parts of Asia (e.g. coal mining in Kalimantan, gold, copper and nickel mining in eastern Indonesia, quarrying in southern India). Mine access roads and temporary settlement by mineworkers can lead to serious indirect impacts. Mining operations sometimes cause river pollution, especially when illegal (and unregulated). In some countries, huge areas are covered by mining concessions or applications.	➤ Make mining concessions and applications subject to strategic and/or environmental impact assessments, with mitigation for any negative impacts on biodiversity, and cancellation of applications that would seriously damage natural habitats and biodiversity. ➤ Review and revise relevant legislation to ensure that mining concessions do not have precedence over gazetted or proposed protected areas. ➤ Ensure that there is adequate mitigation for any loss of natural forest during mining operations, e.g. through improved protection of surrounding forest areas. ➤ Introduce a regional moratorium on further open-cast coal mining in forest areas.
Conversion to aquaculture (Wetland)	
Wetlands are rapidly being converted to shrimp- and fish-ponds in many coastal areas of tropical Asia, e.g. the Philippines, and also inland, e.g. in India. This often affects intertidal flats and mangroves, and reduces habitat for specialised species, although traditionally managed non-intensive aquaculture can provide valuable habitat for many waterbirds.	➤ Make aquacultural projects subject to strategic and/or environmental impact assessments. ➤ Incorporate habitat protection into coastal development plans. ➤ Promote traditional non-intensive management of shrimp- and fish-ponds, to maximise their value to waterbirds, possibly with certification schemes for sustainably produced seafoods.
Fisheries (Wetland)	
Many coastal and freshwater wetlands in Asia are subject to intense fishing, including of fish and shrimp fry, which is reducing the food supply of many waterbirds, as well as people. In eastern Russia, some salmon fisheries are close to collapse because of overharvesting, and industrial-scale collection of eggs from their spawning grounds to supply salmon hatcheries.	➤ Promote the sustainable management of fisheries, with controls on excessive exploitation, local fish sanctuaries and closed seasons to maintain stocks, and reduced use of fishing gear which captures fry. ➤ Strictly protect the spawning grounds of commercially exploited species, e.g. salmon in eastern Russia. ➤ Ban the use of chemicals and dynamite for fishing.

Table 2. Policy-level recommendations to address hunting, the wild bird trade and invasive species in the Asia region.

Overview of issues	Policy recommendations
Exploitation: Hunting	
More than 150 threatened bird species in the Asia region are hunted for food and sport, or captured for the wild bird trade. However, the impact of this exploitation on their populations is generally poorly understood. In particular, very few studies have so far been made to investigate hunting pressure on birds in Asia, and to judge whether it is sustainable or is causing population declines. For some highly threatened birds, commercial hunting is judged to be the primary threat (see p.39). The habitats (and hence populations) of many forest birds are now fragmented, and it is likely that hunting is progressively causing the extinction of some of their subpopulations.	➤ Ensure that all globally threatened species that are hunted are adequately covered by national legislation in all countries in the Asia region. ➤ Where required, improve hunting legislation and its enforcement, including through education and awareness campaigns for hunters and the public, and training for law enforcers. ➤ Control gun, trap and mistnet ownership, especially near to key areas for threatened birds. ➤ Ban the use of poisoned baits for hunting, especially in China.
Exploitation: Wild bird trade	
Capture for the wild bird trade is a major problem for some threatened birds in Indonesia, the Philippines, and parts of several other countries in the Asia region. As with hunting, the impact of this exploitation on their populations is generally poorly understood, although for some highly threatened birds commercial trading is judged to be the primary threat (see p.39). CITES is designed to control international trade in wildlife (see pp.32–33).	➤ See pp.32–33 for recommendations under CITES. ➤ Ensure that all globally threatened species that are traded are adequately covered by national legislation in all countries in the Asia region. ➤ Where required, improve legislation relevant to the wild bird trade and its enforcement, including through education and awareness campaigns for traders and the public, and training for law enforcers.
Invasive species	
Introduced predators and competitors (both mammals and birds) appear to be causing declines in the populations of several endemic forest bird species in the Nansei Shoto and Izu islands in southern Japan, are believed to be a problem for some endemic species in the Andaman and Nicobar islands in India, and could be affecting threatened birds elsewhere, e.g. eastern Indonesia and Sri Lanka. Moreover, invasive exotic plant species are degrading the habitats of threatened birds in many parts of Asia.	➤ Ban any further deliberate releases of alien mammals and birds for 'pest' control or other reasons, especially on islands with endemic bird species. ➤ Take measures to prevent the accidental release of alien species. ➤ Study the impact of invasive species on threatened birds, e.g. on Sulawesi and Halmahera, eastern Indonesia. ➤ Promote eradication programmes to control invasive plants, particularly at key sites for threatened birds.

Underlying and indirect causes of biodiversity loss

The previous sections have analysed the main causes of endangerment to birds in the Asia region, and presented an overview of policy-level approaches towards tackling the direct causes of those threats. However, it is clear that the conservation community in Asia and globally also has to face up to the reality of tackling the underlying and indirect causes of biodiversity loss. This may appear a frustratingly difficult and long-term process, but it is essential if the unequal political and economic structures that underlie environmental problems are to change. Conservation and development workers at a local and national level are often overwhelmed by the reality that powers beyond their reach are the prime determinants of success and failure, and sense that there is little, if anything, they can do to influence factors like policy on land tenure, or resettlement. Nonetheless, if such underlying causes are not addressed, the best that 'on-the-ground' conservation action can hope to achieve is to deflect pressure from areas of high biodiversity to those of more marginal biodiversity value.

Underlying causes are deep-rooted and complex. A detailed analysis is beyond the scope of this review, but it is important for conservation efforts to give them serious attention. Most have their origins in the global economy, in issues like increasing consumption, undervaluation, and perverse subsidies, or in common global trends like participatory democracy or land tenure patterns. Hanging over all of these is the growing and uncertain nature of climate change. The following paragraphs illustrate some of these issues.

INCREASING CONSUMPTION

Economic growth and ever-increasing consumption, notably in some western countries, are arguably the main underlying causes of habitat loss and degradation in Asian countries. Industrialised Asian economies, in Japan for example, have fuelled regional trade in timber. Growing and more affluent nations in Asia are resulting in an increasing demand for paper (e.g. China) and palm oil (e.g. in South Asia, where it is the preferred choice for cooking oil). A promise of strong economic growth and pressure from external donors have caused a number of countries to pursue export-led development strategies, notably in critical sectors for biodiversity conservation like forestry and agriculture. As a consequence, investment capital has poured in over the past 20 years.

POVERTY AND THE ENVIRONMENT

In common with other regions, powerful links also exist between poverty, human population and the environment throughout Asia. In many areas, large number of people live below thresholds accepted nationally, both economically, and in a wide range of other ways (such as access to fresh drinking water, sanitation and primary health care). For such poor communities, natural areas like forests, grasslands and wetlands form a critical component of livelihood strategies, especially in the absence of reliable economic alternatives and in times of stress and poor food security.

LAND TENURE

Land tenure is an important consideration in people's attitude to land use and significant in terms of habitat loss, especially deforestation. Weak or unresolved land tenure systems outside forests often result in spontaneous or government-backed migration and colonisation of forested areas. A lack of recognition of legal rights for indigenous peoples or local communities means other people can easily enter areas and take control. In many cases, it also results in the loss of sustainable land uses by forest dwellers when forest resources are being taken over by large enterprises.

The World Bank's 1991 Forest Policy paper accepts that weak property rights are behind forest loss and degradation in many areas (Contreras-Hermosilla 2000).

■ *UNDERVALUATION*

Conservationists believe that biodiversity has important cultural, spiritual, recreational, and personal values, but invariably official policies tend to recognise natural resources only for their commercial value. The environment, including 'wild nature' or biodiversity, is severely undervalued, despite the fact that it is clear we depend on a full range of complex ecological functions to provide essential physical services, like clean air, pure water, and fertile soils. A recent study concluded that the irreplaceable value of wild nature is about $20 trillion a year (Balmford *et al.* 2002). Forests are particularly undervalued when their full environmental and social value is not taken into account (e.g. nutrient cycling, climate regulation, erosion control and recreation).

■ *PERVERSE SUBSIDIES*

So-called perverse subsidies are defined as those that cause habitat loss, but have no lasting positive impact on sustainable development (Sizer 2000). In the forest sector, perverse subsidies may be direct, such as tax write-offs, grants or low-interests loans, or indirect, such as low or uncollected forest rents, low labour costs, or the construction of 'free' access roads. Government assistance to plantations is supplied through low land and labour costs, weak pollution laws, and financial incentives.

Recent and past adjustment programmes of the IMF and structural loans by the World Bank and Asian Development Bank have encouraged and required trade liberalisation. This has promoted exploitation of a number of products linked to forest loss, including raw timber and agricultural crops from plantations such as oil palm. In Indonesia, for example, recent conditions for loans included a reduction in the levy on log exports, at a time of weakening regulations and rampant illegal trade, and cutting the levy on trade in crude palm oil. Indonesia has also seen resettlement or 'transmigration' programmes and plantations developed together to provide a workforce and employment.

■ *POOR GOVERNANCE*

Many underlying threats arise in situations where governance and administrative systems are weak or in a state of flux. This has been and continues to be a problem in several Asian countries, where it has allowed corruption and collusion to flourish, in part the result of cronically low salary levels and state social security provisions. Frequently downplayed or ignored in the past, corruption often lies behind inadequate regulation of companies, illegal land allocation, and encroachment of natural habitats including protected areas. For example, combined with corruption, poor governance and regulation of Indonesia's banking system has led to misguided investment decisions in forestry, including financing for illegal land clearance, and investment in sectors that were already severely over-capacitated such as pulp and paper.

Another factor behind a poor conservation record in some countries is the low level of investment in regulatory authorities (staff, budgets, training, etc.), which are regarded as a comparatively low priority. Over many years, protected areas have faced a number of basic management problems ranging from poor staff morale and discipline, lack of incentives for good performance, limited capacity, an emphasis on administration rather than field duties for reserve managers, and inappropriate and inflexible budget allocations. Communities that depend on natural resources are rarely, if ever, consulted over the decisions that radically influence local forests. In Indonesia (and many other countries), management of forests, forest land allocation, and protected areas management have been centralised until recently, with little or no authority extending to provincial and local government levels. As a consequence, local administrations have had little incentive to support conservation. This is likely to change, but improvements in terms of forest protection are unlikely to result over the short term.

■ *CLIMATE CHANGE*

An underlying threat which stalks the conservation debate in Asia and globally (and could significantly undermine existing and future conservation efforts) is climate change. In Asia, the most tangible effect has been an increase in the frequency, severity and geographical extent of droughts (and associated forest fires) in the Sundaland region (F07) since the 1960s. It is likely that climate change will have other impacts on Asia's habitats and biodiversity in the medium to long term, but in ways that are far from being fully understood, including the shift of habitats with their characteristic species composition along altitudinal and latitudinal gradients.

Towards sustainable forest management

The problems with forest management in many parts of the Asia region testify to the difficulties in addressing the direct and indirect causes of habitat loss. Ideally, government policy should allow forests (and other natural habitats) to be sustainably managed. However, it should be noted that the real impacts (certainly in terms of environmental and ecological elements) of 'sustainable forest management'

Forest mismanagement in many parts of the Asia region is causing rapid loss of habitat suitable for threatened forest birds.

PHOTO: MARCO LAMBERTINI/BIRDLIFE

(SFM) are still not fully understood. As yey, SFM has not been widely adopted by Asian countries. Ideally, economically, environmentally and socially viable SFM should be pursued in already modified forest, and, in parallel, there should be a total ban on commercial logging of primary areas.

Working towards sustainably managed systems means converting weak forest institutions into strong institutions with a genuine interest in administration and enforcement. It means tackling widespread corruption, and operating within a strong and clear regulatory framework. It means enabling strong public participation in decision-making about natural resources, and making sure that people, particularly indigenous and local communities, have adequate and enforceable rights. These are all aspects of politically stable, well-regulated economies that are resistant to external and internal upheavals. Such conditions will result if many of the underlying causes are tackled.

For forests, another factor is building a constituency of educated and active consumers whose purchasing demonstrates a preference for sustainable forest products. It is the opposite of consumer indifference, and in many markets, consumer choice is beginning to be channelled as a positive force. Increasingly, people are willing to pay a premium—typically 10–20%—for forest products from sustainably managed sources, and the need for an independent system of recognising genuine sources has led to certification. The Forest Stewardship Council (FSC) runs the largest and most advanced scheme. It sets out a series of principles and criteria which apply to all forests, and which are closely allied to concepts of SFM. These are then used to agree national standards, with the participation of all stakeholders, against which forest management units are assessed, certified and labelled. Products carrying an FSC label definitely come from sustainably managed forests (and plantations), and build the confidence of consumers.

Securing an area-based approach to conservation policy in Asia

A continuum exists from undisturbed primary forests, grasslands and wetlands that are rich in biodiversity to habitat that has been converted for monocultivation and from which all but the most determined and resistant species have been excluded. In the case of forest, it is a continuum that sees a sliding scale of endangerment, with species lost as 'management' moves from protection through low-impact logging, to clear felling. It shadows a continuum that starts with a full set of characteristic birds and sees a sequential loss of species as their tolerance of more than limited levels of habitat degradation declines. These species are the barometers, the indicators, of changing forest condition.

For conservation agencies, the continuum starts with the challenge to maintain adequate core areas of natural habitat that shelter a full complement of biodiversity. From a bird conservation angle, this begins with protecting a network of sites that conserves all species of concern. For BirdLife, Important Bird Areas (IBAs: see p.21), are the foundation of this network, but there is a growing trend to adopt similar approaches for other taxa, assembling sites that together form genuine ecological networks. The importance of such areas is central to the objectives of the Convention on Biological Diversity (see p.31), where Article 8 calls on parties to establish a system of protected areas. More recently, Target 9 of the Millennium Development Goals calls upon the global community to: 'Integrate the principles of sustainable development into country policies and programmes and reverse the loss of environmental resources'. Indicators for this target include both the 'proportion of land area covered by forest' (no 25), and 'land area protected to maintain biological diversity' (no 26).

Important Bird Areas can be a major component of the networks, and with the development of other taxa-based evaluations, and the parallel evolution of consensus about priorities, the conservation community needs to lobby for long-term financial resources to maintain these areas. IBAs are chosen for biological reasons, entirely independently of any protection status, and this is a key policy objective, to make it clear that a global ecological network needs to take account of all biodiversity. Protected areas have been the cornerstone of conservation for over a century and good coverage has been achieved, but important gaps remain; the most significant gaps in coverage of globally threatened bird species in Asia, all of them IBAs, are discussed in the relevant forest, grassland and wetland accounts. National protected areas systems in Asia and elsewhere have well-documented weaknesses, linked to insufficient financial and human resources, the failure to integrate protected areas in national and/or local land-use planning, and a lack of participation in management by local government and other stakeholders.

With IBAs now identified in an increasingly large number of countries worldwide, a site-level constituency is growing. Regional directories have been published in Europe, the Middle East, and Africa, and will be available for Asia in 2004, and national directories have been published in more than 50 countries. A key outcome of the IBA process has been high levels of national and local ownership. At many sites this results in individuals and groups emerging who have a strong interest in IBA conservation. Generically known as Site Support Groups (SSGs), these groups are engaging in an increasingly wide range of activities. These include advancing wise resource-use at the local level, engaging in land-use planning and local decision making, building partnerships with other stakeholders including the private sector, and promoting the importance of conserving biodiversity. Often this means helping to strengthen or establish environmental education programmes in schools around sites.

Members of SSGs can also play a major role in monitoring the status of key species and habitats, and particularly the activities of people. Such community-based monitoring of sites leads to the reporting of illegal or destructive activities to the relevant authorities. For site-adjacent communities at many IBAs, marginalisation has meant poor access to work opportunities. As a result, an increasing number of SSGs are starting environmentally friendly conservation-linked projects that help communities generate income (for example, bee-keeping, tree nurseries, ecotourism). At some sites this can extend to providing services, such as assisting researchers and tour-guiding. SSGs are also working with NGOs and government agencies to rehabilitate degraded habitats, for example by tree planting.

In many ways, the growth of this networked site-based constituency has good prospects for conservation success in Asia, both in the short term through tackling direct threats such as illegal logging, and in the longer term by demanding change in those policies that drive the underlying causes of habitat loss. With this constituency increasingly active, SSGs are in a position to provide grassroots communities an opportunity to participate in negotiations and decisions that affect the sites that are their concern.

CONVENTIONS AND RELATED MECHANISMS

A range of international agreements has been established to support the conservation of biological diversity and sustainability in the use of natural resources. The information on threatened birds and their habitats presented in this publication is designed to provide a resource for governments and other decision-makers to implement effectively their obligations under these agreements. The following section introduces the relevant conventions and some related mechanisms, and makes recommendations on how they can benefit the conservation of threatened species in the Asia region. It focuses on agreements that offer direct opportunities to support the conservation of birds and their habitats (see Table 3 for country participation in these agreements). However, note that other conventions, such as the United Nations Framework Convention on Climate Change, will have impacts on species and habitats in the longer term.

Table 3. Participation in international agreements and other mechanisms by countries in the Asia region.

	Convention						
Country	CBD	Ramsar	CITES	CMS	WHC	UNCCD	MAB
Bangladesh	CP–p	CP (2)	CP		CP (3)	CP	NC
Bhutan	CP–C		CP		CP		
Brunei			CP			CP	
Cambodia	CP–C	CP (3)	CP		CP (1)	CP	NC (1)
China	CP–C	CP (21)	CP		CP (28)	CP	NC (22)
India	CP–p	CP (19)	CP	CP	CP (23)	CP	NC (3)
Indonesia	CP–C	CP (2)	CP		CP (6)	CP	NC (6)
Japan	CP–C	CP (13)	CP		CP (11)	CP	NC (4)
North Korea	CP				CP		NC (1)
South Korea	CP–C	CP (2)	CP		CP (7)	CP	NC (2)
Laos	CP–p				CP (2)	CP	
Malaysia	CP–C	CP (4)	CP		CP (2)	CP	NC
Maldives	CP–C				CP	CP	NC
Mongolia	CP–C	CP (6)	CP	CP	CP	CP	NC (4)
Myanmar	CP		CP		CP	CP	NC
Nepal	CP–C	CP (1)	CP		CP (4)	CP	NC
Pakistan	CP–C	CP (19)	CP	CP	CP (6)	CP	NC (1)
Philippines	CP–C	CP (4)	CP	CP	CP (5)	CP	NC (2)
Russia[1]	CP–C	CP (14)	CP		CP (4)	CP	
Singapore	CP–C		CP			CP	
Sri Lanka	CP–C	CP (2)	CP	CP	CP (7)	CP	NC (2)
Thailand	S	CP (10)	CP		CP (4)	CP	NC (4)
Timor-Leste							
Vietnam	CP–C	CP (1)	CP		CP (4)	CP	NC (2)

Key: CBD: Convention on Biological Diversity (CP = Contracting Party; CP–C = Contracting Party, National Biodiversity Strategy and Action Plan (NBSAP) completed; CP–p = NBSAP in preparation; S = signed, but not a Party); Ramsar Convention on Wetlands (CP = Contracting Party; figures in brackets are the number of Ramsar sites at July 2003); CITES: Convention on International Trade in Endangered Species (CP = Contracting Party); CMS: Convention on Migratory Species (CP = Contracting Party); WHC: World Heritage Convention (CP = Contracting Party Committee; figures in brackets are the number of World Heritage Sites at July 2002).; UNCCD: United Nations Convention to Combat Desertification (CP = Contracting Party); MAB: Man and the Biosphere Programme (NC = National Committee; figures in brackets are the number of Biosphere Reserves at November 2002).

[1] The numbers of Ramsar and World Heritage sites given for Russia only cover those in the Asia region (14 of the 35 Ramsar sites and four of the 16 World Heritage Sites).

International agreements and other mechanisms

CONVENTION ON BIOLOGICAL DIVERSITY (CBD)

The Convention on Biological Diversity was adopted in 1992 and came into force in 1993. It currently (July 2003) has 187 parties, amongst them 21 in the Asia region. The objectives of the convention are the conservation of biological diversity, the sustainable use of its components, and the fair and equitable sharing of the benefits arising out of the utilisation of genetic resources. The primary approach of the convention is conservation in the wild. Parties have to identify components of biodiversity, such as threatened species, and ecosystems and habitats containing high diversity, large numbers of endemic or threatened species, or wilderness. Article 8 of the CBD requests parties to establish a system of protected areas, to restore degraded ecosystems, to maintain viable populations of species in natural surroundings, and to develop or maintain necessary legislation and/or other regulatory provisions for the protection of threatened species and populations. The convention has established thematic work programmes on the major ecosystems: marine and coastal biodiversity; forests; inland water ecosystems; dry and sub-humid lands; and agricultural biodiversity. The main tools for the CBD's national implementation are National Biodiversity Strategies and Action Plans (NBSAPs) which in many cases have been established with support from the Global Environment Facility (GEF), which hosts the CBD's financial mechanism. Most Asian countries have developed NBSAPs and are in the process of implementing them, while some are revising their first NBSAPs.

Website: http://www.biodiv.org/

Recommendations

- Countries in the Asia region that are not a party to the CBD (Brunei, Thailand and Timor-Leste) should accede to it.
- Include national lists of (and information on) threatened species and Important Bird Areas (IBAs: see p.21) in NBSAPs.
- Develop national and regional action plans for threatened species within the framework of NBSAPs.
- Review national protected area systems to ensure that they adequately cover threatened species and their habitats.
- Review, and if necessary strengthen, national legislation and/or other regulatory provisions for the protection of threatened species.
- Develop national policies and programmes for the conservation and sustainable use of important habitats for threatened species, supporting the implementation of the CBD's thematic work programmes on the major ecosystems, e.g. activities relevant to forest policy identified by the CBD Work Programme on Forest Biological Diversity.
- Identify gaps in the conservation of threatened species and their habitats and key sites, and hence funding opportunities through the GEF and other international donors (note that only a small proportion of the 311 outstanding IBAs for threatened bird species in Asia are currently receiving funding from these sources).

■ *RAMSAR CONVENTION ON WETLANDS*

The Ramsar Convention on Wetlands of International Importance especially as Waterfowl Habitat, adopted in 1971, entered into force in 1975 and currently (July 2003) has 137 parties, 16 of which are from the Asia region. The convention provides a framework for international cooperation for the conservation and wise use of wetlands. Parties are to designate suitable wetlands for inclusion in the List of Wetlands of International Importance, also known as Ramsar sites, to formulate and implement their planning so as to promote the conservation of wetlands included in the List and the wise use of all wetlands in their territory. At July 2003, 123 Ramsar sites had been designated in the Asia region. As many threatened birds heavily depend on wetlands, the Ramsar Convention is an excellent tool for their protection. A number of Asian wetlands of significance to threatened birds have already been designated as Ramsar sites, and 'wise use' approaches, for example in river basin management, are crucial to the survival of many species and communities. For a comprehensive approach to the national implementation of the convention, many countries have developed National Wetland Policies.

Website: http://www.ramsar.org/

Recommendations

- Countries in the Asia region that are not a party to the Ramsar Convention (Bhutan, Brunei, North Korea, Laos, the Maldives, Myanmar, Singapore and Timor-Leste) should accede to it.
- This review has identified 159 IBAs as outstanding for threatened waterbirds, of which 40 (25%) are currently designated as Ramsar sites (see regions F01, F07, G02, G03, W02, W03, W04, W05, W06, W08, W09, W10, W11, W12, W13, W14, W15, W18, W19 and W20). The remaining 119 wetland IBAs, all of which are likely to qualify under the Ramsar criteria, should be considered for designation as Ramsar sites (other than 11 in countries that are not yet a party to the Convention). Note that BirdLife International will publish a comprehensive Asia IBA Inventory in early 2004; many of the wetland sites included are likely to qualify for designation as Ramsar sites.
- National legislation and institutions should be reviewed for compatibility with good practice in respect of wetlands (using Ramsar guidelines).
- All countries should have a national wetland policy or equivalent, which includes activities for the protection of threatened waterbird species.
- Each significant wetland site should have a management plan, with defined conservation objectives and targets.
- Projects, programmes and policies likely to have significant negative effects on important wetland interests should be subject to environmental and/or strategic impact assessment.
- Wetlands in general should be managed in accordance with the 'wise use' principles of the Ramsar Convention, including environmentally sustainable management of their catchment/watershed/river basin context and of water resources, and including coordinated approaches in transboundary areas.
- All countries should have a national system for monitoring the ecological character of their wetlands, including defined policy responses to potential change.

CONVENTION ON INTERNATIONAL TRADE IN ENDANGERED SPECIES (CITES)

The Convention on International Trade in Endangered Species of Wild Fauna and Flora (CITES), adopted in 1973, came into force in 1975 and currently (July 2003) has 161 parties, amongst them 20 from the Asia region. The convention was established to ensure that trade in wildlife and wildlife products is managed sustainably. It aims to regulate international trade in wildlife products through international cooperation, while recognising national sovereignty over wildlife resources. Species on Appendix I are considered to be threatened with extinction and cannot be traded commercially, while those on Appendix II can only enter international trade under specific controlled circumstances. Many of the 303 Asian threatened species included in the current review are covered by the appendices (62 on Appendix I and 62 on Appendix II: see threatened species list in appendix, pp.241–245), including all raptors, parrots and owls. Trade (either international and/or domestic) is known to be an issue for four threatened species in the Asia region which are not listed on the CITES appendices (see Table 4).

Website: http://www.cites.org/

Recommendations

- Countries in the Asia region that are not a party to CITES (North Korea, Laos, the Maldives and Timor-Leste) should accede to it.
- Conduct a review of those threatened bird species in Asia where international trade is considered possibly to be significant, and which are not at present listed on any CITES appendix (see Table 4), to identify those which might benefit from listing. The situation of Black-winged Starling needs to be urgently assessed, as trade is believed to be the primary threat to this Endangered species.
- Conduct (as a matter of extreme urgency) a review of all threatened bird species on Appendix I where international trade is, or may be, continuing (see Table 5), in order to identify ways to strengthen the enforcement of CITES and to improve national bird trade legislation and its enforcement in the species' range states.
- Conduct a review (possibly including a 'Significant Trade Review') of threatened bird species on Appendix II where trade is considered to be significant (see Table 6), in order to determine appropriate actions such as the establishment of monitoring programmes, the reduction of quotas, or consideration for listing on Appendix I.

Trapping for the bird trade is reducing wild populations of many threatened birds in Asia, and needs to be controlled by improved enforcement of CITES and national bird trade legislation.

PHOTO: MARCO LAMBERTINI/BIRDLIFE

Table 4. Threatened bird species in Asia not listed on a CITES appendix but where trade is considered to be a significant issue.

Species	Status
Storm's Stork *Ciconia stormi*	EN
Greater Adjutant *Leptoptilos dubius*	EN
Timor Sparrow *Padda fuscata*	VU
Black-winged Starling *Sturnus melanopterus*	EN

Table 5. Threatened bird species in Asia on CITES Appendix I in which trade is considered to be continuing.

Species	Status
Palawan Peacock-pheasant *Polyplectron emphanum*	VU
Red-and-blue Lory *Eos histrio*	EN
Salmon-crested Cockatoo *Cacatua moluccensis*	VU
Philippine Cockatoo *Cacatua haematuropygia*	CR

Table 6. Threatened bird species in Asia on CITES Appendix II in which trade is considered to be continuing in significant numbers.

Species	Status
Bornean Peacock-pheasant *Polyplectron schleiermacheri*	EN
Green Peafowl *Pavo muticus*	VU
Chattering Lory *Lorius garrulus*	EN
Yellow-crested Cockatoo *Cacatua sulphurea*	CR
White Cockatoo *Cacatua alba*	VU
Blue-headed Racquet-tail *Prioniturus platenae*	VU
Green Racquet-tail *Prioniturus luconensis*	VU
Straw-headed Bulbul *Pycnonotus zeylanicus*	VU
Omei Shan Liocichla *Liocichla omeiensis*	VU
Green Avadavat *Amandava formosa*	VU
Java Sparrow *Padda oryzivora*	VU

- Review national legislation to ensure that it is adequate to address international and domestic trade in threatened birds, particularly the species listed in Tables 4, 5 and 6.
- Develop mechanisms to effectively control and monitor trade in wild fauna and flora at the national level by, amongst others, strengthening CITES Scientific and Management Authorities and training of customs officers and police. In particular, controls on the trading of Philippine Cockatoo need to be greatly improved in the Philippines (see region F09).
- Raise public awareness of the relevant legislation and the detrimental consequences of non-sustainable taking of and trading in wildlife.

CONVENTION ON MIGRATORY SPECIES (CMS)

The Convention on the Conservation of Migratory Species of Wild Animals (CMS), often known as the Bonn Convention, was adopted in 1979 and entered into force in 1983. At July 2003, it has 83 parties, five of which are in the Asia region. The objective of the convention is to protect migratory species, recognising that protection is needed throughout their migratory ranges, and that this requires international cooperation and action. Two appendices are attached to the Convention. Appendix I lists species which are in danger of extinction throughout all or a significant portion of their range. Appendix II species need not be threatened, but they must have a conservation status which requires, or would benefit from, an international agreement for their conservation. If the circumstances so require, a species can appear on both appendices. Parties are required to prohibit the taking of species on Appendix I, and to conclude agreements with other range states for the conservation and management of species on Appendix II. Parties must also endeavour to conserve and restore important habitats, eliminate impeding activities or obstacles to migration, and tackle other factors that endanger Appendix I species. For Appendix II species, among other things agreements should provide for a network of suitable areas of habitat along their migration routes. Some Asian countries participate in CMS agreements, but are not yet parties to the convention, including Russia and China in an agreement between several Siberian Crane range states. Twenty-seven threatened bird species in the Asia region are on Appendix I of CMS (including 18 which are also on Appendix II), and 15 are on Appendix II (see species list in appendix).

Website: http://www.wcmc.org.uk/cms/

Recommendations

- Only five countries in the Asia region are a party to the CMS (India, Mongolia, Pakistan, the Philippines and Sri Lanka), and more should accede to it.
- Ensure that the appendices correctly reflect the status and needs of Asian migratory bird species by adding species to one or both appendices as appropriate, notably the five species in Table 7. For some species, the accession of one or more relevant Range States will be needed first, in particular to ensure the development of multinational agreements for these species.
- Ensure that all Appendix I species are fully legally protected in all range states that are a party to the CMS.
- Conserve key habitats and sites for Appendix I species, using information on IBAs.
- Develop agreements between range states for the conservation and management of single species or groups of species on CMS Appendix II (e.g. Asian populations of Great Bustard).

Table 7. Threatened migratory birds in the Asia region not listed on a CMS appendix.

Species	Asia region range states
Swinhoe's Rail *Coturnicops exquisitus*	Russia, **Mongolia**, Japan, South Korea, China
Pale-backed Pigeon *Columba eversmanni*	Russia, China, **Pakistan**, **India** (and Central Asia)
Fairy Pitta *Pitta nympha*	Japan, North Korea, South Korea, China (mainland China, Hong Kong, Taiwan), Vietnam, Malaysia, Brunei, Indonesia
Yellow Bunting *Emberiza sulphurata*	Japan, South Korea, China (mainland China, Hong Kong, Taiwan), **Philippines**
Silver Oriole *Oriolus mellianus*	China (mainland China), Thailand, Cambodia

Range states in bold are a party to the CMS.

- Promote existing agreements in the region, in particular the Agreement on the Conservation of Albatrosses and Petrels (ACAP).
- Promote the conservation of South Asian waterbirds by either the extension of the African-Eurasian Waterbird Agreement (AEWA), or the negotiation of a separate agreement.

WORLD HERITAGE CONVENTION

The Convention Concerning the Protection of the World Cultural and Natural Heritage (World Heritage Convention) was adopted in 1972 and entered into force in 1975. At July 2003, it has 176 contracting parties, including 21 in the Asia region. The United Nations Educational, Scientific and Cultural Organisation (UNESCO) provides a secretariat for the convention in the World Heritage Centre in Paris. The aim of the convention is to identify and conserve cultural and natural monuments and sites of outstanding universal value, and this is implemented through the nomination of suitable sites by national governments; as of July 2002, 117 World Heritage Sites had been designated in the Asia region. The great majority of these sites are nominated for their cultural value, as is the case in other regions of the world; the convention has recognised this imbalance, and wishes to see more outstanding natural sitess added. The World Heritage Committee selects World Heritage Sites from among the national nominations, allocates resources for their management, and identifies threatened sites for the List of World Heritage in Danger.

Website: http://whc.unesco.org

Recommendations

- Countries in the Asia region that are not a party to the World Heritage Convention (Brunei, Singapore and Timor-Leste) should accede to it.
- This review has identified 311 IBAs as outstanding for threatened birds, of which 14 (4.5%) are in whole or in part World Heritage Sites (see regions F01, F03, F04, F05, F07, F08, F09, G02, W12 and W15). Many of the other IBAs undoubtedly qualify as World Heritage Sites but are not so designated. Realistic, but challenging, targets should be set gradually to add these sites, and help to redress the imbalance between cultural and natural sites.
- Use information collated during the IBA Programme to help identify threatened World Heritage Sites, for inclusion on the List of World Heritage in Danger, and to help counter the threats.

UNITED NATIONS CONVENTION TO COMBAT DESERTIFICATION (UNCCD)

The United Nations Convention to Combat Desertification (UNCCD) was adopted in 1994 and came into force in 1996. Alongside the Convention on Biological Diversity (CBD) and the Climate Change Convention (UNFCCC), it is one of the Rio conventions that were endorsed by the Rio Earth Summit in 1992. It seeks active cooperation with other conventions, such as the CBD, UNFCCC, Ramsar and CMS. The convention currently (July 2003) has 187 parties, of which 21 are in the Asia region. The objective of the convention is to combat desertification and to mitigate the effects of drought. Desertification is understood as land degradation in arid, semi-arid and sub-humid areas due to climate change and human activities such as deforestation, inadequate agriculture, overgrazing and others; the convention is therefore relevant to many parts of the Asia region, not only deserts. Regional annexes outline the specific obligations for countries to address desertification, with Annex II covering regional implementation in Asia. The main implementation tools are regional and national action programmes. It is expected that GEF's designation of land degradation as a focal area in 2002 will result in a stronger national implementation of the Convention.

Deforestation in the tropics can cause local climate change and water shortage, and should be addressed under the UNCCD through regional and national action programmes.

PHOTO: MICHAEL POULSEN/BIRDLIFE

Website: http://www.unccd.int

Recommendations

- Countries in the Asia region that are not a party to the UNCCD (Bhutan, North Korea and Timor-Leste) should accede to it.
- Develop regional and national action programmes in parts of Asia where habitat clearance and degradation has been sufficiently extensive to cause local climate change and water shortage, including parts of many forest regions (F01–F09) and grassland regions (G01–G03).
- Fully address biodiversity conservation in the development and implementation of action programmes, using information on threatened species and their habitats.
- Ensure that activities to combat desertification, such as reforestation and afforestation, do not negatively impact on threatened species and their habitats, with strategic and/or environmental impact assessments of any large schemes.

MAN AND THE BIOSPHERE PROGRAMME

UNESCO's Man and the Biosphere (MAB) Programme was launched in 1971 and develops the basis, within the natural and social sciences, for the conservation and sustainable use of biological diversity, and for the improvement of the relationship between people and their environment. The programme encourages interdisciplinary research, demonstration and training in natural resource management. An essential tool for the MAB programme is the network of Biosphere Reserves, which are areas of terrestrial and coastal ecosystems where solutions are promoted to reconcile biodiversity conservation with its sustainable use. MAB operates through National Committees and Focal Points amongst the UNESCO member states, and there are National Committees in 18 countries in the Asia region; at November 2002, 54 Biosphere Reserves had been designated in these countries (see Table 3).

Website: http://www.unesco.org/mab/

Recommendations
- Establish National MAB Committees in additional countries in the Asia region (Bhutan, Brunei, Laos, Russia, Singapore and Timor-Leste).
- This review has identified 311 IBAs as outstanding for threatened birds, of which 22 (7.1%) are in whole or in part Biosphere Reserves (see regions F03, F04, F05, F06, F07, F08, F09, W05, W06, W11, W15 and W18). Many of the other IBAs are undoubtedly appropriate for listing as Biosphere Reserves, and more of them should be designated.

WORLD COMMISSION ON DAMS
The World Commission on Dams was initiated by the World Bank and the World Conservation Union (IUCN). The Commission brought together governments, industry, investors, non-governmental organisations and indigenous peoples to thoroughly analyse the economic, social and ecological aspects of large dams, resulting in the publication of a report in 2000 which contains recommendations for their future planning. The United Nations Environment Programme is administrating the Dams and Development Project to implement these recommendations. Of the 45,000 large dams that have been constructed globally, more than 25,000 are in the Asia region. Clearly, threatened species have already been affected, and will continue to be affected, by the flooding of their habitats, the activities associated with the construction of large dams, and by changes in ecology downstream from the dams.

Website: http://www.dams.org/

Recommendations
- Carefully assess the environmental impact of proposed dams (and associated irrigation schemes), using the recommendations of the World Commission on Dams and information on threatened birds and their habitats. Some examples of where proposed dams could negatively affect threatened birds are: the Mekong catchment, especially in Yunnan (China) and Laos (see regions F06 and W18); the lower Ganges and Brahmaputra floodplains in north-east India and Bangladesh (see region W14); the Amur valley in Russia and China (see region W03); and semi-deserts in north-west India (see region G03).
- Operate existing dams to take into account the ecological needs of threatened birds and other biodiversity, particularly species which inhabit river corridors and floodplains downstream. Some examples of where the management of existing dams should take into account the needs of threatened birds are: the Indus valley in Pakistan (see regions G02 and W11); the Gangetic plains of northern India (see regions G02 and W12); and the Yangtze basin in China (especially the Three Gorges Dam: see regions W06 and W08).

Regional agreements and other mechanisms

ASIA-PACIFIC MIGRATORY WATERBIRD CONSERVATION STRATEGY
The Asia-Pacific Migratory Waterbird Conservation Strategy promotes the conservation of migratory waterbirds and wetlands in the Asia-Pacific region. It is supported by the governments of Japan and Australia and coordinated by Wetlands International, with an international conservation committee to oversee its promotion and implementation. The Strategy has greatly increased awareness in the region of the need to protect migratory waterbirds and their habitats, and has led to a series of initiatives undertaken with the active support and involvement of governments, conventions, national and international NGOs, development agencies, the corporate sector and local communities. Regional conservation Action Plans have been developed and are being implemented for shorebirds, cranes and Anatidae. Networks of sites of international importance have been established (under the 'Brisbane Initiative') for these three waterbird groups (Anatidae Site Network in the East Asian Flyway, East Asian Australasian Shorebird Site Network and North East Asian Crane Site Network) that presently involves 74 sites in 12 countries (at June 2003: see regions F01, W01, W02, W03, W04, W05, W06, W08, W09 and W10); new sites are being added each year. Numerous activities have been undertaken at network sites, including public awareness and education, surveys and training courses in wetland management. International and national meetings have been held to share information and skills relevant to wetland management.

The Yellow River delta, a waterbird network site established under the Asia-Pacific Migratory Waterbird Conservation Strategy.

PHOTO: SIMBA CHAN

Website: http://www.wetlands.org/IWC/awc/waterbirdstrategy/Strat.htm

Recommendations
- Ensure that all IBAs which support important populations of threatened shorebirds, cranes and Anatidae are included in the regional site networks.
- Conduct further education and training activities at network sites, to improve awareness of threatened birds and their conservation needs, and to strengthen management of their wetland habitats.

CONSERVATION OF ARCTIC FAUNA AND FLORA (CAFF)
The Conservation of Arctic Fauna and Flora (CAFF) is a working group of the Arctic Council, with a mission to conserve Arctic biodiversity and to ensure that the use of Arctic living resources is sustainable. In the Asia region, Russia is a member of the Arctic Council. Of particular relevance to the conservation of threatened birds is CAFF's Circumpolar Protected Areas Network (CPAN) Strategy and Action Plan that was adopted by the Arctic Countries in 1996, which aims to develop and link the national systems of protected areas into a comprehensive circumpolar protected areas network.

Website: CAFF: http://www.caff.is/

Recommendations

- ➤ Use information on key sites for threatened bird species and IBAs to help develop the Circumpolar Protected Areas Network.

ASEAN AGREEMENT ON THE CONSERVATION OF NATURE AND NATURAL RESOURCES

The ASEAN Agreement on the Conservation of Nature and Natural Resources was adopted in 1985 and has been signed by Brunei, Indonesia, Malaysia, the Philippines, Singapore and Thailand, although it has not yet entered into force. It covers a broad range of conservation and development issues, including the conservation of threatened and endemic species and their habitats.

Website: ASEAN agreement: http://www.aseansec.org/6080.htm

Recommendations

- ➤ Review and ratify the ASEAN Agreement on the Conservation of Nature and Natural Resources, and use information on threatened bird species and IBAs to help in its implementation.

Bilateral migratory bird agreements/treaties

Ten bilateral migratory bird agreements/treaties involving territories in the Asia region have already been established (Table 8), and others are under discussion. These agreements provide mechanisms for the promotion of bilateral and international actions for the conservation of migratory birds.

Website: http://www.wetlands.org/IWC/awc/waterbirdstrategy/Strat.htm

Recommendations

- ➤ Implement international action programmes for selected threatened migratory waterbirds, including using information and recommendations from the *Threatened birds of Asia: the BirdLife International Red Data Book* and the current review.
- ➤ Consider further bilateral agreements where these would benefit the conservation of migratory birds.

Table 8. Bilateral agreements/treaties on the conservation of migratory birds in the Asia-Pacific region.

	Australia	China	India	Japan	North Korea	South Korea	Russia	U.S.A.
Australia								
China	Yes							
India								
Japan	Yes	Yes						
North Korea								
South Korea								
Russia			Yes	Yes	Yes	Yes		
U.S.A.		Yes		Yes			Yes	

Table reproduced with permission from the Asia-Pacific Migratory Waterbird Conservation Strategy: 2001–2005.

Okinawa Woodpecker is confined to a single IBA in southern Japan, Yambaru on Okinawa, where its habitat is under pressure from forestry and development.

PHOTO: TAKUKI HANASHIRO

PRIORITIES TO PREVENT THE EXTINCTION OF ASIAN BIRDS

The 324 globally threatened bird species that occur in the Asia region include some that have a very high risk of extinction, the 41 that are Critical and the 66 that are Endangered[1]. The conservation actions required for the protection of these birds and their habitats are therefore of the highest priority, and require immediate action. The following analysis focuses on these highly threatened species, to identify the pressures that could lead to their extinction in the near future[2]. Limited information is given here on the conservation measures required to address these issues; for more details go to the *Conservation issues and strategic solutions* sections of the relevant accounts (the codes for these are given in the Tables below: F01–F09 refer to forests, G01–G03 to grasslands, W01–W20 to wetlands and S01 to seabirds).

[1] According to the *IUCN Red List Categories and Criteria* (IUCN 2001), Critical species face an 'extremely high risk of extinction in the wild' (e.g. population viability analysis giving a probability of extinction of at least 50% within 10 years or three generations, whichever is the longer) and Endangered species a 'very high risk' (probability of extinction at least 20% within 20 years or five generations, whichever is the longer). Note that five of these species are excluded from the following analysis, two Critical seabirds because they nest (and face threats) outside the Asia region on Christmas Island, and three Endangered species which are endemic to Biak Island in Irian Jaya (which is not covered in this book).

[2] Note that many of the pressures on Critical and Endangered birds covered below are also affecting groups of Vulnerable species, and in the long-term could lead to them being upgraded to a higher category of threat.

Critical habitats and sites

Habitat loss is the major cause of endangerment in birds in the Asia region. Most Critical and Endangered bird species have restricted ranges, and/or are specialised to a particular habitat type, and they often occur in the areas and habitats that are being most rapidly cleared and modified by human activities. Table 9 includes all forest, grassland and wetland regions that support significant populations of one or more Critical and/or Endangered species. These birds are listed for each region, together with the main land-use issues that are impacting on their habitats. The tropical forests of continental South-East Asia, Indonesia and the Philippines are clearly of primary importance for highly threatened bird species, and immediate and effective conservation actions are needed if mass extinctions of species are to be averted. However, there are species and conservation issues in many other parts of the Asia region that require urgent attention.

Table 9. Key habitat conservation issues for Critical and Endangered bird species in Asia.

Habitat region	Issues	Species affected (Critical and Endangered only)
F01 Boreal and northern temperate forests	Forestry; Development	Scaly-sided Merganser, Blakiston's Fish-owl
F02 Japanese forests	Development; Forestry; Introduced predators	Japanese Night-heron; Okinawa Rail, Okinawa Woodpecker, Amami Thrush
F03 South-east Chinese forests	Forestry and illegal logging	White-eared Night-heron, Sichuan Partridge
F04 Sino-Himalayan mountain forests	Shifting cultivation; Exploitation of forest products	White-browed Nuthatch
F05 Indian peninsula and Sri Lankan forests	Forestry and illegal logging; Conversion to agriculture and plantations; Development	Forest Owlet, Sri Lanka Whistling-thrush, Rufous-breasted Laughingthrush
F06 Indo-Burmese forests	Forestry and illegal logging; Conversion to agriculture and plantations; Shifting agriculture; Exploitation of forest products; Development	White-bellied Heron, White-winged Duck, Orange-necked Partridge, Chestnut-headed Partridge, Edwards's Pheasant, Vietnamese Pheasant, Collared Laughingthrush, Grey-crowned Crocias
F07 Sundaland forests	Forestry and illegal logging; Pulp and paper industry; Conversion to plantations (especially oil palm); Conversion to agriculture; Forest fires; Development; Mining; Transmigration	Storm's Stork, White-shouldered Ibis, White-winged Duck, Javan Hawk-eagle, Bornean Peacock-pheasant, Silvery Wood-pigeon, Sumatran Ground-cuckoo, Gurney's Pitta, Rueck's Blue-flycatcher, Black-winged Starling, Bali Starling
F08 Wallacea	Forestry and illegal logging; Exploitation of forest products; Conversion to agriculture and plantations; Livestock grazing and fire; Transmigration; Development (including mining)	Maleo, Talaud Rail, Moluccan Woodcock, Wetar Ground-dove, Timor Green-pigeon, Timor Imperial-pigeon, Red-and-blue Lory, Chattering Lory, Blue-fronted Lorikeet, Yellow-crested Cockatoo, Sangihe Hanging-parrot, Flores Hanging-parrot, Taliabu Masked-owl, Siau Scops-owl, Flores Scops-owl, Lompobatang Flycatcher, Matinan Flycatcher, Caerulean Paradise-flycatcher, Flores Monarch, White-tipped Monarch, Black-chinned Monarch, Sangihe Shrike-thrush, Elegant Sunbird, Sangihe White-eye, Rufous-throated White-eye, Banggai Crow, Flores Crow
F09 Philippine forests	Conversion to agriculture; Forestry and illegal logging; Mining; Development	Japanese Night-heron (nb), Philippine Eagle, Mindoro Bleeding-heart, Negros Bleeding-heart, Mindanao Bleeding-heart, Sulu Bleeding-heart, Tawitawi Brown-dove, Negros Fruit-dove, Philippine Cockatoo, Blue-winged Racquet-tail, Black-hooded Coucal, Sulu Hornbill, Mindoro Tarictic, Visayan Tarictic, Visayan Wrinkled Hornbill, Streak-breasted Bulbul, Black Shama, Flame-templed Babbler, Negros Striped-babbler, White-throated Jungle-flycatcher, Cebu Flowerpecker, Isabela Oriole
G02 Indo-Gangetic grasslands	Conversion to agriculture and plantations; Grass harvesting; Livestock grazing; burning	Bengal Florican
G03 South Asian arid habitats	Irrigation and conversion to agriculture; Agricultural intensification; Livestock grazing; Grassland management	Great Indian Bustard, Lesser Florican, Jerdon's Courser
W02 Sea of Okhotsk and Sea of Japan coasts	Development	Crested Shelduck, Red-crowned Crane, Spotted Greenshank
W03 Amur, Ussuri and Sungari river basins	Wetland drainage; Development; Cutting of nesting trees	Oriental Stork, Red-crowned Crane
W06 Yellow Sea coast	Coastal reclamation; Potential development of the Demilitarised Zone (DMZ) in Korea	Black-faced Spoonbill, Swan Goose (nb), Red-crowned Crane (nb), Spotted Greenshank (nb)
W07 Central Chinese wetlands	Wetland conversion and agricultural change; Pollution/pesticides	Crested Ibis
W08 Lower Yangtze basin	Wetland conversion and agricultural change; Changes in river flow (caused by the Three Gorges Dam)	Oriental Stork (nb), Swan Goose (nb), Siberian Crane (nb)
W10 China Sea coast	Coastal reclamation; Conversion to aquaculture	Black-faced Spoonbill (nb), Spotted Greenshank (nb)
W14 Assam and Sylhet plains	Conversion to agriculture; Cutting of nesting trees	Greater Adjutant
W15 Bay of Bengal coast	Coastal reclamation; Conversion to aquaculture	Spotted Greenshank (nb)
W18 Lower Mekong basin	Conversion to agriculture; Cutting of nesting trees; Development (including dams)	Greater Adjutant, White-shouldered Ibis, Giant Ibis, Bengal Florican

Key: Species that occur in key areas only as non-breeding visitors are marked '(nb)'.

Table 10. Key sites for Critical and Endangered bird species in Asia.

Important Bird Area	Status	Species affected (Critical and Endangered only)
Bikin river basin, Primorye, Russia (F01)	—	Scaly-sided Merganser, Blakiston's Fish-owl
Central Amami forests, Japan (F02)	(PA)	**Amami Thrush**
Yambaru, northern Okinawa, Japan (F02)	(PA)	**Okinawa Rail**, **Okinawa Woodpecker**
Natma Taung National Park, Myanmar (F04)	PA	White-browed Nuthatch
Toranmal Reserve Forest (Shahada), Maharashtra, India (F05)	—	Forest Owlet
Ke Go and Khe Net, Vietnam (F06)	(PA)	**Vietnamese Pheasant**
Phong Dien and Dakrong, Vietnam (F06)	(PA)	Edwards's Pheasant
Chu Yang Sin National Park, Vietnam (F06)	PA	Collared Laughingthrush, Grey-crowned Crocias
Cat Tien National Park, Vietnam (F06)	PA	Orange-necked Partridge
Khao Nor Chuchi, Thailand (F07)	(PA)	Gurney's Pitta
Southern Tanintharyi Division, Myanmar (F07)	—	Gurney's Pitta
Bali Barat National Park, Bali, Indonesia (F07)	PA	**Bali Starling**
Mbeliling, Flores, Indonesia (F08)	—	Flores Hanging-parrot, Flores Monarch, Flores Crow
Ruteng Nature Recreation Park, Indonesia (F08)	PA	Flores Scops-owl
Paitchau-Iralalora, Timor-Leste (F08)	—	Yellow-crested Cockatoo, Timor Green-pigeon, Timor Imperial-pigeon, probably Wetar Ground-dove
Karakelang Hunting Reserve, Talaud, Indonesia (F08)	PA	**Talaud Rail**, Red-and-blue Lory
Gunung Sahendaruman, Sangihe, Indonesia (F08)	—	Sangihe Hanging-parrot, **Caerulean Paradise-flycatcher**, **Sangihe Shrike-thrush**, Elegant Sunbird, **Sangihe White-eye**
Pulau Siau, Indonesia (F08)	—	**Siau Scops-owl**
Bogani Nani Watabone National Park, Sulawesi, Indonesia (F08)	PA	Maleo, Matinan Flycatcher
Lompobatang Protection Forest, Sulawesi, Indonesia (F08)	PA	**Lompobatang Flycatcher**
Taliabu proposed nature reserve, Indonesia (F08)	—	Taliabu Masked-owl
Tanahjampea, Indonesia (F08)	—	**White-tipped Monarch**
Pulau Boano, Indonesia (F08)	—	**Black-chinned Monarch**
Kapalat Mada, Buru, Indonesia (F08)	—	**Rufous-throated White-eye**, possibly Blue-fronted Lorikeet
Siburan, Mindoro, Philippines (F09)	—	Mindoro Bleeding-heart, Black-hooded Coucal, Mindoro Tarictic
Northern Sierra Madre Nature Park, Luzon, Philippines (F09)	PA	Japanese Night-heron (nb), Philippine Eagle, Isabela Oriole
North-west Panay (Pandan peninsula), Philippines (F09)	—	Negros Bleeding-heart, Visayan Tarictic, White-throated Jungle-flycatcher
Central Panay Mountains, Philippines (F09)	—	Visayan Tarictic, Visayan Wrinkled Hornbill, Flame-templed Babbler, White-throated Jungle-flycatcher
Mt Canlaon National Park, Negros, Philippines (F09)	PA	**Negros Fruit-dove**, Visayan Tarictic, Flame-templed Babbler
Cuernos de Negros, Negros, Philippines (F09)	—	Visayan Tarictic, Visayan Wrinkled Hornbill, Flame-templed Babbler, **Negros Striped-babbler**
Tabunan, Cebu, Philippines (F09)	PA	Streak-breasted Bulbul, Black Shama, Cebu Flowerpecker
Mt Kaluayan-Mt Kinabalian complex, Mindanao, Philippines (F09)	—	Japanese Night-heron (nb), Philippine Eagle, Mindanao Bleeding-heart
Tawitawi island, Philippines (F09)	—	**Sulu Bleeding-heart**, **Tawitawi Brown-dove**, Philippine Cockatoo, Blue-winged Racquet-tail, Sulu Hornbill
Sri Lankamalleswara Wildlife Sanctuary, Andhra Pradesh, India (G03)	PA	Jerdon's Courser
Yancheng Nature Reserve, Jiangsu, China (W06)	PA	Swan Goose (nb), Red-crowned Crane (nb)
Yang Xian county, Shaanxi, China (W07)	(PA)	**Crested Ibis**
Poyang Hu lake, Jiangxi, China (W08)	(PA)	Swan Goose (nb), **Siberian Crane** (nb)
Tsengwen estuary, Taiwan (W10)	—	Black-faced Spoonbill (nb)
Chhep, Cambodia (W18)	(PA)	Greater Adjutant, Giant Ibis
Western Siem Pang, Cambodia (W18)	—	White-shouldered Ibis, Giant Ibis
Mazu (Matzu) Dao islands (S01)	PA	**Chinese Crested-tern**

Key: PA = IBA is a protected area; (PA) = IBA partially protected; — = IBA unprotected.
Species which are entirely or largely confined to a single IBA are shown in bold, and those which occur in IBAs only as non-breeding visitors are marked '(nb)'

The ranges of some Critical or Endangered birds are naturally very small, or their habitats have been reduced to just a few fragments, and they are now confined to a small number of IBAs. Table 10 identifies 41 IBAs that are crucial for the survival of at least 66 highly threatened bird species[1]. Only 22 of these are protected areas, or partially protected, and the establishment of new reserves should be considered at the other IBAs, or other mechanisms such as protection through land-use planning. Some of the Critical and Endangered species are entirely or largely confined to a single IBA, meaning that habitat loss or degradation at a single site could cause their extinction. Particularly notable IBAs include Gunung Sahendaruman on Sangihe in Indonesia, which supports five highly threatened species of which three are unique to the site, and Yambaru (northern Okinawa) in Japan and Tawitawi in the Philippines, both of which have two unique birds.

[1] Note that additional IBAs for some of these species are given in the forest, grassland and wetland accounts in this book, and more will be identified in the directory of *Important Bird Areas in Asia* which the BirdLife Asia network is scheduled to publish in early 2004.

Bali Starling is on the brink of extinction because of illegal trapping for the wild bird trade.

PHOTO: MARK EDWARDS/BIRDLIFE

Exploitation of birds

Exploitation is considered to be a significant threat to the 10 Critical and 25 Endangered species listed in Table 11. Of these, 23 species are heavily hunted for food and/or sport, but in most cases there is insufficient information available on capture rates and bird populations to judge whether such exploitation might be sustainable, or whether it is causing population declines. Two species are affected by both commercial collection of their eggs and hunting. Seven species are captured in significant numbers (in relation to the size of their global populations) for the wild bird trade, and three species are affecting by both hunting and capture for the wild bird trade.

Unsustainable exploitation is believed to be the principal cause of endangerment of several Critical and Endangered species. Large-scale commercial hunting of Swan Goose in the lower Yangtze basin in China is almost certainly the main reason for the species's rapid decline. Commercial egg collection is the most serious threat to Maleo on Sulawesi, and has caused abandonment of, or serious declines at, many of its nesting colonies, although habitat loss and hunting are also important factors. The eggs and chicks of Greater Adjutant and other waterbirds are subject to large-scale commercial collection at Tonle Sap lake in Cambodia, but projects are now underway to try to control this activity. Trapping for the wild bird trade is the main pressure on Philippine Cockatoo, and several Indonesian species, including Red-and-blue Lory, Chattering Lory, Yellow-crested Cockatoo, Black-winged Starling and Bali Starling. All of these species are listed on Appendix I or Appendix II of CITES, other than Black-winged Starling (see Table 11).

Gaps in knowledge

The distributions and habitat requirements of some of Asia's Critical and Endangered bird species are very poorly known, meaning that it is not possible to precisely define the habitat protection and other measures that are required to ensure their survival. Eleven Critical and Endangered Asian birds (and four Vulnerable and two Data Deficient species) that have not been recorded in recent decades are listed in Table 12. Some of them may already be extinct (e.g. Pink-headed Duck), but others probably still survive (e.g. Sulu Bleeding-heart) and need to be searched for to identify key sites and the most appropriate conservation actions. There are critical gaps in knowledge of the distributions and/or ecological requirements of the 16 highly threatened species in Table 13, which make it difficult to devise appropriate measures for their conservation. Research projects must be developed to improve understanding of their status and ecological requirements, key sites for their conservation, and the threats that they face.

Table 11. Critical and Endangered bird species in Asia that are seriously affected by exploitation.

Species	Status	CITES	Issue
White-eared Night-heron *Gorsachius magnificus*	EN		H
Oriental Stork *Ciconia boyciana*	EN	I	H
Greater Adjutant *Leptoptilos dubius*	EN		H, E
White-shouldered Ibis *Pseudibis davisoni*	CR		H
Giant Ibis *Thaumatibis gigantea*	CR		H
Swan Goose *Anser cygnoides*	EN	I	H
White-winged Duck *Cairina scutulata*	EN	I	H
Javan Hawk-eagle *Spizaetus bartelsi*	EN	II	W
Maleo *Macrocephalon maleo*	EN	I	E, H
Orange-necked Partridge *Arborophila davidi*	EN		H
Chestnut-headed Partridge *Arborophila cambodiana*	EN		H
Edwards's Pheasant *Lophura edwardsi*	EN	I	H
Vietnamese Pheasant *Lophura hatinhensis*	EN		H
Bornean Peacock-pheasant *Polyplectron schleiermacheri*	EN	II	H, W
Great Indian Bustard *Ardeotis nigriceps*	EN	I	H
Bengal Florican *Houbaropsis bengalensis*	EN	I	H
Lesser Florican *Sypheotides indica*	EN	II	H
Mindoro Bleeding-heart *Gallicolumba platenae*	CR		H
Negros Bleeding-heart *Gallicolumba keayi*	CR		H
Mindanao Bleeding-heart *Gallicolumba criniger*	EN		H
Wetar Ground-dove *Gallicolumba hoedtii*	EN		H
Tawitawi Brown-dove *Phapitreron cinereiceps*	CR		H
Timor Green-pigeon *Treron psittacea*	EN		H
Timor Imperial-pigeon *Ducula cineracea*	EN		H
Red-and-blue Lory *Eos histrio*	EN	I	W
Chattering Lory *Lorius garrulus*	EN	II	W
Yellow-crested Cockatoo *Cacatua sulphurea*	CR	II	W
Philippine Cockatoo *Cacatua haematuropygia*	CR	I	W
Blue-winged Racquet-tail *Prioniturus verticalis*	EN	II	H
Sulu Hornbill *Anthracoceros montani*	CR	II	H
Mindoro Tarictic *Penelopides mindorensis*	EN	II	H
Visayan Tarictic *Penelopides panini*	EN	II	H, W
Visayan Wrinkled Hornbill *Aceros waldeni*	CR	II	H, W
Black-winged Starling *Sturnus melanopterus*	EN		W
Bali Starling *Leucopsar rothschildi*	CR	I	W

Key: CITES: Species listed on CITES Apendix I or II (see pp.32–33); Issue: H = hunting; E = egg collection; W = wild bird trade.

Table 12. Asia's 'lost species': Critical and Endangered bird species not recorded in recent decades.

Species (Critical and Endangered only)	Last record	Areas to search
Crested Shelduck *Tadorna cristata*	1964	Wetlands in eastern **Russia**, **North Korea** and probably north-east China (F01, W02, W03), including forested rivers in mountains
Pink-headed Duck *Rhodonessa caryophyllacea*	1949	Wetlands in northern **India** (W12, W13), especially in Bihar and Assam, and northern **Myanmar** (W16)
Himalayan Quail *Ophrysia superciliosa*	1876	Mountain grasslands and forests in the Western Himalayas in **India** (F04), including following up several unconfirmed reports
Javanese Lapwing *Vanellus macropterus*	1940	Coastal grasslands and wetlands on Java (W20), and possibly elsewhere in **Indonesia**
Silvery Wood-pigeon *Columba argentina*	1931	Small islands off Sumatra and other Greater Sunda islands, **Malaysia** and **Indonesia** (F07), and the coasts of the larger islands, including following up several recent unconfirmed sightings
Sulu Bleeding-heart *Gallicolumba menagei*	1891	Forests on Tawitawi and other islands in the Sulu archipelago, **Philippines** (F09)
Negros Fruit-dove *Ptilinopus arcanus*	1953	Forest on Mt Canlaon, where the single known specimen was collected, and elsewhere on Negros and Panay, **Philippines** (F09)
Siau Scops-owl *Otus siaoensis*	1866	Forest on Siau, where the single known specimen was collected, and possibly on other small islands off northern Sulawesi, **Indonesia** (F08)
White-eyed River-martin *Eurychelidon sirintarae*	1978	Riverine habitats in **Thailand** (W17) and elsewhere in South-East Asia
Rueck's Blue-flycatcher *Cyornis ruckii*	1918	Lowland forest in northern Sumatra, **Indonesia**, and possibly elsewhere in the Sundaland forests (F07)
Banggai Crow *Corvus unicolor*	1880s	Banggai and other islands in the Banggai and Sula island groups, **Indonesia** (F08)

In addition to the Critical and Endangered species in the table, four Vulnerable species (found in areas where their habitats are currently under relatively low pressure) have not been recorded in recent decades: Nicobar Sparrowhawk *Accipiter butleri* (F06: last definite record in the Nicobar islands in 1901, but possible sightings in the 1990s); Manipur Bush-quail *Perdicula manipurensis* (G02: no confirmed records in the grasslands of north-east India and Bangladesh since 1932); Black-browed Babbler *Malacocincla perspicillata* (F07: known by a single specimen collected in Kalimantan, Indonesia in the 1840s); Rusty-throated Wren-babbler *Spelaeornis badeigularis* (F04: known by a single specimen collected in the eastern Himalayas of India in 1947), plus two Data Deficient species (found in areas where their habitats might not be under any pressure): Vaurie's Nightjar *Caprimulgus centralasicus* (G01: known by a single specimen collected in the Taklimakan desert, Xinjiang, China in 1929), and Sillem's Mountain-finch *Leucosticte sillemi* (F04: known by two specimens collected at high altitude in the Western Himalayas in 1929).

Table 13. Key gaps in knowledge of Critical and Endangered bird species in Asia.

Species (Critical and Endangered only)	Actions to address main gaps in knowledge
White-eared Night-heron *Gorsachius magnificus*	Surveys in south-east **China** and northern **Vietnam** (F03) to investigate the species's breeding distribution and status, and the key sites for its conservation
Japanese Night-heron *Gorsachius goisagi*	National survey in **Japan** (F02) to update information on the species's breeding distribution and status, and the key sites for its conservation
White-rumped Vulture *Gyps bengalensis*	Research on the factors currently affecting its population in **South Asia** (G03), to determine the emergency action required for its protection
Indian Vulture *Gyps indicus*	Research on the factors currently affecting its population in **South Asia** (G03), to determine the emergency action required for its protection
Slender-billed Vulture *Gyps tenuirostris*	Research on the factors currently affecting its population in **South Asia** (G03), to determine the emergency action required for its protection
Sumba Buttonquail *Turnix everetti*	Surveys on Sumba in Nusa Tenggara, **Indonesia** (F08), to investigate the species's distribution and status, ecological requirements, and the key sites for its conservation
Moluccan Woodcock *Scolopax rochussenii*	Surveys in Northern Maluku, **Indonesia** (F08), to investigate the species's distribution and status, ecological requirements, and the key sites for its conservation
Jerdon's Courser *Rhinoptilus bitorquatus*	Surveys in arid parts of eastern **India** (G03), including thorough analysis of satellite images, to investigate the species's distribution and status, and the key sites for its conservation
Chinese Crested-tern *Sterna bernsteini*	Surveys on islands off the coast of eastern **China** (S01) and **Vietnam**, including at presumed former colonies in Shandong, to investigate the species's status and the key sites for its conservation
Blue-fronted Lorikeet *Charmosyna toxopei*	Surveys on Buru in Maluku, **Indonesia** (F08), to investigate the species's distribution and status, ecological requirements, and the key sites for its conservation
Sumatran Ground-cuckoo *Carpococcyx viridis*	Surveys on Sumatra, **Indonesia** (F07), to investigate the species's distribution and status, ecological requirements, and the key sites for its conservation
Taliabu Masked-owl *Tyto nigrobrunnea*	Surveys on the Sulu islands, **Indonesia** (F08), to investigate the species's distribution and status, ecological requirements, and the key sites for its conservation
Flores Scops-owl *Otus alfredi*	Surveys on Flores and other islands in northern Nusa Tenggara, **Indonesia** (F08), to investigate the species's distribution and status, and the key sites for its conservation
Forest Owlet *Heteroglaux blewitti*	Surveys in the forests of central **India** (F05), to investigate the species's distribution and status, and the key sites for its conservation
Gurney's Pitta *Pitta gurneyi*	Surveys of the forests of southern **Myanmar** (F07), to investigate the species's distribution and status, and the key sites for its conservation
Isabela Oriole *Oriolus isabellae*	Surveys on Luzon, **Philippines** (F09), to investigate the species's distribution and status, ecological requirements, and the key sites for its conservation

DATA PRESENTATION

HABITAT OVERVIEW

The habitat region accounts which form the major part of this book have been prepared in a standardised way. The opening page of each account summarises the number of globally threatened species that occur there, the habitat requirements and altitudinal ranges of these birds, and the countries and territories which the region covers. The special features and global importance of the region are highlighted in the summary text, including whether it overlaps with one of Conservation International's Global Hotspots (see pp.20–21).

SUMMARY TABLE

The number of species that qualify for the IUCN Red List Categories (see p.18) of CR (Critically Endangered), EN (Endangered) and VU (Vulnerable) are given, and are subdivided according to their occurrence status within the habitat region. The symbols and species categories used in this table cross-refer to those used in Table 2.

MAP

The maps are colour coded according to habitat type (e.g. forest regions are in a variety of shades of green), and show the geographical extent of each habitat region. In most forest and grassland regions, further colour coding is used to show sub-regions (usually Endemic Bird Areas: see p.20–21), each of which has its own group of threatened birds. The locations of the outstanding Important Bird Areas for threatened birds given in Table 1 are shown on the maps. Only those countries and territories where the habitat region's threatened birds have been recorded are labelled on the maps. The seabirds map (S01) differs in that it shows all known breeding localities for the five threatened seabirds that breed in Asia.

OUTSTANDING IBAs FOR THREATENED BIRDS

The most outstanding sites for threatened birds in the habitat region, generally those with the most extensive and highest quality natural habitats, or wetlands that regularly support globally outstanding congregations of threatened waterbirds.

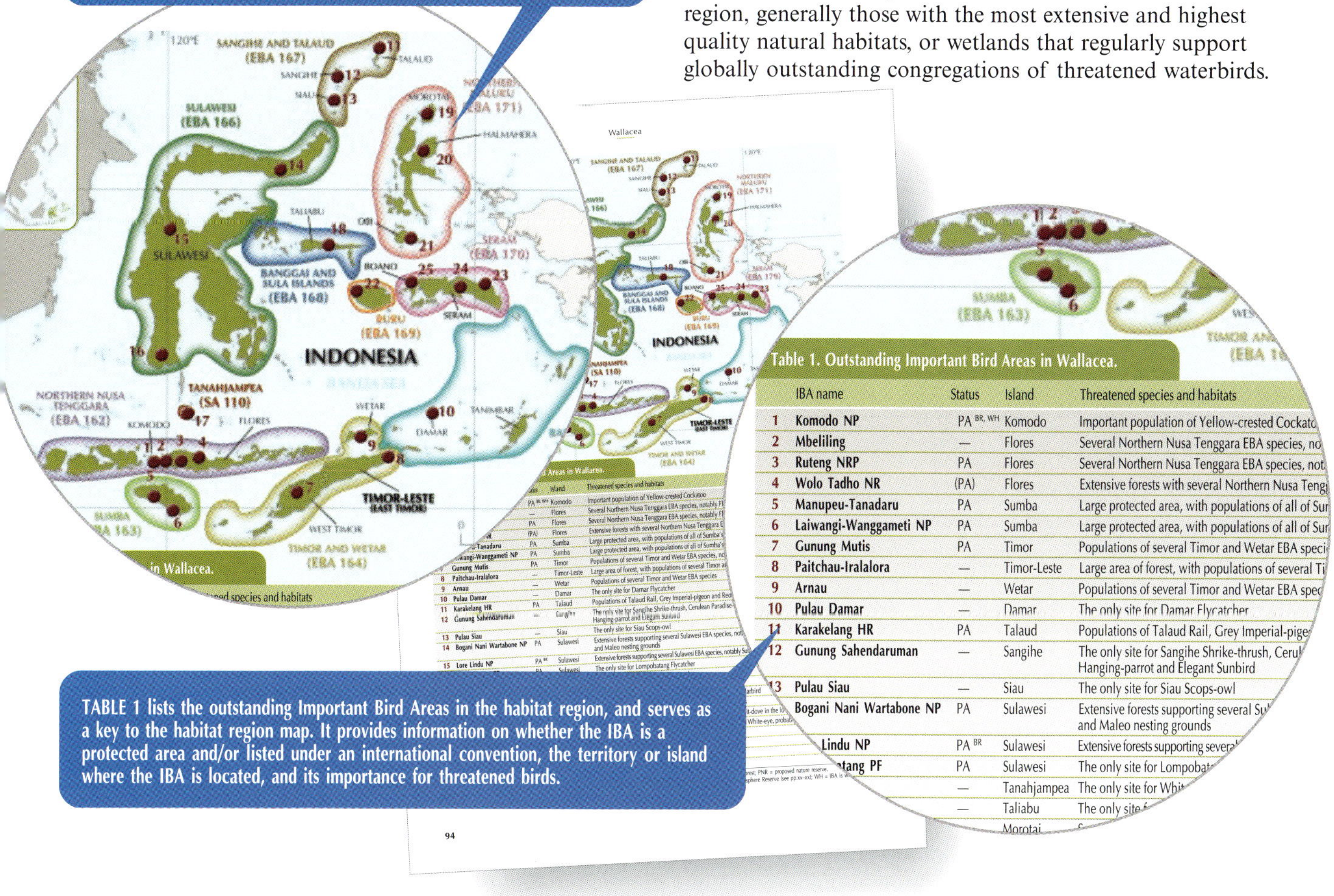

TABLE 1 lists the outstanding Important Bird Areas in the habitat region, and serves as a key to the habitat region map. It provides information on whether the IBA is a protected area and/or listed under an international convention, the territory or island where the IBA is located, and its importance for threatened birds.

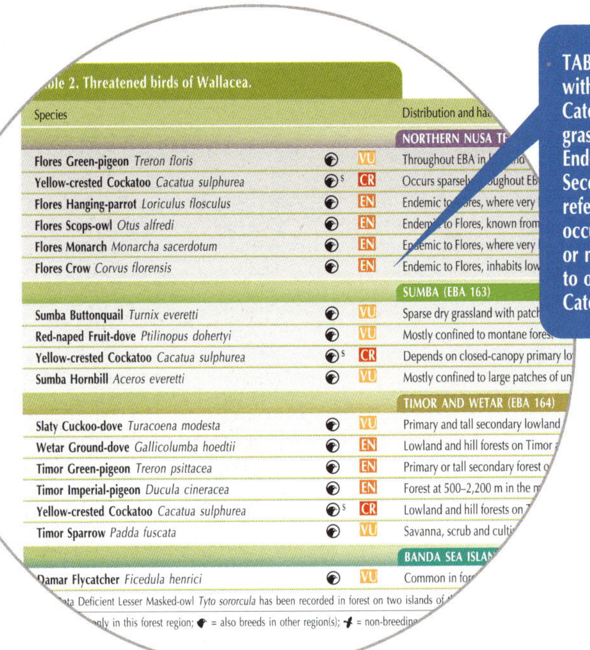
...ble 2. Threatened birds of Wallacea.

Species			Distribution and ha...
			NORTHERN NUSA TE...
Flores Green-pigeon *Treron floris*		VU	Throughout EBA in l...
Yellow-crested Cockatoo *Cacatua sulphurea*	s	CR	Occurs sparsely throughout EB...
Flores Hanging-parrot *Loriculus flosculus*		EN	Endemic to Flores, where very...
Flores Scops-owl *Otus alfredi*		EN	Endemic to Flores, known from...
Flores Monarch *Monarcha sacerdotum*		EN	Endemic to Flores, where very...
Flores Crow *Corvus florensis*		EN	Endemic to Flores, inhabits low...
			SUMBA (EBA 163)
Sumba Buttonquail *Turnix everetti*		VU	Sparse dry grassland with patch...
Red-naped Fruit-dove *Ptilinopus dohertyi*		VU	Mostly confined to montane fore...
Yellow-crested Cockatoo *Cacatua sulphurea*	s	CR	Depends on closed-canopy primary lo...
Sumba Hornbill *Aceros everetti*		VU	Mostly confined to large patches of un...
			TIMOR AND WETAR (EBA 164)
Slaty Cuckoo-dove *Turacoena modesta*		VU	Primary and tall secondary lowland...
Wetar Ground-dove *Gallicolumba hoedtii*		EN	Lowland and hill forests on Timor...
Timor Green-pigeon *Treron psittacea*		EN	Primary or tall secondary forest o...
Timor Imperial-pigeon *Ducula cineracea*		EN	Forest at 500–2,200 m in the m...
Yellow-crested Cockatoo *Cacatua sulphurea*	s	CR	Lowland and hill forests on T...
Timor Sparrow *Padda fuscata*		VU	Savanna, scrub and culti...
			BANDA SEA ISLAN...
...amar Flycatcher *Ficedula henrici*		VU	Common in for...

...ta Deficient Lesser Masked-owl *Tyto sororcula* has been recorded in forest on two islands of...
...nly in this forest region; = also breeds in other region(s); = non-breedin...

TABLE 2 lists the globally threatened species found in the habitat region, together with a symbol to represent their occurrence status there, their IUCN Red List Category, and notes on their distribution and/or habitats. In most forest and grassland regions, the table is divided into subgroups of threatened birds (usually by Endemic Bird Areas: see pp.20–21; the codes used for Endemic Bird Areas and Secondary Areas are from Stattersfield *et al.* 1998), which are colour coded to cross-refer visually to the coloured sub-regions on the map. In wetland regions, the occurrence status symbols are used to show whether a species is a breeding, passage or non-breeding visitor, and the proportion of its global population that is estimated to occur in the wetland region. The occurrence status symbols and IUCN Red List Category cross-refer to the summary table and the species list in the Appendix.

Wallacea

Table 2. Threatened birds of Wallacea.

Species			Distribution and habitat
			NORTHERN NUSA TENGGARA (EBA 162)
Flores Green-pigeon *Treron floris*		VU	Throughout EBA in lowland forest
Yellow-crested Cockatoo *Cacatua sulphurea*	s	CR	Occurs sparsely throughout EBA in lowland forest
Flores Hanging-parrot *Loriculus flosculus*		EN	Endemic to Flores, where very local in mid-elevation semi-evergreen rainforest
Flores Scops-owl *Otus alfredi*		EN	Endemic to Flores, known from two localities in montane forest above 1,000 m
Flores Monarch *Monarcha sacerdotum*		EN	Endemic to Flores, where very local in mid-elevation semi-evergreen rainforest
Flores Crow *Corvus florensis*		EN	Endemic to Flores, inhabits lowland forest below 950 m
			SUMBA (EBA 163)
Sumba Buttonquail *Turnix everetti*		VU	Sparse dry grassland with patches of bushes in the lowlands
Red-naped Fruit-dove *Ptilinopus dohertyi*		VU	Mostly confined to montane forest
Yellow-crested Cockatoo *Cacatua sulphurea*	s	CR	Depends on closed-canopy primary lowland forest with tall trees
Sumba Hornbill *Aceros everetti*		VU	Mostly confined to large patches of undisturbed lowland forest
			TIMOR AND WETAR (EBA 164)
Slaty Cuckoo-dove *Turacoena modesta*		VU	Primary and tall secondary lowland and hill forests on Timor and Wetar
Wetar Ground-dove *Gallicolumba hoedtii*		EN	Lowland and hill forests on Timor and Wetar
Timor Green-pigeon *Treron psittacea*		EN	Primary or tall secondary forest on Timor, chiefly in the extreme lowlands
Timor Imperial-pigeon *Ducula cineracea*		EN	Forest at 500–2,200 m in the mountains of Timor and Wetar
Yellow-crested Cockatoo *Cacatua sulphurea*	s	CR	Lowland and hill forests on Timor
Timor Sparrow *Padda fuscata*		VU	Savanna, scrub and cultivation in the lowlands of Timor
			BANDA SEA ISLANDS (EBA 165)
Damar Flycatcher *Ficedula henrici*		VU	Common in forest on the tiny island of Damar

The Data Deficient Lesser Masked-owl *Tyto sororcula* has been recorded in forest on two islands of the Tanimbar group
= breeds only in this forest region; = also breeds in other region(s); = non-breeding visitor from another region; s = also occurs in other EBA(s) and/or SA(s) in Wallacea

continued

Flores Green-pigeon is widespread but localised in the fragmented lowland forests of northern Nusa Tenggara.

PHOTO: COLIN TRAINOR/BIRDLIFE

Yellow-crested Cockatoo has declined rapidly in many parts of Wallacea because of capture for the wild bird trade.

PHOTO: BIRDLIFE

F08

95

CURRENT STATUS OF HABITATS AND THREATENED SPECIES

This section gives an overview of the current conditions of the natural and semi-natural habitats in the region, with information on past and current threats, and conservation measures that are already in place. Where species are directly threatened by exploitation or other issues that do not relate to their habitats, information is given on the current status and rates of decline of the birds affected.

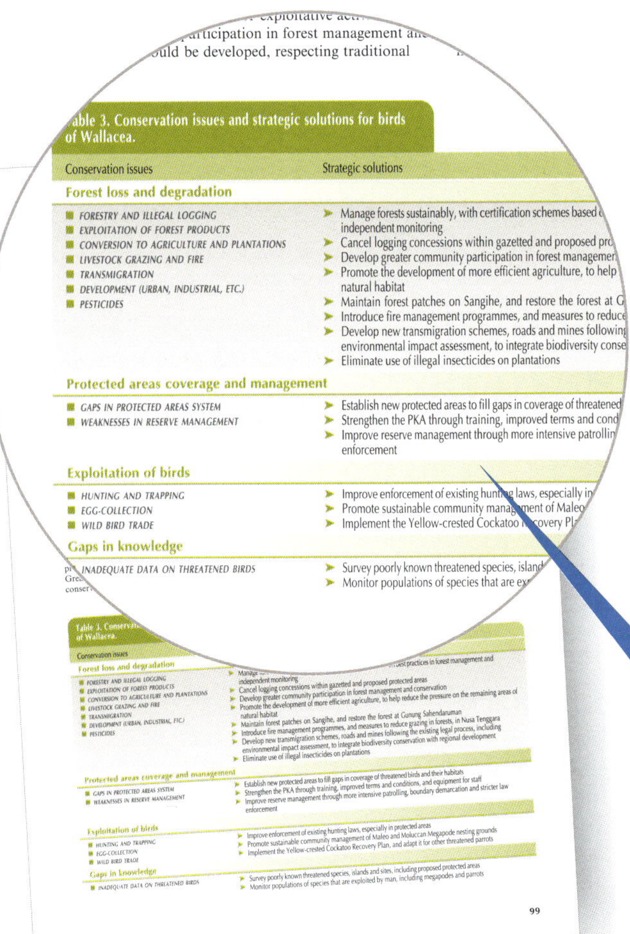
...exploitative act...
...articipation in forest management an...
...ould be developed, respecting traditional...

...able 3. Conservation issues and strategic solutions for birds of Wallacea.

Conservation issues	Strategic solutions
Forest loss and degradation	
■ FORESTRY AND ILLEGAL LOGGING ■ EXPLOITATION OF FOREST PRODUCTS ■ CONVERSION TO AGRICULTURE AND PLANTATIONS ■ LIVESTOCK GRAZING AND FIRE ■ TRANSMIGRATION ■ DEVELOPMENT (URBAN, INDUSTRIAL, ETC.) ■ PESTICIDES	➤ Manage forests sustainably, with certification schemes based o... independent monitoring ➤ Cancel logging concessions within gazetted and proposed pro... ➤ Develop greater community participation in forest manageme... ➤ Promote the development of more efficient agriculture, to help... natural habitat ➤ Maintain forest patches on Sangihe, and restore the forest at G... ➤ Introduce fire management programmes, and measures to reduce... ➤ Develop new transmigration schemes, roads and mines following... environmental impact assessment, to integrate biodiversity conse... ➤ Eliminate use of illegal insecticides on plantations
Protected areas coverage and management	
■ GAPS IN PROTECTED AREAS SYSTEM ■ WEAKNESSES IN RESERVE MANAGEMENT	➤ Establish new protected areas to fill gaps in coverage of threatene... ➤ Strengthen the PKA through training, improved terms and cond... ➤ Improve reserve management through more intensive patrollin... enforcement
Exploitation of birds	
■ HUNTING AND TRAPPING ■ EGG-COLLECTION ■ WILD BIRD TRADE	➤ Improve enforcement of existing hunting laws, especially in... ➤ Promote sustainable community management of Maleo... ➤ Implement the Yellow-crested Cockatoo Recovery Pl...
Gaps in knowledge	
■ INADEQUATE DATA ON THREATENED BIRDS	➤ Survey poorly known threatened species, island... ➤ Monitor populations of species that are ex...

99

CONSERVATION ISSUES AND STRATEGIC SOLUTIONS

This section is usually subdivided under four main headings: Habitat loss and degradation; Protected areas coverage and management; Exploitation of birds: and Gaps in knowledge. Within each of these subsections, there are overviews of the main threats to habitats and birds in the region, with recommendations for conservation actions to address these issues. These conservation issues, and the 'strategic solutions' proposed to address them, are summarised in Table 3.

TABLE 3 summarises the main conservation issues affecting the threatened birds in the habitat region, and the 'strategic solutions' proposed to address these threats. Each of the conservation issues listed in the table corresponds to a subheading in the *Conservation issues and strategic solutions* section.

BOREAL and NORTHERN TEMPERATE FORESTS

THIS region includes the eastern part of the vast belt of boreal (or taiga) forest which extends from northern Europe to north-east Asia, together with the northern temperate forests of south-east Russia, North Korea, and north-east China. Six threatened bird species breed in these forests and the associated wetlands, including three which are confined to the relatively developed south and east of the region (the riverine Scaly-sided Merganser and Blakiston's Fish-owl, and Rufous-backed Bunting). The other three are relatively widespread within the region, with Baikal Teal also nesting in tundra wetlands (see W01) and Greater Spotted Eagle ranging westwards from Asia to eastern Europe. An additional seven threatened species (Oriental Stork, Crested Shelduck, Baer's Pochard, Steller's Sea-eagle, Red-crowned Crane, Swinhoe's Rail and Spotted Greenshank) are found in wetlands on the south-eastern and eastern edges of this forest zone, and their conservation is covered in regions W02 and W03.

- **Key habitats** Boreal and temperate forest, and associated wetlands.
- **Altitude** Lowlands to c.2,000 m.
- **Countries and territories** **Russia** (Krasnoyarsk, Irkutsk, Buryatia, Chita, Yakutia, Koryakia, Kamchatka, Magadan, Khabarovsk, Amur, Jewish Autonomous Region, Primorye, Sakhalin); **Mongolia**; **Japan** (Hokkaido); **North Korea**; **China** (Heilongjiang, Jilin, Inner Mongolia).

Threatened species

	CR	EN	VU	Total
(breeds only in this forest region)[1]	—	2	2	4
(also breeds in other region(s))	—	—	2	2
Total	—	2	4	6

Key: (symbol) = breeds only in this forest region.
[1] Two species which nest only in this forest region migrate to other regions outside the breeding season, Scaly-sided Merganser and Hooded Crane.
(symbol) = also breeds in other region(s).

Some extensive tracts of pristine boreal forest remain in eastern Russia.
PHOTO: OTTO PFISTER

OUTSTANDING IBAs FOR THREATENED BIRDS (see Table 1)

Most threatened birds in this region breed at low densities, and their conservation principally depends on broad-scale habitat protection measures rather than the management of key sites. However, the three large IBAs selected together support significant proportions of the global populations of Scaly-sided Merganser, Blakiston's Fish-owl and Rufous-backed Bunting.

CURRENT STATUS OF HABITATS AND THREATENED SPECIES

Large areas of the region remain pristine and still relatively unaffected by intensive economic activities, but forests in some areas have been much reduced and fragmented by logging and development, notably in south-east Russia, Hokkaido and north-east China. From the 1960s to early 1980s, logging in the Ussuri river drainage (including the lower Bikin and Iman valleys) greatly reduced the habitat of

Table 1. Outstanding Important Bird Areas in the boreal and northern temperate forests.

	IBA name	Status	Territory	Threatened species and habitats
1	**Bikin river basin**[W03]	— [WH]	Primorye	Important breeding populations of Scaly-sided Merganser and Blakiston's Fish-owl, also Hooded Crane
2	**Iman river basin**[W03]	—	Primorye	Important breeding populations of Scaly-sided Merganser and Blakiston's Fish-owl, also Hooded Crane
3	**Xianghai NNR**[W03]	PA [AP,R]	Jilin	Important breeding population of Rufous-backed Bunting

Several of the forest birds of this region breed in or near to some IBAs listed in regions W02 (Blakiston's Fish-owl) and W03.
Note that more IBAs in this region will be included in the *Important Bird Areas in Asia*, due to be published in early 2004.

Key *IBA name*: NNR = National Nature Reserve.
Status: PA = IBA is a protected area; (PA) = IBA partially protected; — = unprotected; AP = IBA is wholly or partially an Asia-Pacific waterbird network site (see p.35); R = IBA is wholly or partially a Ramsar Site (see pp.31–32); WH = IBA is wholly or partially a World Heritage Site (see p.34); W03 = supports some threatened wetland birds of region W03.

Scaly-sided Merganser and Blakiston's Fish-owl. However, large-scale deforestation in river valleys was then prohibited, and plans to log the Bikin basin were dropped. Wetlands in the taiga zone are also relatively secure, but there have been localised losses through development. On Hokkaido, the human population has increased greatly since the 1950s, and much of the habitat of Blakiston's Fish-owl has been lost to farmland, urban development, and logging. Fish stocks in the rivers have been reduced by the construction of dams. The protection afforded to the forests where the species nests is still inadequate, and many of these areas remain under threat. However, immediately to the north of Hokkaido, the southern Kuril islands are relatively undeveloped, and retain large areas of undisturbed natural forest. In north-east China, much original boreal and northern temperate forest has been cleared, and much remaining forest degraded; the Chinese breeding population of Scaly-sided Merganser is greatly reduced, and Blakiston's Fish-owl may be near to national extinction.

Blakiston's Fish-owl requires stretches of mature riverine forest, a habitat that has been much reduced in north-east China, Hokkaido and parts of Russia.

PHOTO: YASUYUKI MAKINO

CONSERVATION ISSUES AND STRATEGIC SOLUTIONS (summarised in Table 3)

Habitat loss and degradation

FORESTRY

Forestry provides almost 10% of total industrial output in the Russian Far East, but timber production fell significantly in the 1990s following the political changes in Russia. Logging concessions have recently been granted in many areas (many to joint ventures between Russian and foreign companies), new logging roads are being constructed, and it is likely that the intensity of logging will accelerate. The largest remaining primary forests in the south-east Russian breeding grounds of Scaly-sided Merganser and Blakiston's Fish-owl are inside concessions, included those in the Bikin and Iman river basins. Extensive areas of primary forest must be protected from logging, through the cancellation of concessions and the establishment of nature reserves. Although all riparian forests are technically protected in Russia, to preserve water supplies and salmon spawning grounds, the upper levels of the riverine terraces and the banks of small streams are often subject to intensive logging; protection of riverine flood-plain forests should be extended to include riverine terraces and small streams on valley sides, as these are an integral part of the ecosystems required by Scaly-sided Merganser and Blakiston's Fish-owl. The logging practices in this part of Russia are generally inefficient, and 40–60% of all timber cut may be lost in the production process; foreign capital and reduced-impact logging equipment and practices could reduce waste, and help create a sustainable industry, with processing at local Russian sawmills rather than exporting raw logs. Very little undisturbed taiga forest remains in north east China, but it was recently protected by a national logging ban; this ban should be strictly enforced (and continued) to prevent any further logging of

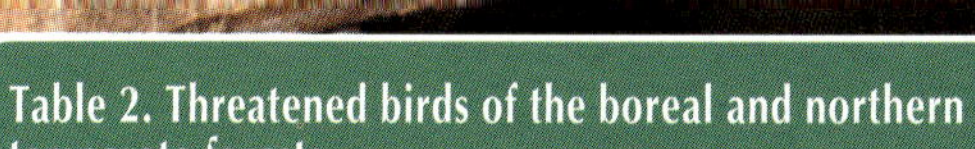

Table 2. Threatened birds of the boreal and northern temperate forests.

Species			Distribution and habitat
Baikal Teal *Anas formosa*	● m	VU	Breeds in wetlands in the northern taiga of Krasnoyarsk, Chita, Yakutia, Kamchatka, Magadan, Khabarovsk and Amur (as well as in tundra: see W01)
Scaly-sided Merganser *Mergus squamatus*	◉ m	EN	Breeds along mountain rivers with tall riverine forest, in Khabarovsk, Amur, Jewish A. R., Primorye, North Korea, Heilongjiang, Jilin and Inner Mongolia
Greater Spotted Eagle *Aquila clanga*	● mo	VU	In the Asia region, breeds in forest and forest steppe, usually near wetlands, in Irkutsk, Buryatia, Khabarovsk, Amur, Jewish A. R., Primorye, Heilongjiang, Jilin and Inner Mongolia, and presumably northern Mongolia
Hooded Crane *Grus monacha*	◉ m	VU	Breeds in wetlands in larch taiga in Krasnoyarsk, Yakutia, Khabarovsk, Amur, Primorye and Heilongjiang, probably in Irkutsk, Khakassia and Sakhalin, and possibly northern Mongolia
Blakiston's Fish-owl *Ketupa blakistoni*	◉	EN	Inhabits densely forested, rather large rivers in Magadan, Khabarovsk, Jewish A. R., Primorye, Sakhalin, Hokkaido and (at least formerly) Heilongjiang, Jilin and Inner Mongolia
Rufous-backed Bunting *Emberiza jankowskii*	◉	VU	Breeds (or formerly bred) in sparse vegetation in semi-humid transitional areas between Manchurian deciduous forest and Mongolian steppe, in Primorye, North Korea, Heilongjiang and Jilin

◉ = breeds only in this forest region; ● = also breeds in other region(s); m = migrates to other region(s); o = also breeds outside the Asia region

Much boreal forest has been cleared or logged in Hokkaido and north-east China, and logging operations are set to expand in eastern Russia.

PHOTO: OTTO PFISTER

mature forest, with special protection given to mature trees near river banks. In logged forests, where nest holes may be limited, nest boxes should be provided for Scaly-sided Mergansers and Blakiston's Fish-owls. Rufous-backed Bunting has been found nesting in young plantations, and there may be potential to influence forestry practices for its benefit, perhaps by introducing a rotational system of planting.

■ *FOREST AND WETLAND CONVERSION*
Natural forests and wetlands continue to be converted to farmland and urban areas in the more densely populated parts of the region. This is affecting the habitat of Greater Spotted Eagle in southern Russia, Blakiston's Fish-owl on Hokkaido and Rufous-backed Bunting in north-east China. On Hokkaido, a recovery plan is needed for the natural river systems and forests, to preserve the remaining areas of habitat for Blakiston's Fish-owl, and to restore these habitats in selected areas.

■ *DEVELOPMENT (URBAN, INDUSTRIAL, ETC.)*
Many ongoing and planned industrial and infrastructure development projects could affect the natural forests and wetlands. New ports and free-trade zones (with associated new roads and rail links), such as the Tumen River Development Project, will create new opportunities to export timber, oil and gas, and will probably affect natural habitats inland, particularly in areas most accessible from the coast. Environmental impact assessments should review and, if necessary, revise these projects, to reconcile nature conservation and economic development. New developments should be avoided near protected areas and other sites of high biodiversity value. The southern Kuril islands retain most of their natural habitats, unlike nearby Hokkaido, which is now highly developed. Should the current dispute between Japan and Russia over ownership of the islands be resolved, there is a danger that rapid development could take place. Measures are needed to protect the richest areas of natural habitat on the islands, perhaps through the designation of an international Peace Park.

■ *POLLUTION*
Mining of minerals is seriously affecting the quality of Scaly-sided Merganser (and presumably also Blakiston's Fish-owl) habitat in the Bikin and Iman river basins, through direct pollution of the water with heavy metals, and increased clay in suspension, which may reduce fish populations and their detectability by birds. Controls on mining activities in environmentally sensitive areas need to be improved.

■ *REDUCED FOOD SUPPLY*
Overharvesting of fish, especially salmonids, is affecting the food supply of Blakiston's Fish-owl (and presumably also Scaly-sided Merganser), for example in the Bikin river basin. On Hokkaido, the collection of salmon and trout at the river mouths for artificial spawning has reduced food availability upstream, and some pairs of the fish-owl have therefore been fed supplementary fish. However, the long-term solution is to improve the management of river fisheries, to allow sufficient fish for both birds and human consumption.

■ *DISTURBANCE*
The riverine habitat of Scaly-sided Merganser and Blakiston's Fish-owl is subject to disturbance, principally from river traffic which has continued to increase since motorboats became widely available in eastern Russia in the 1970s. Another source of disturbance is the rafting of logged timber down rivers. These activities need to be managed, to minimise disturbance in the richest areas for riverine threatened birds, particularly during the breeding season when the mergansers have young. Human disturbance is a problem for Blakiston's Fish-owl on Hokkaido, including by bird photographers, and measures

need to be put in place to address this problem. The Russian breeding range of Greater Spotted Eagle is in a relatively densely populated region, and human disturbance causes some pairs to abandon their nests, and increased predation of nests by crows; improved protection of nests is required (see *Gaps in protected areas system* below).

■ *INCREASED MORTALITY*
Many Blakiston's Fish-owls are killed on Hokkaido as a result of human activities, including drowning in fishponds (usually when entangled in nets), traffic accidents (often on bridges, where the owls perch) and electrocution by power-lines. Measures are therefore needed to prevent fish-owls being caught in nets and other equipment at fish farms, to discourage the birds from perching on road bridges, and to prevent them coming into contact with power-lines. Many young Scaly-sided Mergansers are drowned in fishing nets in their Russian breeding grounds, and fishing activities must be scaled back on these stretches of river in the breeding season.

Protected areas coverage and management

■ *GAPS IN PROTECTED AREAS SYSTEM*
All countries in this region have well-developed protected areas systems, with many important areas of taiga forest within nature reserves, but there are some significant gaps. The only relatively large protected area in the Russian breeding range of Scaly-sided Merganser is the Sikhote-Alin' State Biosphere Reserve, with the other nature reserves only protecting small, isolated blocks of primary forest. It is therefore a high priority to give formal protection (e.g. national park status) to the Bikin river basin, which has the largest remaining area of primary forest in this part of Russia and large breeding populations of several threatened bird species. Proposals for other new reserves in Russia include in the Iman river basin (possibly in the largely undeveloped Armu river basin), along the Anyuy and Khor rivers in Khabarovsk (for Blakiston's Fish-owl), in Yakutia (for Baikal Teal and Hooded Crane) and in the Barguzinskaya valley and the Uoyan–Kumora extension to the Verkhnyaya Angara riverine floodplain in Buryatia (for Greater Spotted Eagle). In China, new protected areas have been proposed at Xiaobei Hu lake in Heilongjiang (for Scaly-sided Merganser) and at Yaotuo and Yushutai in Jilin (for Rufous-backed Bunting).

■ *WEAKNESSES IN RESERVE MANAGEMENT*
A number of problems are affecting the protected areas system in eastern Russia, linked to the ongoing decentralisation of administrative structures within the country, a recent steep decline in funding for reserves, the lack of a unified management and planning structure, and poor public support. In Buryatia, the habitats of Greater Spotted Eagle are under pressure from recreational and agricultural activities inside several national parks; improved protection of nests of this species is needed, by establishing 'protecting zones' around the nest sites. Although Sikhote-Alin' State Biosphere Reserve is a major stronghold for Scaly-sided Merganser, ongoing threats to the surrounding areas must be addressed if the current population of this species is to be maintained. In China, there is a need to prepare and implement explicit and transparent management plans for nature reserves. In general, many reserves are overstaffed, and need fewer but better trained management staff. In Xianghai National Nature Reserve in Jilin, Mongolian oak *Quercus mongolica*, the Rufous-backed Bunting's favoured songpost tree, is being cut; the reserve's existing management plan needs to be implemented, including measures designed to maintain the present numbers of the bunting.

Table 3. Conservation priorities for birds of the boreal and northern temperate forests.

Conservation issues	Strategic solutions
Habitat loss and degradation	
■ *FORESTRY* ■ *FOREST AND WETLAND CONVERSION* ■ *DEVELOPMENT (URBAN, INDUSTRIAL, ETC.)* ■ *POLLUTION* ■ *REDUCED FOOD SUPPLY* ■ *DISTURBANCE* ■ *INCREASED MORTALITY*	➤ Protect key areas of primary forest in Russia from logging, and maintain the logging ban in north-east China ➤ Enforce the laws protecting riverine forests in Russia from logging, and extend them to cover forests on the valley sides ➤ Promote reduced-impact logging practices in Russia ➤ Provide nest boxes for Scaly-sided Merganser and Blakiston's Fish-owl in logged forest ➤ Manage plantations for the benefit of Rufous-backed Bunting ➤ Prepare a recovery plan for key natural river systems and forests on Hokkaido ➤ Assess the environmental impact of development projects in Russia ➤ Develop measures to protect key sites in the southern Kuril islands ➤ Prevent pollution of key rivers in Russia by mining activities ➤ Improve management of river fisheries ➤ Control human activities along rivers to minimise disturbance of threatened birds, particularly in the breeding season ➤ Develop methods to reduce accidental mortality of Blakiston's Fish-owls on Hokkaido
Protected areas coverage and management	
■ *GAPS IN PROTECTED AREAS SYSTEM* ■ *WEAKNESSES IN RESERVE MANAGEMENT*	➤ Establish new protected areas in Russia and China, notably in the Bikin and Iman river basins in Primorye ➤ Address the current problems with reserve funding and management in eastern Russia ➤ Prepare and implement management plans for nature reserves in north-east China
Exploitation of birds	
■ *HUNTING*	➤ Improve enforcement of existing hunting legislation in Russia and China, including through education and awareness programmes
Gaps in knowledge	
■ *INADEQUATE DATA ON THREATENED BIRDS*	➤ Survey Scaly-sided Merganser and Blakiston's Fish-owl in eastern Russia and north-east China, to identify further key sites for their conservation ➤ Search for Rufous-backed Bunting in Russia and North Korea

The Bikin and Iman river basins in Primorye support the most important known breeding populations of Scaly-sided Merganser.

PHOTO: FRANK TODD

Exploitation of birds

■ *HUNTING*

Shooting of birds for food and sport is widespread. It is unlikely that hunting on the breeding grounds is a major problem for Baikal Teal and Hooded Crane, which nest at low densities in the more remote parts of the taiga, but it could be reducing the populations (at least locally) of Greater Spotted Eagle and Blakiston's Fish-owl in Russia and probably China. Shooting of Scaly-sided Merganser is a problem in Russia, with 80–100 birds killed annually in the Bikin river basin alone, mainly for sport and shooting practice (as they have an unpleasant taste and smell) by local people, who were recently given licences to shoot any species of duck at any time of year (despite the fact that the species is officially protected in Russia). Education and awareness programmes are required; however, Scaly-sided is difficult to distinguish from Common Merganser *Mergus merganser*, so the shooting of all merganser species probably needs to be banned. Measures to enforce existing hunting legislation are required, backed up by education programmes.

Gaps in knowledge

■ *INADEQUATE DATA ON THREATENED BIRDS*

The distribution and numbers of all of threatened species in this region are incompletely known. Further surveys are a priority to clarify the most important sites for Scaly-sided Merganser and Blakiston's Fish-owl, for example in the river basins along the Okhotsk Sea coast and in the lower Amur river valley, along the rivers flowing into the Sea of Japan, and in the Changbai, Xiao Hinggan and Da Hinggan mountains in China. Synchronised surveys are required to obtain more accurate information on the size of the Scaly-sided Merganser population in the Bikin basin; an international research centre has been proposed there, to develop research projects for the improved management of threatened species. There are no recent records of Rufous-backed Bunting from Russia, and very little information on its status in North Korea, so surveys are required, together with investigations to try to establish how changes to the habitats in its former breeding range caused it to decline so rapidly.

JAPANESE FORESTS

THREE threatened bird species breed in the forests on the main southern islands of Japan: Japanese Night-heron and Fairy Pitta are mainly confined to lowland forests in southern Honshu, Shikoku and Kyushu, and Yellow Bunting nests only in mid-altitude forests in the mountains of Honshu. Another eight threatened species breed on small islands in the three Japanese Endemic Bird Areas: the Izu islands, Ogasawara islands and Nansei Shoto. These include the highly threatened Okinawa Rail and Okinawa Woodpecker, which are endemic to the forests of Okinawa, and Amami Thrush, which is found only on Amami and nearby small islands.

- **Key habitats** Subtropical and temperate forest.
- **Altitude** Lowlands to 1,500 m.
- **Countries and territories** **Japan** (Honshu, Izu islands, Ogasawara islands, Shikoku, Kyushu, Nansei Shoto).

Threatened species

	CR	EN	VU	Total
[1]	2	2	6	10
[2]	—	—	1	1
Total	2	2	7	11

Key: = breeds only in this forest region.
[1] Three species which nest only in this forest region migrate to other regions outside the breeding season, Japanese Night-heron, Izu Leaf-warbler and Yellow Bunting.
= also breeds in other region(s).
[2] Fairy Pitta nests in this forest region and the South-east Chinese forests (F03), and migrates to another region outside the breeding season.

The Yambaru forests in northern Okinawa support many unique animals and plants, including two endemic bird species. PHOTO: MARTIN HALE

F02

OUTSTANDING IBAs FOR THREATENED BIRDS (see Table 1)

Four IBAs have been selected in the Japanese forests, which together support populations of most threatened forest birds of this region. However, the three mainland Japanese species, Japanese Night-heron, Fairy Pitta and Yellow Bunting, are all relatively widespread with low population densities, and no key sites have been selected for them. The two key sites in the Nansei Shoto islands are both globally outstanding, Yambaru in Northern Okinawa because it holds the entire world populations of Okinawa Rail and Okinawa Woodpecker, and Central Amami because it supports the most important populations of Amami Thrush, Amami Jay and Ryukyu Woodcock. Several of the Izu islands support important populations of the endemic Izu Thrush and Izu Leaf-warbler, and Miyake-jima has been selected as the outstanding site to represent these islands.

Table 1. Outstanding Important Bird Areas in the Japanese forests.

	IBA name	Status	Territory	Threatened species and habitats
1	**Miyake-jima**	(PA)	Izu islands	Important populations of Izu Thrush and Izu Leaf-warbler
2	**Haha-jima**	PA	Ogasawara islands	Supports the largest population of Bonin White-eye
3	**Central Amami forests**	(PA)	Nansei Shoto	Important populations of Amami Thrush, Amami Jay and Ryukyu Woodcock
4	**Yambaru, northern Okinawa**	(PA)	Nansei Shoto	Supports the entire global populations of Okinawa Rail and Okinawa Woodpecker, also Ryukyu Woodcock

Note that more IBAs in this region will be included in the *Important Bird Areas in Asia*, due to be published in early 2004.

Key *Status*: PA = IBA is a protected area; (PA) = IBA partially protected; — = unprotected.

CURRENT STATUS OF HABITATS AND THREATENED SPECIES

The habitat available to many threatened forest birds has been greatly reduced in the densely populated and highly developed lowlands of Japan, but deforestation has now almost ceased in the lowlands and extensive forest cover remains in the hills and mountains. Recent declines in the numbers of Japanese Night-heron, Izu Leaf-warbler and other migratory species cannot be accounted for by habitat loss in their breeding ranges in Japan, prompting speculation that the cause may lie with deforestation in their winter quarters. On Honshu, Shikoku and Kyushu the lower-altitude forests favoured by Japanese Night-heron and Fairy Pitta have been affected in the past by urban and industrial development, logging and firewood collection, but in recent decades forest cover has gradually increased and many of the remaining forests are now in protected areas. The higher-altitude forests where Yellow Bunting breeds on Honshu are not seriously affected by development, and it is unclear if changes to these habitats could account for its decline.

On the Izu islands, much natural forest has been lost to timber plantations and urban and infrastructural development, and powerful volcanic eruptions in 2000 caused serious damage to forests on Miyake-jima; however, sufficient habitat remains to support substantial populations of the threatened birds on several islands. Virtually all original subtropical forest has already been cleared from the Ogasawara islands, and this presumably caused the extinction of three endemic species in the nineteenth century (Bonin Wood-pigeon *Columba versicolor*, Bonin Thrush *Zoothera terrestris* and Bonin Grosbeak *Chaunoproctus ferreorostris*), as well as the extinction of Bonin White-eye on several islands; however, the white-eye survives in secondary and man-modified habitats on the Haha-jima island group, and an active conservation programme is underway.

In the Nansei Shoto, forest on Amami and Okinawa has been greatly reduced in recent decades, with only small areas of forest officially protected. On Amami, large areas of mature forest have been replaced by young secondary growth, although this logging is only economically feasible through government subsidy; mature forest now only covers 10–15 km^2, less than 5% of the island. Okinawa has suffered substantial deforestation, particularly since 1945, through wide-scale clear-cutting (often followed by afforestation with conifers that are unsuitable to Okinawa Woodpecker), dam construction and associated road-

Eight threatened bird species are endemic to small islands in southern Japan, including Okinawa Rail.

PHOTO: TAKUKI HANASHIRO

Table 2. Threatened birds of the Japanese forests.

Species			Distribution and habitat
			JAPANESE MAINLAND (HONSHU, SHIKOKU AND KYUSHU)
Japanese Night-heron *Gorsachius goisagi*	m	EN	Watercourses and damp areas on forested hills and lower mountain slopes, on Honshu, Shikoku, Kyushu and the Izu islands
Fairy Pitta *Pitta nympha*	m	VU	Broadleaf evergreen forest, mainly near the coast, in southern Honshu, Shikoku and Kyushu
Yellow Bunting *Emberiza sulphurata*	m	VU	Deciduous and mixed forest at c.600–1,500 m in central Honshu, on the forest edge and in park-like areas
			IZU ISLANDS (EBA 146)
Izu Thrush *Turdus celaenops* [1]		VU	Subtropical forests on the islands between O-shima and Aoga-shima
Izu Leaf-warbler *Phylloscopus ijimae* [1]	m	VU	Subtropical forests on the islands between O-shima and Aoga-shima
			OGASAWARA ISLANDS (EBA 147)
Bonin White-eye *Apalopteron familiare*		VU	Secondary subtropical evergreen forest, and man-modified habitats
			NANSEI SHOTO (EBA 148)
Okinawa Rail *Gallirallus okinawae*		EN	Subtropical evergreen forest in northern Okinawa
Ryukyu Woodcock *Scolopax mira*		VU	Subtropical evergreen forest on several islands in the central Nansei Shoto islands
Okinawa Woodpecker *Sapheopipo noguchii*		CR	Mature subtropical evergreen forest in northern Okinawa
Amami Thrush *Zoothera major*		CR	Mature subtropical evergreen forest on Amami and adjacent small islands
Amami Jay *Garrulus lidthi*		VU	Subtropical evergreen forest and woodland on Amami and adjacent small islands

= breeds only in this forest region; = also breeds in other region(s); m = migrates to other region(s)
1 = both Izu Thrush and Izu Leaf-warbler also breed very locally in the northern Nansei Shoto islands

building, agricultural development and golf course construction. Introduced predators and competitors are known or expected to be causing rapid declines in the populations of several threatened species (and even local extinctions) on the Izu islands, Ogasawara islands and Nansei Shoto; effective control of these introduced species is probably the highest and most immediate priority for conservation in the region.

CONSERVATION ISSUES AND STRATEGIC SOLUTIONS (summarised in Table 3)

Habitat loss and degradation

■ *DEVELOPMENT (URBAN, INDUSTRIAL, ETC.)*

Clearance and disturbance of forest for urban, industrial and infrastructural developments still affects some of the threatened forest birds of Japan. On Honshu, Shikoku and Kyushu, the habitat of Yellow Bunting appears secure, and forest suitable for nesting Fairy Pitta is believed to be increasing, but many Japanese Night-heron breeding sites are unprotected and vulnerable. Road construction and/or tourist developments are damaging natural habitats on some of the Izu islands. For example, in southern Mikura-jima there is a long-term plan to relocate villagers, and on Miyake-jima the Tokyo prefecture government was planning (at least before the recent volcanic eruption) to construct either a camp ground or a marine park at Toga Point. On the Ogasawara islands, although the Bonin White-eye occurs in a variety of man-modified habitats, it is at risk from economic development (including for tourism) and a consequent reduction of forest cover. In the Nansei Shoto, clearance of forest is a major threat to the threatened endemic birds. In northern Okinawa, small-scale development projects are gradually reducing the forest area, and a potentially serious new threat is a plan jointly agreed by the government of Japan and the US Army to move the latter's existing base from the south to an area near Nago adjacent to Yambaru, with a further intention to build several helicopter pads in the centre of Yambaru. One effect of the helicopter pads will be the fragmentation and opening up of forest by their access roads. These roads will facilitate access by introduced predators (see below), and increase roadkills in Okinawa Rails.

These threats need to be addressed through improved site protection (see below) and management. All new development projects in habitats of threatened birds need to be carefully assessed, particularly those that could affect the critical sites, and if necessary plans should be modified or even abandoned. For example, the proposal to construct helicopter pads in northern Okinawa should be reviewed and revised. On the Ogasawara islands, further efforts are needed to restore the habitats of Bonin White-eye. Given the potential conflicts between development and the needs of the threatened species, public awareness should be raised of the conservation importance of these birds and their habitats.

■ *FORESTRY*

Logging has reduced the habitat of the threatened species on Amami and Okinawa in the Nansei Shoto. The cutting of mature forest should cease (and subsidies that support such activities withdrawn), with areas of medium-aged forest allowed to develop. On Amami, logging should be restricted to a regime allowing a permanent mosaic of cut-over and mature stands, and forest road construction should be halted; government subsidies previously used in creating employment through logging and farming should be redirected into work that assists the conservation of native forest and wildlife. Nest-boxes should be provided for Amami Jay and Okinawa Woodpecker, to enable these species to nest in relatively young secondary and logged forest where the trees are not yet mature enough to provide suitable nest holes.

Protected areas coverage and management

■ *GAPS IN PROTECTED AREAS SYSTEM*

Japanese Night-heron, Fairy Pitta and Yellow Bunting occur in many protected areas on Honshu, Shikoku and Kyushu, but some important sites are unprotected, notably of Japanese Night-heron. The protection of the most important breeding localities of Japanese Night-heron is

Table 3. Conservation issues and strategic solutions for birds of the Japanese forests.

Conservation issues	Strategic solutions
Habitat loss and degradation	
■ *DEVELOPMENT (URBAN, INDUSTRIAL, ETC.)* ■ *FORESTRY*	➤ Assess the environmental impact of development projects, particularly at critical sites ➤ Review and revise plans for the construction of military facilities in northern Okinawa ➤ Restore natural habitats on the Ogasawara islands ➤ Cease logging of mature forest on Amami and Okinawa ➤ Provide nest-boxes for Amami Jay and Okinawa Woodpecker in logged forest
Protected areas coverage and management	
■ *GAPS IN PROTECTED AREAS SYSTEM* ■ *WEAKNESSES IN RESERVE MANAGEMENT*	➤ Protect important Japanese Night-heron breeding grounds ➤ Create new protected areas embracing all remaining natural forest in northern Okinawa and on Amami ➤ Strengthen the infrastructure and manpower of the national park in the Izu islands
Gaps in knowledge	
■ *INADEQUATE DATA ON THREATENED BIRDS*	➤ Conduct national surveys of Japanese Night-heron and Yellow Bunting ➤ Continue monitoring threatened birds population on the Izu islands, Ogasawara islands, Okinawa and Amami ➤ Search for Bonin White-eye on the smaller Ogasawara islands, and survey Kakeroma-jima for Amami Thrush and Amami Jay
Other conservation issues	
■ *INTRODUCED PREDATORS* ■ *SMALL RANGE AND POPULATION*	➤ Control introduced predators (and garbage dumping) on the Izu islands and the Nansei Shoto, and establish predator-free safe havens ➤ Consider the reintroduction of Bonin White-eye to some of the smaller Ogasawara islands ➤ Establish a captive population of Okinawa Rail

Yellow Bunting nests only in mid-altitude forests in the mountains of Honshu, where surveys are required to determine whether it is still declining.

PHOTO: PETE MORRIS/BIRDQUEST

crucial but must be based on surveys (see *Inadequate data on threatened birds* below). In the Nansei Shoto, there are several small protected areas in northern Okinawa and on Amami, but most forests are not officially protected. In 1996, the Environment Agency of Japan decided to designate a national park in northern Okinawa, but this has yet to be established and efforts should be continued until a major protected area is created embracing all the area's remaining natural forest. A large new reserve should also be established to protect the remaining natural forests on Amami.

■ *WEAKNESSES IN RESERVE MANAGEMENT*

Although the Izu archipelago is a national park, with several sites designated as 'special protected areas', there are few rangers and loss of habitat continues on many islands. Infrastructure and manpower of the park need to be strengthened.

Gaps in knowledge

■ *INADEQUATE DATA ON THREATENED BIRDS*

The distribution and population size of Japanese Night-heron are not fully understood, and a coordinated survey is required at known sites and in other areas of apparently suitable forest, recording habitat condition as well as the birds. A national survey is also required for Yellow Bunting, including a comparison with historical data to help determine the extent and persistence of its decline. On the Izu islands, the populations of Izu Thrush and Izu Leaf-warbler should be monitored, particularly to determine the impact of introduced predators on thrush numbers on Miyake-jima. The health of the largest population of Bonin White-eye on Haha-jima island should continue to be monitored. This species's current status on many of the smaller Ogasawara islands is poorly understood, and surveys are needed to determine the existence, size and conservation needs of any populations. Information should also be gathered on the islands where it is extinct, to try to determine the causes of its disappearance. In the Nansei Shoto, the status of the population of Okinawa Rail should continue to be closely monitored, and ecological studies continued, particularly to determine the impact of introduced predators. A ringing and radio-tracking programme would yield a great deal of information about survival and movements, and help to define territory sizes, breeding densities and therefore total population size. Populations of Ryukyu Woodcock on Amami and Okinawa should be monitored in a carefully planned, replicable manner, particularly to determine the impact of introduced predators on its numbers, and surveys could also be conducted on the other islands where it occurs. Populations of Amami Thrush and Amami Jay should continue to be monitored on Amami in a replicable manner, and surveys conducted for both species on Kakeroma-jima to develop appropriate conservation measures there.

Other conservation issues

■ *INTRODUCED PREDATORS*

Several threatened species are known or expected to be negatively affected by the introduction of predators (primarily for snake control) to small islands in Japan. On the Izu islands, the introduction of the Siberian weasel *Mustela sibirica* to Miyake-jima in the 1970s and 1980s appears to have caused significant declines of Japanese Night-herons and Izu Thrushes. The weasels tend to concentrate near human developments where food is available, and the increased availability of raw garbage on the Izu islands has also led to a large increase in Large-billed Crows *Corvus macrorhynchos*, which predate many Izu Thrush nests. On the Ogasawara islands, introduced cats and rats presumably played a part in the extinction of several endemic bird species in the nineteenth century, and these predators may affect Bonin White-eye, particularly on some smaller islands. On Okinawa, feral dogs and cats and the introduced Javan mongoose *Herpestes javanicus* and weasel *Mustela itatsi* are possible predators of Okinawa Rail, Ryukyu Woodcock and Okinawa Woodpecker, and feral pigs damage potential ground-foraging sites for Okinawa Woodpecker (and presumably the other two species). The brown tree snake *Boiga irregularis*, which is responsible for the almost complete elimination of the native terrestrial avifauna of Guam, has also been observed on Okinawa. On Amami, the Javan mongoose *Herpestes javanicus* (or *H. edwardsi*) has become very numerous, and Ryukyu Woodcock has declined steeply in areas where the mongoose is common (but where the forest is still in good condition), suggesting that the mongoose is causing a high level of predation; however, feral dogs and cats were encountered during the survey, which are also potential predators of the woodcock. The mongoose may prey on young Amami Jays, and on Amami Thrush. The eggs or chicks of Amami Jays are also predated by Large-billed Crow, which has recently increased on Amami, probably because of increased garbage on the island.

On the Izu islands, control of both Siberian weasel and Large-billed Crow is needed, and the spread of weasels to new islands must be prevented. New controls on the dumping of garbage should be introduced throughout the islands to reduce the numbers of crows and perhaps of weasels (if they feed on rodents which typically concentrate

The population of Ryukyu Woodcock has declined on several islands, apparently because of predation by mammals introduced to control snakes.

PHOTO: YASUYUKI MAKINO

at refuse). On the Ogasawara islands, the factors affecting Bonin White-eye numbers need study, and any predators that are found to be limiting its numbers on the smaller islands should be removed. On Okinawa, control of some or all introduced species is essential for the survival of the near-flightless Okinawa Rail, and important for Ryukyu Woodcock and Okinawa Woodpecker. As a precautionary measure, predator exclusion zones should be established and maintained in core areas of Yambaru, to provide a safe haven for the rail and woodcock. On Amami, control of introduced predators is essential to ensure the survival of Ryukyu Woodcock and of several other threatened endemic species, including the Amami rabbit *Pentalagus furnessi*, and may also be beneficial to Amami Thrush and Amami Jay. The feasibility and cost of eradicating the mongoose from Amami (or at least the key forests on the island) needs assessment, with similar work to determine how to control feral cats and dogs. No new introduction of alien predators for biological control purposes should be allowed on the Nansei Shoto and Izu islands, and in particular no weasels and no mongooses (the pressure to introduce the latter to control the highly poisonous snakes in the Nansei Shoto must be resisted at all costs) should be permitted to be released on any other islands.

■ *SMALL RANGE AND POPULATION*

Bonin White-eye has a tiny range, with most of its known population confined to the island of Haha-jima, and Okinawa Rail and Okinawa Woodpecker are both confined to a small area of northern Okinawa, meaning that they are potentially vulnerable to chance events, including disease and natural disasters. On the Ogasawara islands, the reintroduction of Bonin White-eye on selected islands is an important precautionary measure; however, before this is done suitable habitat may need to be restored and introduced predators removed. It may also prove necessary to remove the introduced Japanese White-eye *Zosterops japonicus* from these islands, as competition between the two has possibly been a factor in the extinction of Bonin White-eye on some smaller islands in its former range. Captive breeding might be appropriate for Okinawa Rail, to establish a secure and healthy reserve population, and any such enterprise could be tied in closely with awareness campaigns which stress the immense biological value of Yambaru.

SOUTH-EAST CHINESE FORESTS

F03

TWELVE threatened bird species breed in the subtropical forests of south-east China and Taiwan, one of which also occurs in northern Vietnam. They are all found in the three Endemic Bird Areas of this region—the Chinese subtropical forests, South-east Chinese mountains and Taiwan—other than Reeves's Pheasant, which inhabits the forests of central China. Most of them occur in forests on hills and lower mountains slopes, and in Sichuan their ranges meet those of the threatened birds of the Sino-Himalayan mountain forests (F04).

- **Key habitats** Subtropical forest.
- **Altitude** Lowlands to 2,600 m.
- **Countries and territories** **China** (*mainland*: Gansu, Sichuan, Chongqing, Yunnan, Guizhou, Shaanxi, Henan, Hubei, Anhui, Jiangsu, Zhejiang, Fujian, Jiangxi, Hunan, Guangxi, Guangdong; *Taiwan*); **Vietnam**.

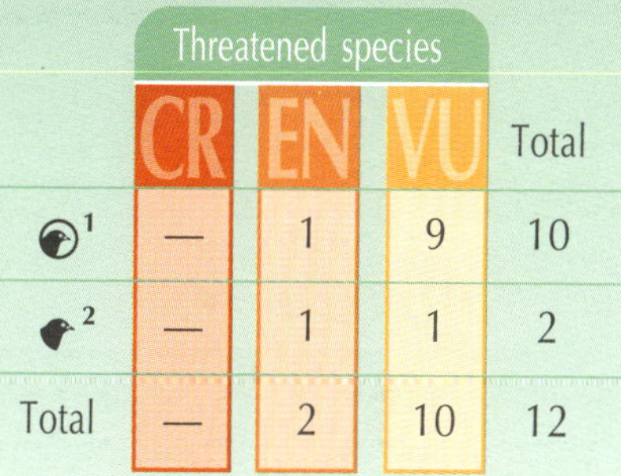

Threatened species

	CR	EN	VU	Total
(breeds only)[1]	—	1	9	10
(also breeds elsewhere)[2]	—	1	1	2
Total	—	2	10	12

Key: (icon) = breeds only in this forest region.
[1] Brown-chested Jungle-flycatcher and Silver Oriole nest only in this forest region but migrate to other regions outside the breeding season.

(icon) = also breeds in other region(s).
[2] Fairy Pitta nests in this forest region and the Japanese forests (F02), and migrates to another region outside the breeding season.

The forests of the Wuyi Shan mountains of Fujian are amongst the most extensive remaining in south-east China. PHOTO: MIKE CROSBY/BIRDLIFE

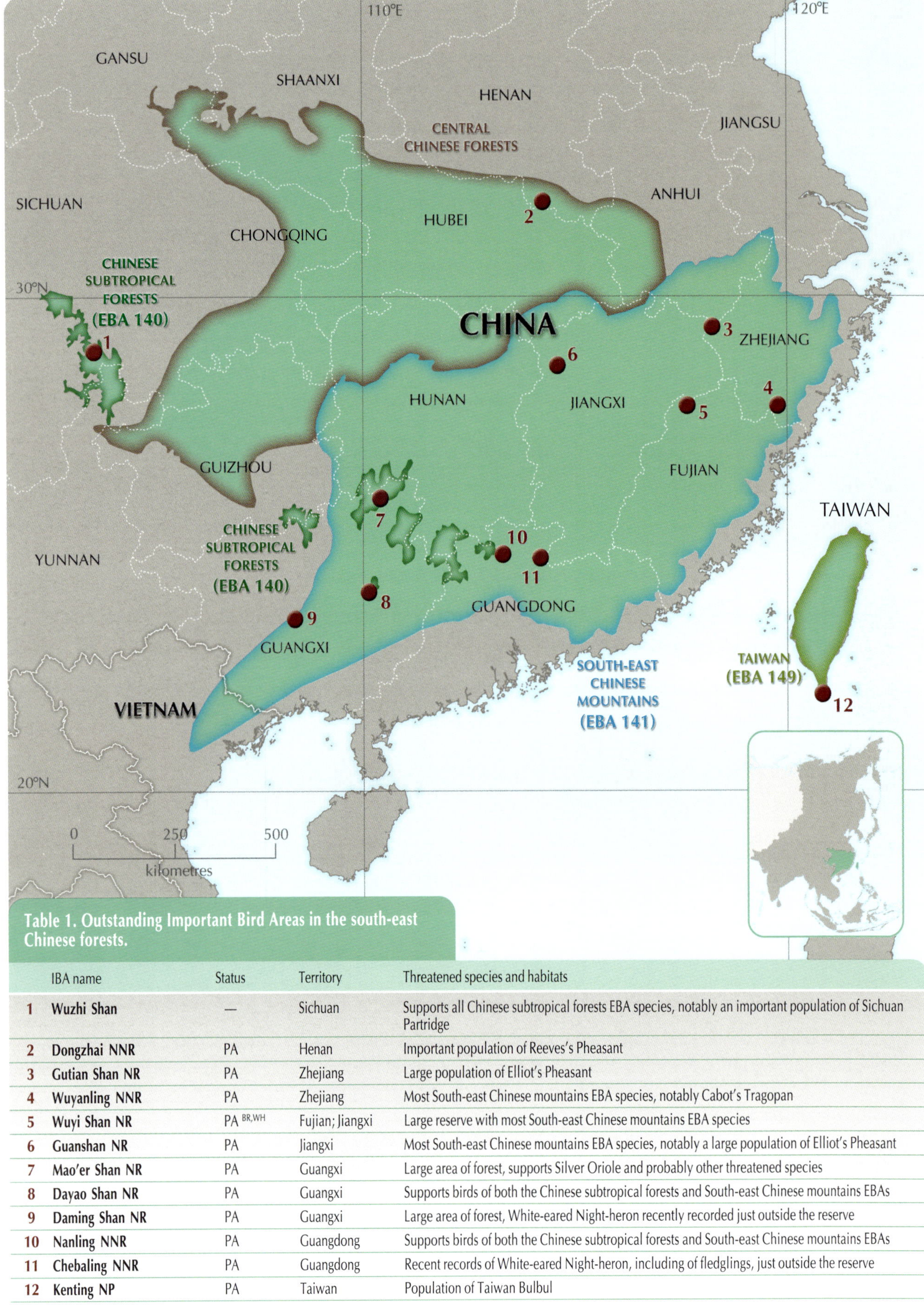

Table 1. Outstanding Important Bird Areas in the south-east Chinese forests.

	IBA name	Status	Territory	Threatened species and habitats
1	**Wuzhi Shan**	—	Sichuan	Supports all Chinese subtropical forests EBA species, notably an important population of Sichuan Partridge
2	**Dongzhai NNR**	PA	Henan	Important population of Reeves's Pheasant
3	**Gutian Shan NR**	PA	Zhejiang	Large population of Elliot's Pheasant
4	**Wuyanling NNR**	PA	Zhejiang	Most South-east Chinese mountains EBA species, notably Cabot's Tragopan
5	**Wuyi Shan NR**	PA [BR,WH]	Fujian; Jiangxi	Large reserve with most South-east Chinese mountains EBA species
6	**Guanshan NR**	PA	Jiangxi	Most South-east Chinese mountains EBA species, notably a large population of Elliot's Pheasant
7	**Mao'er Shan NR**	PA	Guangxi	Large area of forest, supports Silver Oriole and probably other threatened species
8	**Dayao Shan NR**	PA	Guangxi	Supports birds of both the Chinese subtropical forests and South-east Chinese mountains EBAs
9	**Daming Shan NR**	PA	Guangxi	Large area of forest, White-eared Night-heron recently recorded just outside the reserve
10	**Nanling NNR**	PA	Guangdong	Supports birds of both the Chinese subtropical forests and South-east Chinese mountains EBAs
11	**Chebaling NNR**	PA	Guangdong	Recent records of White-eared Night-heron, including of fledglings, just outside the reserve
12	**Kenting NP**	PA	Taiwan	Population of Taiwan Bulbul

Note that more IBAs in this region will be included in the *Important Bird Areas in Asia*, due to be published in early 2004.

Key *IBA name*: NP = National Park; NR = Nature Reserve; NNR = National Nature Reserve.
Status: PA = IBA is a protected area; (PA) = IBA partially protected; — = unprotected; BR = IBA is wholly or partially a Biosphere Reserve (see pp.34–35); WH = IBA is wholly or partially a World Heritage Site (see p.34).

OUTSTANDING IBAs FOR THREATENED BIRDS (see Table 1)

Twelve IBAs have been selected in south-east China, which together support populations of all of the threatened forest birds of this region, and include some of the largest and richest forests remaining in this part of China. Many other sites in the region with significant populations of threatened forest birds will be documented during BirdLife's ongoing regional IBA Project.

CURRENT STATUS OF HABITATS AND THREATENED SPECIES

South-east China has a long history of human habitation, and extensive deforestation had already taken place there by the nineteenth century. Rapid forest loss has continued in most provinces in the region over the past 50 years: in Fujian timber reserves declined by 50% between 1949 and 1980, and in Sichuan forest cover was estimated to have been reduced from 19% to 12.6% between the early 1950s and 1988. The

Fairy Pitta inhabits mid-altitude forests, and some key breeding areas on Taiwan and elsewhere are under pressure for development.

PHOTO: WEN-HSIN HUANG

Table 2. Threatened birds of the south-east Chinese forests.

Species			Distribution and habitat
			CENTRAL CHINESE FORESTS
Reeve's Pheasant *Syrmaticus reevesii*	◐	VU	Forest at 400–2,600 m (optimum elevation c.1,000 m)
			CHINESE SUBTROPICAL FORESTS (EBA 140)
Sichuan Partridge *Arborophila rufipectus*	◐	EN	Subtropical broadleaf forest at c.1,100–2,250 m in Sichuan and probably extreme north-east Yunnan
Omei Shan Liocichla *Liocichla omeiensis*	◐	VU	Subtropical broadleaf forest, scrub and bamboo at c.1,000–2,400 m in Sichuan and north-east Yunnan
Gold-fronted Fulvetta *Alcippe variegaticeps*	◐	VU	Subtropical broadleaf forest, usually with bamboo, at c.700–2,000 m in Sichuan and Guangxi
Silver Oriole *Oriolus mellianus*	◐ m	VU	Subtropical broadleaf forest at c.600–1,700 m in Sichuan, Yunnan, Guizhou, Guangxi and Guangdong
			SOUTH-EAST CHINESE MOUNTAINS (EBA 141)
White-eared Night-heron *Gorsachius magnificus*	◖	EN	Forest on lower and middle mountain slopes, usually near water
White-necklaced Partridge *Arborophila gingica*	◐	VU	Broadleaf and mixed broadleaf-coniferous forest, bamboo and scrub at c.500–1,900 m
Cabot's Tragopan *Tragopan caboti*	◐	VU	Evergreen broadleaf and mixed broadleaf-coniferous forest at c.600–1,800 m
Elliot's Pheasant *Syrmaticus ellioti*	◐	VU	Broadleaf and mixed broadleaf-coniferous forest, bamboo and scrub at c.200–1,900 m
Fairy Pitta *Pitta nympha*	◖ ms	VU	Broadleaf and mixed broadleaf-coniferous forest at c.400–1,900 m
Brown-chested Jungle-flycatcher *Rhinomyias brunneata*	◐ m	VU	Bamboo undergrowth in evergreen broadleaf forest at c.600–1,600 m
			TAIWAN (EBA 149)
Fairy Pitta *Pitta nympha*	◖ ms	VU	Subtropical forest in the lowlands and foothills
Taiwan Bulbul *Pycnonotus taivanus*	◐	VU	Secondary forest, scrub, agricultural land and gardens in the lowlands of southern and eastern Taiwan

◐ = breeds only in this forest region; ◖ = also breeds in other region(s); m = migrates to other region(s); s = also occurs in another EBA in the south-east Chinese forests

Much of lowland south-east China has long been cleared for cultivation, principally of rice.

PHOTO: MIKE CROSBY/BIRDLIFE

relatively accessible, low- to mid-altitude subtropical forests have been disproportionately badly affected, so the remaining habitat of the threatened forest birds of this region is severely fragmented, and their populations have been subdivided into ever smaller and more isolated groups of birds. In the past, the main causes of deforestation were conversion of forest to agricultural land, and clear-felling for timber. In many areas, particularly rapid forest loss took place to fuel steel furnaces during the 'Great Leap Forward' in the 1950s, and at a few sites uncontrolled fires have destroyed large blocks of forest. In recent decades, the rate of forest loss has slowed considerably, although some localised losses of natural forest still occur.

The national and provincial governments of mainland China have recently designated many new protected areas, and there are now several hundred in south-east China which officially protect many of the largest and richest areas of natural forest in the region. On Taiwan, the lowlands have long been cleared for agriculture and habitations, but the hills and mountains retain extensive forest cover with large areas protected inside nature reserves.

CONSERVATION ISSUES AND STRATEGIC SOLUTIONS (summarised in Table 3)

Forest loss and degradation

■ *FORESTRY AND ILLEGAL LOGGING*

Until recently, commercial logging by state-run enterprises was a major reason for the diminution of natural forests in south-east China. However, in 1998 a national logging ban (following catastrophic flooding in the lower Yangtze basin, and linked to concerns over siltation of the Three Gorges Dam) was enacted under the National Forest Protection Program (NFPP), which limits logging to local subsistence needs. Although this measure has been largely effective, a significant level of illegal logging has been reported, and subsistence logging can have a major local impact. The ban provides an opportunity for the forestry authorities in China to work together with local communities, NGOs and conservationists to develop sustainable forestry under the NFPP; this would protect primary forest and forests inside nature reserves, whilst allowing sustainable exploitation of forest resources by local people. Since the ban, forestry workers in southern Sichuan have been redeployed in replanting some of the steeper denuded slopes, mainly using seeds from local broadleaf trees. However, similar schemes elsewhere in south-east China often use monocultures and exotic species; instead, reforestation using appropriate mixtures of native species is needed.

■ *CONVERSION TO AGRICULTURE AND PLANTATIONS*

Forestland conversion for agriculture has greatly reduced, fragmented and degraded natural habitats in much of south-east China. Although deforestation for agriculture is now illegal, small-scale encroachment still occurs. Clearance of forest is often linked to uncertainty over land tenure, which makes resource management difficult. Boundary disputes therefore need to be urgently resolved, and land ownership and land-use rights clarified, to help strengthen forest management and prevent illegal conversion. Although plantations may be locally important where natural forest has been destroyed, for example at a site in southern Guangxi where suspected White-eared Night-heron nests have been found, they are rarely as rich as natural forest and are highly susceptible to invasive diseases, e.g. pinewood nematode.

■ *EXPLOITATION OF FOREST PRODUCTS*

The collection of fuelwood from collective forests by local people is permitted under the NFPP, but poses a threat to forest ecosystems, particularly as logging waste is no longer available for fuel. Sustainable alternatives to fuelwood should be promoted, including the use of biogas (produced from livestock manure) and solar power.

■ *DEVELOPMENT (URBAN, INDUSTRIAL, ETC.)*

New roads, dams, power grids, tourism facilities and other developments strongly contribute to habitat loss in south-east China and Vietnam, and the impact of these projects on biodiversity is not usually evaluated properly. Habitat loss linked to economic development is also a threat on Taiwan: an area of forest in Yünlin county supporting a large population of Fairy Pittas was recently scheduled for clearance to allow gravel extraction, although this was suspended following a high-profile media campaign. At Na Hang Nature Reserve, the only locality where White-eared Night-heron has recently been recorded in Vietnam, the government has proceeded with a major dam project which will flood some of the last primary forest in this area, despite the existence of an ongoing, full-size GEF project at the site. Environmental impact assessments should be conducted for development projects that have the potential to damage forested areas; these need to address the potential impact on the ecology of the area, and make appropriate mitigation plans. New developments should be avoided near protected areas and other sites of high biodiversity value. Information on important areas of forest for threatened birds should be provided to local governments, so that protection measures can be incorporated into county development plans.

■ *DISTURBANCE*

The collection of forest products is a major source of disturbance, for example by large numbers of bamboo-shoot collectors in the Daliang Shan in southern Sichuan, where livestock passing through forests to grazing areas is also a problem. Quotas are needed to control such movements, and to limit and regulate bamboo-shoot collecting.

Protected areas coverage and management

■ *GAPS IN PROTECTED AREAS SYSTEM*

Since 1990, China has rapidly expanded its protected areas system, and by 2002 a total of 1,757 nature reserves covered 13.2% of the nation's land. Several hundred of these are in south-east China, many harbouring important populations of threatened forest birds. However, it is important to note that many of these are county reserves with little funding or management infrastructure. Moreover, some significant gaps remain, which need to be addressed by establishing new reserves or modifying the boundaries of existing reserves, to improve coverage of key forest types or create 'green corridors' to link isolated reserves. County reserves with important populations of threatened species should be upgraded to provincial or national reserves, so that they receive more resources.

The most important gap in coverage of the threatened birds of this region is in the Daliang Shan mountains in Sichuan, where very little of the known range of Sichuan Partridge is protected inside nature reserves. The Wildlife Division of the Sichuan Forestry Department is now implementing a long-term plan to improve the protection of its habitat, by extending Mabian Dafengding Nature Reserve (the only reserve known to support the species) and by establishing several new protected areas. On Taiwan, montane habitats are relatively well protected, but more reserves are needed to protect the lower-altitude forests inhabited by the threatened birds. A special refuge is needed

Table 3. Conservation issues and strategic solutions for birds of the south-east Chinese forests.

Conservation issues	Strategic solutions
Forest loss and degradation	
■ *FORESTRY AND ILLEGAL LOGGING* ■ *CONVERSION TO AGRICULTURE AND PLANTATIONS* ■ *EXPLOITATION OF FOREST PRODUCTS* ■ *DEVELOPMENT (URBAN, INDUSTRIAL, ETC.)* ■ *DISTURBANCE*	➤ Develop a sustainable forestry strategy under the Chinese National Forest Protection Program ➤ Promote reforestation using mixtures of native tree species ➤ Reduce illegal forest conversion and cutting by clarifying administrative boundaries, land ownership and land-use rights ➤ Develop sustainable alternatives to wood as a source of fuel ➤ Assess the environmental impact of development projects in forest areas, and incorporate forest protection in local development plans ➤ Control human disturbance at key sites for threatened birds
Protected areas coverage and management	
■ *GAPS IN PROTECTED AREAS SYSTEM* ■ *WEAKNESSES IN RESERVE MANAGEMENT*	➤ Improve coverage of threatened birds by extending the boundaries or upgrading the status of existing reserves, or establishing new reserves ➤ Establish new reserves in the Daliang Shan in Sichuan, and extend Mabian Dafengding Nature Reserve ➤ Create a special refuge to protect a genetically pure population of Taiwan Bulbuls ➤ Revise inadequate or outdated protected area management plans ➤ Strengthen reserve management through improved funding, infrastructure and staff training
Exploitation of birds	
■ *HUNTING* ■ *WILD BIRD TRADE*	➤ Strengthen enforcement of hunting laws in China, including through improved training for law enforcers and conservation education programmes ➤ List Omei Shan Liocichla as a nationally protected species in China
Gaps in knowledge	
■ *INADEQUATE DATA ON THREATENED BIRDS*	➤ Conduct a review of forest cover in south-east China, to help identify sites for survey and protection ➤ Survey the poorly known birds of the Chinese subtropical forests EBA
Other conservation issues	
■ *HYBRIDIZATION OF TAIWAN BULBUL*	➤ Study the distribution and spread of hybrid bulbuls on Taiwan ➤ Establish a genetically pure captive population of Taiwan Bulbuls ➤ Conduct awareness campaigns to discourage releases of Chinese Bulbuls for religious purposes

for the Taiwan Bulbul where a genetically pure population survives; the feasibility of strictly enforcing buffer zones from which all Chinese Bulbuls *Pycnonotus sinensis* and hybrids are removed could be tested, although such a drastic measure could prove difficult to implement.

WEAKNESSES IN RESERVE MANAGEMENT

Although many new protected areas have recently been designated in China, there is no stable financial mechanism in the national budget to support the provincial- and county-level reserves, and many reserves have to generate income for their own operating budget. Weak institutional capabilities and poor staff morale often hamper reserve management, and some reserves are managed by several different government departments, undermining administrative coherence. As a result, illegal activities are widespread in many protected areas, including logging, mining, forest grazing, agricultural encroachment and hunting, and tourism is often a problem. The preparation of management plans is the responsibility of nature reserve staff, subject to government approval, but many existing plans are inadequate or outdated, and require revision. The National Endangered Plant and Wildlife Protection and Nature Reserve Construction Program is a new government initiative to improve the existing protected area system, mainly focused on national (rather than provincial and county) reserves. It could fund the establishment of new reserves to address gaps in coverage, and help resolve management problems by providing stable funding for reserves to improve their infrastructure, staff training and working conditions, and the livelihood of local communities.

Exploitation of birds

HUNTING

Although a ban on gun ownership in China in 1995 may have helped to reduce the pressure, the hunting of birds for food is widespread, despite being illegal in protected areas and for certain protected species. This is believed to be negatively affecting several threatened birds, especially in areas where their habitats have already been fragmented. The larger forest species are probably most seriously affected, notably the partridges and pheasants, but even small birds such as Fairy Pitta are hunted. Stronger enforcement of existing hunting laws is needed both inside and outside protected areas. Forest conservation education programmes are required in south-east China, possibly using threatened forest birds as flagships, with the aim of reducing hunting pressure.

The Sichuan Partridge is virtually confined to the Daliang Shan mountains of south-central Sichuan.

PHOTO: DAI BO

WILD BIRD TRADE

Trapping for the national and international cagebird trade may be a significant threat to Omei Shan Liocichla (and possibly other threatened species), which has appeared in trade in Europe, and which should now be listed as a nationally protected (i.e. untradeable) species in China. There is also a need for improved training and capacity-building for inspectors and enforcers of wildlife legislation.

Gaps in knowledge

INADEQUATE DATA ON THREATENED BIRDS

Apart from the pheasants and partridges, the distributions of most threatened forest birds of south-east China are poorly known, notably the passerines. A regional forest review, using forest-cover maps and possibly satellite images and GIS data, should be conducted to select potentially suitable areas for threatened birds, to guide surveys to identify key sites for their conservation. This will help determine the location of new nature reserves, and the modifications needed to the boundaries of existing reserves. The distributions of the four species of the Chinese subtropical forests EBA are particularly poorly known: they all occur in south-central Sichuan, with breeding records of Gold-fronted Fulvetta and Silver Oriole several hundred kilometres to the south-east in Guangxi and Guangdong, and it is possible that some of them also occur in the intervening areas. Surveys targeting these species are required in southern Sichuan, north-west Yunnan, Guizhou, southern Hunan, Guangxi and Guangdong. White-eared Night Heron was rediscovered in northern Vietnam during the only recent ornithological surveys of the Na Hang area, and it is likely that further surveys in that area would locate other sites for the heron, or possibly other threatened birds of the South-East Chinese mountains EBA. Several threatened species appear to be very local, presumably reflecting specialised habitat requirements, and ecological studies are required to improve understanding of their needs and future management.

Other conservation issues

HYBRIDISATION OF TAIWAN BULBUL

The ranges of Taiwan Bulbul and the closely related Chinese Bulbul have changed on Taiwan, presumably because of alterations to the original habitats there, and they now overlap and hybridise in several areas. The problem has been exacerbated by releases of Chinese Bulbuls for religious purposes. Hybrids are now widespread, with genetically pure populations of Taiwan Bulbul remaining in only a few isolated parts of its range. More field studies are required to clarify the distribution and spread of hybrids within the wild population of the Taiwan Bulbul, to help develop the most appropriate conservation measures. If genetically pure Taiwan Bulbuls can be obtained, a rigorously managed captive breeding programme should be initiated by Taiwan's zoos to maintain pure stocks while the problem in the wild is being resolved. An awareness campaign should be used to alert the Taiwan government and public to the plight of one of the island's endemic species, particularly to discourage any further releases of Chinese Bulbuls for religious purposes.

SINO-HIMALAYAN MOUNTAIN FORESTS

F04

THIS region is made up of the mid- and high-elevation forests, scrub and grasslands that cloak the southern slopes of the Himalayas and the mountains of south-west China and northern Indochina. A total of 28 threatened species are confined (as breeding birds) to the Sino-Himalayan mountain forests, including six which are relatively widespread in distribution, and 22 which inhabit one of the region's six Endemic Bird Areas: Western Himalayas, Eastern Himalayas, Shanxi mountains, Central Sichuan mountains, West Sichuan mountains and Yunnan mountains. All 28 taxa are considered Vulnerable, apart from the Endangered White-browed Nuthatch, which has a tiny range, and the Critically Endangered Himalayan Quail, which has not been seen for over a century and may be extinct.

- **Key habitats** Montane temperate, subtropical and subalpine forest, and associated grassland and scrub.
- **Altitude** 350–4,500 m.
- **Countries and territories** **China** (Tibet, Qinghai, Gansu, Sichuan, Yunnan, Guizhou, Shaanxi, Shanxi, Hebei, Beijing, Guangxi); **Pakistan**; **India** (Jammu and Kashmir, Himachal Pradesh, Uttaranchal, West Bengal, Sikkim, Arunachal Pradesh, Assam, Meghalaya, Nagaland, Manipur, Mizoram); **Nepal**; **Bhutan**; **Myanmar**; **Thailand**; **Laos**; **Vietnam**.

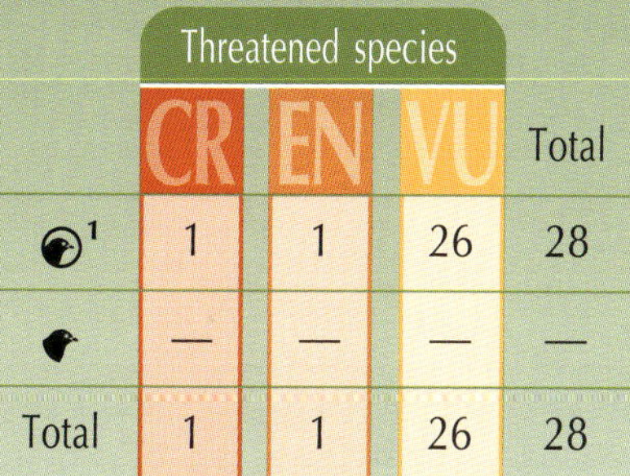

	Threatened species			
	CR	EN	VU	Total
[1]	1	1	26	28
	—	—	—	—
Total	1	1	26	28

Key: = breeds only in this forest region.
[1] Three species which nest only in this forest region migrate to other regions outside the breeding season, Wood Snipe, Rufous-headed Robin and Kashmir Flycatcher.
= also breeds in other region(s).

The Sino-Himalayan mountain forests region includes Conservation International's South-central China mountains Hotspot (see pp.20–21).

The forests of the Palas valley in northern Pakistan support the largest known population of Western Tragopan. PHOTO: BIRDLIFE

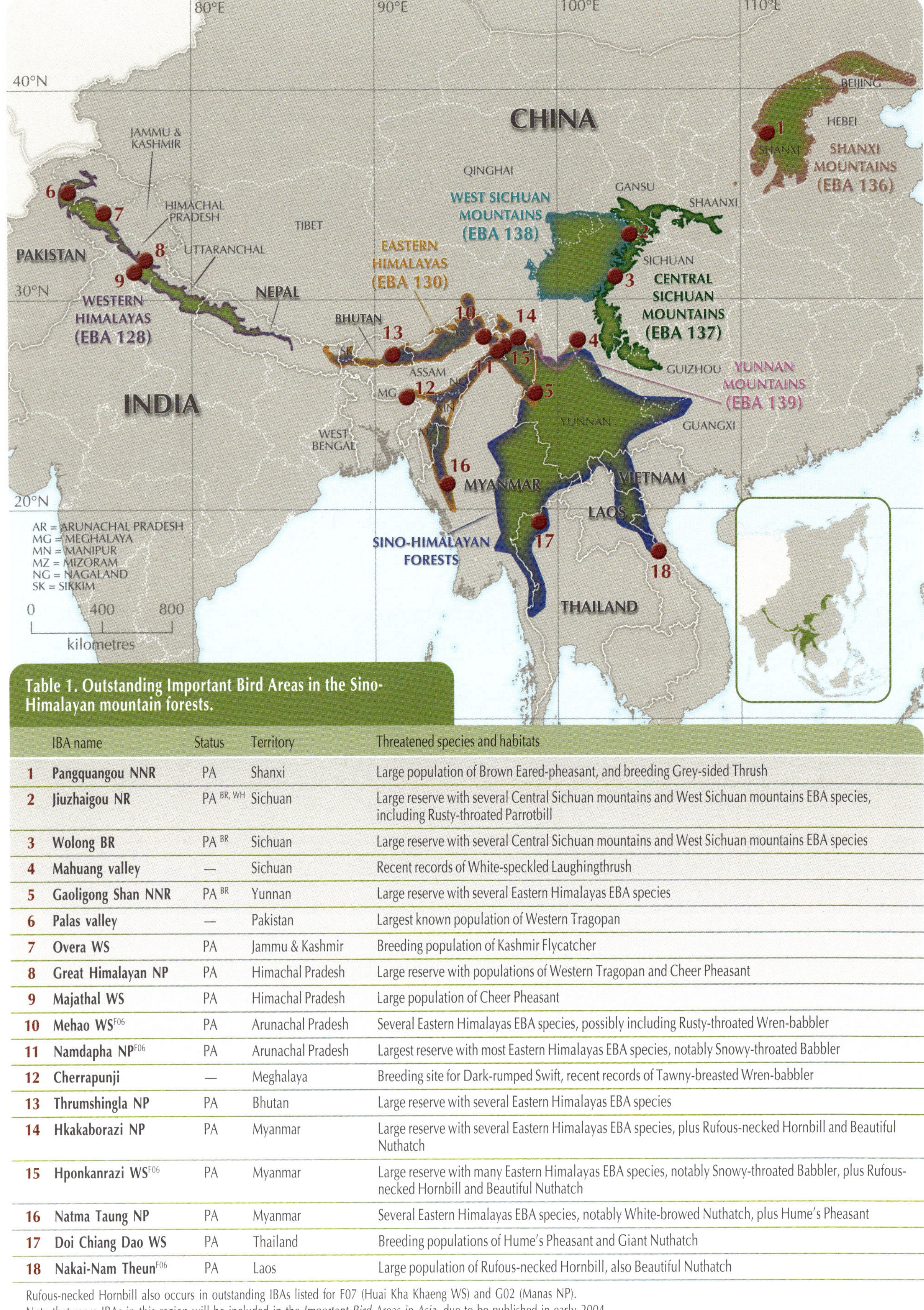

Table 1. Outstanding Important Bird Areas in the Sino-Himalayan mountain forests.

	IBA name	Status	Territory	Threatened species and habitats
1	**Pangquangou NNR**	PA	Shanxi	Large population of Brown Eared-pheasant, and breeding Grey-sided Thrush
2	**Jiuzhaigou NR**	PA BR, WH	Sichuan	Large reserve with several Central Sichuan mountains and West Sichuan mountains EBA species, including Rusty-throated Parrotbill
3	**Wolong BR**	PA BR	Sichuan	Large reserve with several Central Sichuan mountains and West Sichuan mountains EBA species
4	**Mahuang valley**	—	Sichuan	Recent records of White-speckled Laughingthrush
5	**Gaoligong Shan NNR**	PA BR	Yunnan	Large reserve with several Eastern Himalayas EBA species
6	**Palas valley**	—	Pakistan	Largest known population of Western Tragopan
7	**Overa WS**	PA	Jammu & Kashmir	Breeding population of Kashmir Flycatcher
8	**Great Himalayan NP**	PA	Himachal Pradesh	Large reserve with populations of Western Tragopan and Cheer Pheasant
9	**Majathal WS**	PA	Himachal Pradesh	Large population of Cheer Pheasant
10	**Mehao WS** F06	PA	Arunachal Pradesh	Several Eastern Himalayas EBA species, possibly including Rusty-throated Wren-babbler
11	**Namdapha NP** F06	PA	Arunachal Pradesh	Largest reserve with most Eastern Himalayas EBA species, notably Snowy-throated Babbler
12	**Cherrapunji**	—	Meghalaya	Breeding site for Dark-rumped Swift, recent records of Tawny-breasted Wren-babbler
13	**Thrumshingla NP**	PA	Bhutan	Large reserve with several Eastern Himalayas EBA species
14	**Hkakaborazi NP**	PA	Myanmar	Large reserve with several Eastern Himalayas EBA species, plus Rufous-necked Hornbill and Beautiful Nuthatch
15	**Hponkanrazi WS** F06	PA	Myanmar	Large reserve with many Eastern Himalayas EBA species, notably Snowy-throated Babbler, plus Rufous-necked Hornbill and Beautiful Nuthatch
16	**Natma Taung NP**	PA	Myanmar	Several Eastern Himalayas EBA species, notably White-browed Nuthatch, plus Hume's Pheasant
17	**Doi Chiang Dao WS**	PA	Thailand	Breeding populations of Hume's Pheasant and Giant Nuthatch
18	**Nakai-Nam Theun** F06	PA	Laos	Large population of Rufous-necked Hornbill, also Beautiful Nuthatch

Rufous-necked Hornbill also occurs in outstanding IBAs listed for F07 (Huai Kha Khaeng WS) and G02 (Manas NP).
Note that more IBAs in this region will be included in the *Important Bird Areas in Asia*, due to be published in early 2004.

Key *IBA name*: BR = Biosphere Reserve; NP = National Park; NR = Nature Reserve; NNR = National Nature Reserve; WS = Wildlife Sanctuary.
Status: PA = IBA is a protected area; (PA) = IBA partially protected; — = unprotected; BR = IBA is wholly or partially a Biosphere Reserve (see pp.34–35); WH = IBA is wholly or partially a World Heritage Site (see p.35); F06 = also supports threatened forest birds of region F06.

Table 2. Threatened birds of the Sino-Himalayan mountain forests.

Species			Distribution and habitat
			SINO-HIMALAYAN FORESTS
Hume's Pheasant *Syrmaticus humiae*	◐	VU	Forest at 750–3,300 m in Myanmar and adjacent parts of north-east India and northern Thailand, and China (Yunnan and Guangxi)
Wood Snipe *Gallinago nemoricola*	◐ m	VU	Breeds in alpine grassland and scrub near the treeline in the Himalayas and south-west China (EBAs 128, 130, 138 and probably 139)
Rufous-necked Hornbill *Aceros nipalensis*	◐	VU	Broadleaf evergreen forest at 700–2,000 m from Bhutan, north-east India, Myanmar and southern China (Tibet and Yunnan) to northern Thailand, Laos and Vietnam
Grey-sided Thrush *Turdus feae*	◐	VU	Breeds in forest above 1,000 m in Shanxi, Hebei and Beijing (EBA 136), and winters in the mountains of north-east India, Myanmar, Thailand and Laos
Giant Nuthatch *Sitta magna*	◐	VU	Pine forests at 1,000–2,000 m in China (Yunnan and adjacent parts of Sichuan and Guizhou), eastern Myanmar and northern Thailand
Beautiful Nuthatch *Sitta formosa*	◐	VU	Broadleaf evergreen forest at 600–2,400 m from Bhutan and north-east India to northern Thailand, Laos and Vietnam
			WESTERN HIMALAYAS (EBA 128)
Himalayan Quail *Ophrysia superciliosa*	◐	CR	Known by nineteenth century records from two localities in Uttar Pradesh, in grassland and scrub on steep hillsides at 1,650–2,400 m
Western Tragopan *Tragopan melanocephalus*	◐	VU	Ranges from northern Pakistan to Uttar Pradesh, where it nests in coniferous and mixed forests at 2,500–3,600 m
Cheer Pheasant *Catreus wallichi*	◐	VU	Ranges from northern Pakistan to western Nepal, favouring steep slopes with scrub, tall grass and stunted trees at 1,200–3,050 m
Kashmir Flycatcher *Ficedula subrubra*	◐ m	VU	Virtually confined as a breeding bird to Kashmir, nesting at 1,800–2,300 m in open broadleaf deciduous forests with scrub
The Data Deficient Sillem's Mountain-finch *Leucosticte sillemi* is known by two specimens collected on a barren plateau at c.5,125 m in southern Xinjiang, China, immediately to the north of the Western Himalayas			
			EASTERN HIMALAYAS (EBA 130)
Chestnut-breasted Partridge *Arborophila mandellii*	◐	VU	Himalayas of north-east India and Bhutan, in broadleaf evergreen forest at 350–2,500 m
Blyth's Tragopan *Tragopan blythii*	◐	VU	Himalayas from Bhutan to Yunnan, and the mountains south of the Brahmaputra in north-east India and north-west Myanmar, in forest at c.1,500–3,300 m
Sclater's Monal *Lophophorus sclateri*	◐	VU	Himalayas from Arunachal Pradesh to Yunnan, in subalpine scrub, grassland and coniferous forest at 2,500–4,200 m
Dark-rumped Swift *Apus acuticauda*	◐	VU	Nests on cliffs near forest, with known breeding colonies in India (Meghalaya and Mizoram) and eastern Bhutan
Rusty-bellied Shortwing *Brachypteryx hyperythra*	◐	VU	Himalayas from West Bengal to northern Myanmar and Yunnan, in forest up to c.2,900 m
Rusty-throated Wren-babbler *Spelaeornis badeigularis*	◐	VU	Known by a single specimen collected in winter in eastern Arunachal Pradesh, in subtropical forest at 1,600 m
Tawny-breasted Wren-babbler *Spelaeornis longicaudatus*	◐	VU	Mountains south of the Brahmaputra river in Meghalaya, Assam and Manipur, in forest at 1,000–2,000 m
Snowy-throated Babbler *Stachyris oglei*	◐	VU	Recorded in forest and scrub in ravines in eastern Arunachal Pradesh and northern Myanmar
White-browed Nuthatch *Sitta victoriae*	◐	EN	Only found on Mt Victoria and in southern Chin State in western Myanmar, in broadleaf forest at c.2,300–3,000 m
			SHANXI MOUNTAINS (EBA 136)
Brown Eared-pheasant *Crossoptilon mantchuricum*	◐	VU	Mixed coniferous-broadleaf forest at 800–2,600 m
			CENTRAL SICHUAN MOUNTAINS (EBA 137)
Rufous-headed Robin *Luscinia ruficeps*	◐ m	VU	Forest and scrub at c.2,400–2,800 m in northern Sichuan and southern Shaanxi
Black-throated Blue Robin *Luscinia obscura*	◐	VU	Forest and bamboo in the mountains of southern Gansu and northern Sichuan
Snowy-cheeked Laughingthrush *Garrulax sukatschewi*	◐	VU	Forest, bamboo and scrub at 2,000–3,500 m in southern Gansu and northern Sichuan
Grey-hooded Parrotbill *Paradoxornis zappeyi*	◐	VU	Bamboo and stunted forest above 2,000 m in southern Sichuan and northern Guizhou
Rusty-throated Parrotbill *Paradoxornis przewalskii*	◐	VU	Forest, bamboo and scrub above 2,000 m in southern Gansu and northern Sichuan
			WEST SICHUAN MOUNTAINS (EBA 138)
Chinese Monal *Lophophorus lhuysii*	◐	VU	Breeds in grassland, scrub and rocky areas above the treeline at c.3,300–4,500 m
Sichuan Jay *Perisoreus internigrans*	◐	VU	Subalpine coniferous forest and mixed fir and rhododendron forest at c.3,000–4,300 m
			YUNNAN MOUNTAINS (EBA 139)
White-speckled Laughingthrush *Garrulax bieti*	◐	VU	Coniferous forest and mixed fir and rhododendron forest at c.2,500–4,300 m

◐ = breeds only in this forest region; m = migrates to other region(s)

OUTSTANDING IBAs FOR THREATENED BIRDS (see Table 1)

Eighteen IBAs have been selected, which together support populations of all threatened forest birds of this region apart from Himalayan Quail and possibly Rusty-throated Wren-babbler, and include some of the largest and richest forests remaining in the Sino-Himalayan mountains. Many other forest sites in this region with significant populations of threatened forest birds will be documented during BirdLife's ongoing IBA Project.

CURRENT STATUS OF HABITATS AND THREATENED SPECIES

The extent of forest loss varies widely within this region. In China, large-scale logging operations have clear-felled huge areas of forest, including subalpine forests up to the treeline. In Sichuan, forest cover declined from 19% to 12.6% between the early 1950s and 1988 and in Yunnan it declined from c.55% to c.30% between the early 1950s and 1975; subsequent loss of habitat in China has, if anything, been even more rapid (see F03: South-east Chinese forests). However, some extensive tracts of forest are protected in reserves (often established for charismatic mammals such as giant panda *Ailuropoda melanoleuca* and Yunnan snub-nosed monkey *Rhinopithecus bieti*) or survive in unlogged areas, and in 1998 a national logging ban restricted logging to local subsistence needs. In the Himalayas, although natural habitat remains relatively secure at high altitudes, forests at mid-elevations tend to be under greater pressure, and are now highly fragmented in many areas. Three-quarters of original forest cover has been lost in the western Himalayas, and this figure is even higher for Nepal, where almost all lower-altitude subtropical forests have been lost or degraded. In north-east India and northern Myanmar the situation is less extreme, as large, almost continuous areas of mid- and high-altitude forests remain. However, forests were being rapidly logged in north-east India until a recent ban, and shifting cultivation has cleared or degraded large areas. In Bhutan an admirable national policy proposes to maintain forests over 60% of the country, although some localised habitat loss has been reported. Some montane forests in Myanmar, Thailand, Laos and Vietnam have been badly affected by clearance for permanent and (especially) shifting agriculture, as well as commercial timber extraction; the situation is particularly acute in Chin State, Myanmar, northern Laos and north-west Vietnam, but some large areas of relatively undisturbed forest remain, for example in northern Myanmar and the Annamite mountains in Laos and Vietnam.

CONSERVATION ISSUES AND STRATEGIC SOLUTIONS (summarised in Table 3)

Forest loss and degradation

■ *SHIFTING CULTIVATION*

Some parts of the region are populated by ethnic groups whose lifestyle is based on shifting or semi-shifting cultivation. Practised on a small scale using long clearing cycles, this has little lasting impact and the forest eventually recovers. However, in many areas increased population pressure and a reduction in the area of forest available have

Table 3. Conservation issues and strategic solutions for birds of the Sino-Himalayan mountain forests.

Conservation issues	Strategic solutions
Forest loss and degradation	
■ *SHIFTING CULTIVATION* ■ *CONVERSION TO AGRICULTURE AND PLANTATIONS* ■ *FORESTRY AND ILLEGAL LOGGING* ■ *EXPLOITATION OF FOREST PRODUCTS* ■ *DEVELOPMENT (URBAN, INDUSTRIAL, ETC.)* ■ *DISTURBANCE AND LIVESTOCK GRAZING*	➤ Promote sustainable forms of upland agriculture that do not result in net loss of natural forest ➤ Prevent conversion to agriculture and plantations within protected areas ➤ Reduce illegal forest conversion and cutting in China by clarifying administrative boundaries, land ownership and land-use rights ➤ Continue logging bans in China and north-east India, and promote similar bans elsewhere in the region ➤ Develop sustainable, community-based forest management in China, northern India and Pakistan ➤ Control illegal logging at key sites for threatened birds ➤ Establish community forests to provide non-timber forest products, and develop sustainable alternatives to wood as a source of fuel ➤ Assess the environmental impact of development projects in forested areas, and minimise development near key sites ➤ Control human disturbance and overgrazing at key sites
Protected areas coverage and management	
■ *GAPS IN PROTECTED AREAS SYSTEM* ■ *WEAKNESSES IN RESERVE MANAGEMENT*	➤ Improve coverage of threatened species by modifying the boundaries of existing reserves, or by establishing new reserves (including village sanctuaries) ➤ Review and revise the protected areas system of Myanmar ➤ Strengthen reserve management through improved funding, infrastructure and staff training
Exploitation of birds	
■ *HUNTING AND TRAPPING*	➤ Improve enforcement of hunting laws, and control gun ownership, particularly near protected areas ➤ Conduct education programmes to improve understanding of forest conservation, threatened species and hunting laws ➤ Establish community-based conservation initiatives to control hunting at key sites
Gaps in knowledge	
■ *INADEQUATE DATA ON THREATENED BIRDS*	➤ Search for Himalayan Quail in the Western Himalayas ➤ Survey north-east India, Bhutan and northern Myanmar for poorly known Eastern Himalayan species ➤ Determine whether White-speckled Laughingthrush has populations in any protected area in Yunnan, or if new reserves are required ➤ Investigate whether the poorly known threatened birds of the Sichuan mountains are adequately protected by giant panda reserves

Logs floating down a river illustrate the huge scale of clear-felling of forests in China, but a recent logging ban provides a great opportunity to develop sustainable forestry under the National Forest Protection Program.

PHOTO: MIKE CROSBY/BIRDLIFE

resulted in a rapid shortening of the cycle, and an increase in pioneer shifting cultivation, which is causing widespread forest loss and degradation, principally in the mountains of north-east India, Myanmar and northern Indochina. Often associated with shifting cultivation, fires have been one of biggest causes of montane forest loss in northern Vietnam and Laos. There is an urgent need to promote sustainable forms of upland agriculture that do not result in net loss of natural forest, including improved agricultural and agro-forestry practices that enable farmers to remain longer on established clearings, and thereby reduce their demand for new land. Efforts are needed to rehabilitate abandoned land, including the development of community forestry plantations. At selected key sites, projects should integrate conservation and land-use development, and work in collaboration with local people. Such projects could prevent shifting cultivation within core areas of reserves, and control it in buffer zones. Simple socio-economic incentives to replace destructive practices by shifting cultivators should be promoted through environmental awareness and training programmes.

■ *CONVERSION TO AGRICULTURE AND PLANTATIONS*

Clearance for permanent agriculture and plantations has contributed to the reduction and fragmentation of natural forests in many areas. Deforestation for agriculture is now illegal in China, but small-scale encroachment still happens. Conversion for tea and other plantations is still taking place in north-east India. These activities need to be effectively banned within core areas of reserves, and controlled in buffer zones. In China (and probably elsewhere), disputes over administrative boundaries need to be resolved and land ownership and land-use rights clarified, to help strengthen forest management and prevent illegal conversion for agriculture.

■ *FORESTRY AND ILLEGAL LOGGING*

Until recently, commercial logging was a major reason for the diminution of natural forests in this region, particularly subtropical forests at mid-altitudes. The clearance and degradation of forests has caused major environmental problems, including erosion, which causes siltation of rivers, and damage to watersheds, which affects water supplies (causing periodic flooding and drought). A national logging ban was enacted in China in 1998, which limits logging to local subsistence needs. Following disastrous floods in 1992, logging was banned in North West Frontier Province, Pakistan; although the ban was lifted in 2001, strict new regulations have so far prevented the resumption of wide-scale logging. Logging and timber export was recently banned in Arunachal Pradesh and other north-east Indian states. These bans have greatly reduced the rate of forest loss, but some illegal logging continues, and in Arunachal Pradesh it is still legal to cut timber to supply plywood and veneer factories. In Laos and Vietnam, logging of the commercially very valuable conifer *Fokienia hodginsii*, mainly for export to Japan, affects montane areas.

The commercial logging bans in China and India should be continued, and, where appropriate, similar restrictions promoted in other parts of the region. The ban in China allows the forestry authorities to work together with local communities, NGOs and conservationists to develop sustainable forestry under the National Forest Protection Program (see *Forestry and illegal logging* under F03).

A huge variety of foods and medicines are harvested from forests in China.

PHOTO: MIKE CROSBY/BIRDLIFE

Measures are needed to minimise illegal logging, particularly in protected areas and other sites of high biodiversity value. Since the ban, forestry workers in Sichuan have been redeployed in replanting some of the steeper denuded slopes, mainly using seeds from local broadleaf trees; similar reforestation with appropriate mixtures of native tree species is needed elsewhere in the region. In Pakistan, the government is currently revising its Forest Working Plan, which provides an opportunity to prescribe extraction rates and logging practices which benefit local people and conserve biodiversity.

■ *EXPLOITATION OF FOREST PRODUCTS*

The collection of wood and other forest vegetation for fuel, construction material and fodder is causing forest degradation in many parts of the region. The collection of fuelwood from collective forests by local people is permitted in China, but poses a threat to forest ecosystems, particularly as logging waste is no longer available for fuel. In some areas of Nepal (and presumably also India), the cutting of fuelwood to provide heating and hot water for tourists is degrading forests near trekking routes. These activities need to be minimised throughout, and controlled within core zones of protected areas. Community or collective forests need to be established more widely, and their management improved and made more sustainable through training in forestry techniques for local people. Sustainable alternatives to fuelwood should be promoted, including the use of biogas (produced from livestock manure) and solar power.

■ *DEVELOPMENT (URBAN, INDUSTRIAL, ETC.)*

New roads, dams, mines, buildings and other developments strongly contribute to habitat loss in this region, both by directly damaging forests, and indirectly by displacing people into forest areas. Construction of roads in highland areas may cause landslips, and provide improved access to remote montane habitats for shifting cultivators, illegal loggers, hunters and harvesters of forest products. Environmental impact assessments should be conducted for development projects (and associated new and upgraded access routes) that have the potential to damage forested areas, with appropriate mitigation plans. New developments should be avoided near protected areas and other sites of high biodiversity value.

■ *DISTURBANCE AND LIVESTOCK GRAZING*

In addition to the permanent residents, there are seasonal influxes of large numbers of people, their livestock and dogs into montane habitats. In summer, flocks of sheep, goats or yaks are taken to the high-altitude forests and alpine grasslands, damaging vegetation structure, particularly the understorey; moreover, production of cheese from yak milk requires considerable fuelwood. The summer peak in both grazing and the (locally large-scale) collection of edible fungi and medicinal plants coincides with the bird breeding season, probably posing a threat to some threatened species (e.g. partridges and pheasants) through disturbance and hunting. New management measures are needed in areas where human disturbance and overgrazing are a problem; for example, in the Daliang Shan in Sichuan control of livestock movements through forests to grazing areas is required, along with restrictions on bamboo-shoot collecting.

Protected areas coverage and management

■ *GAPS IN PROTECTED AREAS SYSTEM*

There are many protected areas in this region, which together support populations of most threatened birds. However, it is unclear whether some of the more poorly known species, such as Himalayan Quail, Dark-rumped Swift, White-speckled Laughingthrush and Rusty-throated Wren-babbler, are adequately covered by existing reserves. The most important populations of Western Tragopan in northern Pakistan are currently unprotected, notably in the Palas valley; conservation measures are needed to protect these forests from high-impact commercial logging. Where surveys locate important populations of poorly known species, or reveal that current levels of protection are inadequate (as in the case of Hume's Pheasant in India), new reserves need to be established for the species in question. Many protected areas are too small to support viable populations of the larger threatened species; to improve coverage of these low-density birds, selected reserves should be enlarged, and their buffer zones expanded. Villagers are often not opposed to the idea of protected areas, and indeed themselves confer some level of protection to montane forests in order to maintain their water supply and populations of prey species; such local protection could be reinforced and supported, by encouraging communities to establish village sanctuaries. The protected areas system in Myanmar is currently being reviewed and revised, and surveys are required in ornithologically undocumented areas to locate suitable sites for new reserves, for example in Shan State to safeguard Hume's Pheasant and Giant Nuthatch.

■ *WEAKNESSES IN RESERVE MANAGEMENT*

There are major problems with the management of protected areas, linked to a lack of resources and the practical difficulties associated with remote mountainous terrain, with forestry departments in most countries understaffed, underfunded and badly equipped. As a consequence, illegal activities are widespread in many reserves, including logging, mining, forest grazing, agricultural encroachment, hunting, etc. Increased resources and improved management planning and training of reserve staff are needed, with a particular focus on the most outstanding sites for threatened biodiversity. Appropriate funding and training of the governmental departments responsible for conservation are also required. In China, the National Endangered Plant and Wildlife Protection and Nature Reserve Construction Program is a new government initiative that has the potential to make these types of improvement.

Exploitation of wildlife

■ *HUNTING AND TRAPPING*

Shooting and snaring of wildlife is common throughout much of this region, which is populated by ethnic groups whose lifestyle is based on shifting or semi-shifting cultivation and hunting, although pressure is less intense in Bhutan and Tibet for cultural reasons. The larger forest species are probably most seriously affected, notably the hornbills, and partridges and pheasants, which have been much reduced in many areas. Improved law enforcement is required to prevent illegal hunting both inside and outside protected areas. Control of gun ownership is meeting with some success in Laos and Vietnam, and could be expanded to other countries in the region, particularly near important protected areas. Education programmes concerned with forest conservation, threatened species and the hunting laws

Blyth's Tragopan and other galliforms are shot and snared in many parts of the eastern Himalayas.

PHOTO: JEAN HOWMAN/W²A

The threatened birds of the Sichuan mountains are poorly known, and studies are required to investigate whether they are adequately protected by the reserves established for giant panda.

PHOTO: JOHN HOLMES

could help reduce hunting pressure, possibly using the most charismatic threatened forest birds such as Blyth's Tragopan or Rufous-necked Hornbill as flagships. Community-based initiatives, including signing of stakeholder agreements with local households and establishment of patrolling groups, have helped control hunting at key sites in Vietnam, and should be tried in other areas of high biodiversity value.

Gaps in knowledge

■ *INADEQUATE DATA ON THREATENED BIRDS*

Knowledge of the distribution and ecology of the threatened birds is very incomplete, including their tolerance of logging and other modifications of their habitats. Two Himalayan species—Himalayan Quail in the west and Rusty-throated Wren-babbler in the east—have gone unreported for years, and need to be searched for. Most other Eastern Himalayan species are poorly known, notably Chestnut-breasted Partridge, Dark-rumped Swift, Snowy-throated Babbler and Rusty-bellied Shortwing, and surveys are required throughout the remaining forests in the highlands of north-east India, Bhutan and northern Myanmar. In China, surveys are required for White-speckled Laughingthrush, to determine whether it has populations in any protected areas and, if not, which of its sites should be established as reserves. The threatened species of the West and Central Sichuan mountains EBAs are poorly known, most notably Rufous-headed Robin, Black-throated Blue Robin and Rusty-throated Parrotbill, and studies are required to investigate whether these birds are sufficiently protected by the network of reserves established for giant panda and other mammals.

INDIAN PENINSULA and SRI LANKAN FORESTS

F05

TWELVE threatened bird species are confined to forests in this region, and two others occur as non-breeding visitors. They are found in three discrete forest areas, two Endemic Bird Areas—the Western Ghats and Sri Lanka—and one Secondary Area—the Central Indian forests. The Central Indian forests support the Critically Endangered Forest Owlet, a species only recently rediscovered after a gap of over 100 years. In the more humid forests of the Western Ghats, three of the four endemic species are confined to the higher altitude habitats in the south, but Nilgiri Wood-pigeon occurs throughout. On Sri Lanka, six of the seven endemics are confined to the moist forests of the wet zone in the south-west of the island, with only Red-faced Malkoha ranging into the dry zone of the north and east. In the wet zone, Green-billed Coucal is confined to rainforests in the lowlands, and Sri Lanka Wood-pigeon and Sri Lanka Whistling-thrush to forests at higher altitudes.

- **Key habitats** Lowland wet evergreen and semi-evergreen rainforest, moist deciduous forest and montane wet temperate forest; montane grasslands.
- **Altitude** 0–2,600 m.
- **Countries and territories** **India** (Madhya Pradesh, Maharashtra, Goa, Karnataka, Kerala, Tamil Nadu, Orissa); **Sri Lanka**.

Threatened species

	CR	EN	VU	Total
breeds only in this forest region	1	2	9	12
also breeds in other region(s)	—	—	—	—
non-breeding visitor from another region	—	—	2	2
Total	1	2	11	14

Key: = breeds only in this forest region. = also breeds in other region(s). = non-breeding visitor from another region.

The Indian peninsula and Sri Lankan forests region corresponds closely to Conservation International's Western Ghats and Sri Lanka Hotspot (see pp.20–21).

Montane grassland interspersed with wet temperate shola forests, a characteristic landscape of the southern Western Ghats. PHOTO: AMEEN AHMED

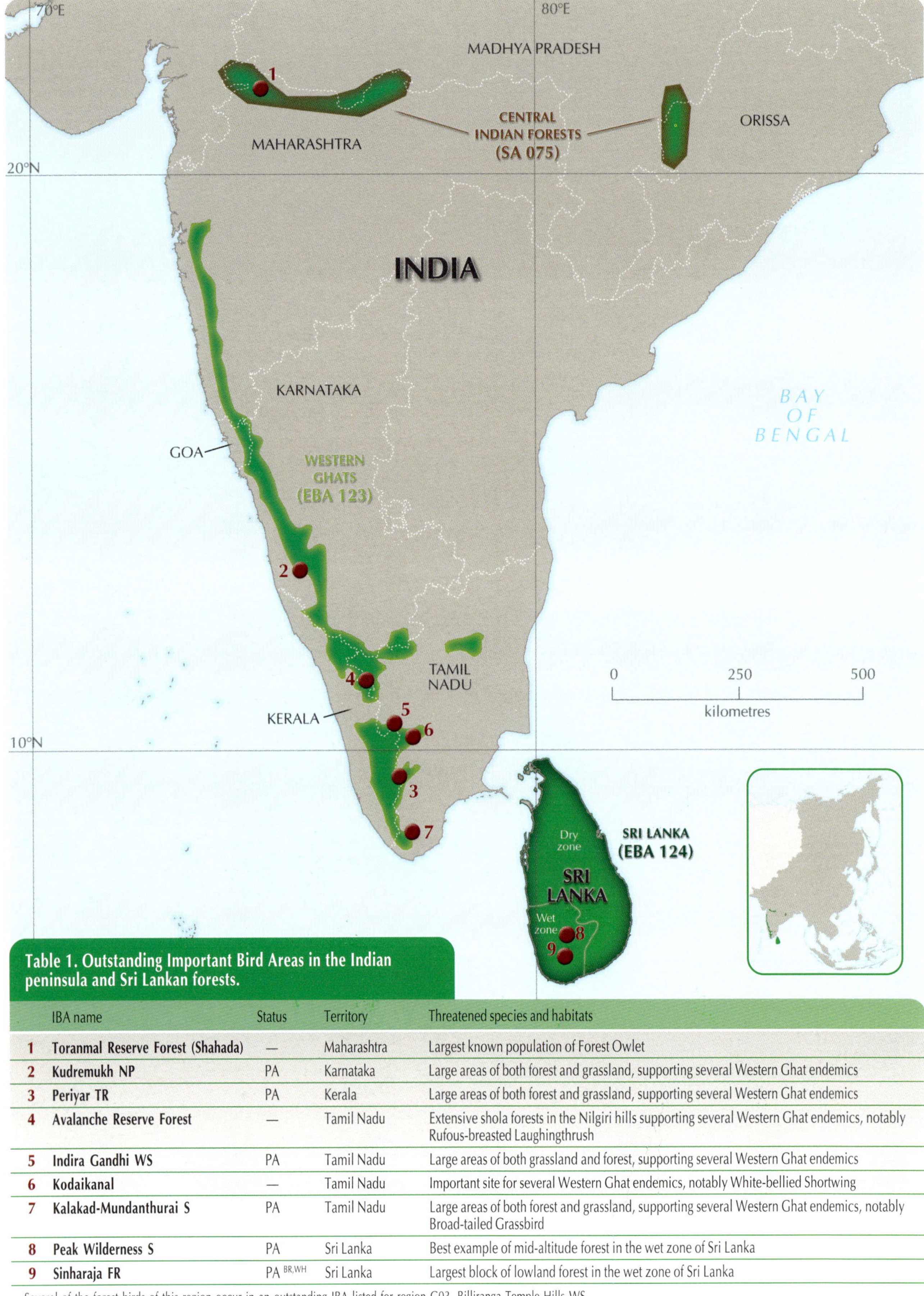

Table 1. Outstanding Important Bird Areas in the Indian peninsula and Sri Lankan forests.

	IBA name	Status	Territory	Threatened species and habitats
1	**Toranmal Reserve Forest (Shahada)**	—	Maharashtra	Largest known population of Forest Owlet
2	**Kudremukh NP**	PA	Karnataka	Large areas of both forest and grassland, supporting several Western Ghat endemics
3	**Periyar TR**	PA	Kerala	Large areas of both forest and grassland, supporting several Western Ghat endemics
4	**Avalanche Reserve Forest**	—	Tamil Nadu	Extensive shola forests in the Nilgiri hills supporting several Western Ghat endemics, notably Rufous-breasted Laughingthrush
5	**Indira Gandhi WS**	PA	Tamil Nadu	Large areas of both grassland and forest, supporting several Western Ghat endemics
6	**Kodaikanal**	—	Tamil Nadu	Important site for several Western Ghat endemics, notably White-bellied Shortwing
7	**Kalakad-Mundanthurai S**	PA	Tamil Nadu	Large areas of both forest and grassland, supporting several Western Ghat endemics, notably Broad-tailed Grassbird
8	**Peak Wilderness S**	PA	Sri Lanka	Best example of mid-altitude forest in the wet zone of Sri Lanka
9	**Sinharaja FR**	PA [BR,WH]	Sri Lanka	Largest block of lowland forest in the wet zone of Sri Lanka

Several of the forest birds of this region occur in an outstanding IBA listed for region G03, Billiranga Temple Hills WS.
Note that more IBAs in this region will be included in the *Important Bird Areas in Asia*, due to be published in early 2004.

Key *IBA name*: FR = Forest Reserve; NP = National Park; S = Sanctuary; TR = Tiger Reserve: WS = Wildlife Sanctuary.
Status: PA = IBA is a protected area; (PA) = IBA partially protected; — = unprotected; BR = IBA is wholly or partially a Biosphere Reserve (see pp.34–35); WH = IBA is wholly or partially a World Heritage Site (see p.34).

White-bellied Shortwing is confined to high-altitude forests of the Western Ghats.

PHOTO: TIM LOSEBY

OUTSTANDING IBAs FOR THREATENED BIRDS (see Table 1)

Eight IBAs have been selected, which together support populations of all of the threatened forest birds of the Indian peninsula and Sri Lanka, and include some of the largest and richest forests that remain there. Many other sites in the region with significant populations of threatened forest birds will be documented during BirdLife's ongoing IBA Project.

CURRENT STATUS OF HABITATS AND THREATENED SPECIES

There has been extensive clearance and degradation of forests (and montane grasslands) in all areas. Relatively low-

Table 2. Threatened birds of the Indian peninsula and Sri Lankan forests.

Species			Distribution and habitat
			CENTRAL INDIAN FORESTS (SA 075)
Forest Owlet *Heteroglaux blewitti*	⊛	CR	Recorded at six localities (recently at four in Madhya Pradesh and Maharashtra), in dry deciduous forest at c.400–500 m
			WESTERN GHATS (EBA 123)
Wood Snipe *Gallinago nemoricola*	✈	VU	Winters in forested areas at high altitudes
Nilgiri Wood-pigeon *Columba elphinstonii*	⊛	VU	Ranges from Maharashtra to Kerala, in evergreen forest at c.50–2,000 m, but mostly at high altitudes
White-bellied Shortwing *Brachypteryx major*	⊛	VU	Found in Karnataka, Kerala and Tamil Nadu, in evergreen forest above 1,000 m
Rufous-breasted Laughingthrush *Garrulax cachinnans*	⊛	EN	Confined to the Nilgiri hills in Kerala and Tamil Nadu, in evergreen forest above 1,200 m
Broad-tailed Grassbird *Schoenicola platyura*	⊛	VU	Found in Karnataka, Kerala and Tamil Nadu, in tall grassland and bracken at 900–2,000 m
Kashmir Flycatcher *Ficedula subrubra*	✈ s	VU	Winters in forest at high altitudes
			SRI LANKA (EBA 124)
Sri Lanka Wood-pigeon *Columba torringtoni*	⊛	VU	Found mainly in hill evergreen forest in the wet zone above c.900 m
Red-faced Malkoha *Phaenicophaeus pyrrhocephalus*	⊛	VU	Tall primary forest, mainly in the wet zone, but also in riverine forests in the dry zone
Green-billed Coucal *Centropus chlororhynchus*	⊛	VU	Virtually confined to lowland rainforest below c.760 m in the wet zone
Sri Lanka Whistling-thrush *Myiophonus blighi*	⊛	EN	Rocky areas along streams in undisturbed hill evergreen forest in the wet zone above c.900 m
Ashy-headed Laughingthrush *Garrulax cinereifrons*	⊛	VU	Lowland and hill forests in the wet zone below c.1,500 m
Kashmir Flycatcher *Ficedula subrubra*	✈ s	VU	Winters in hill evergreen forest in the wet zone above c.760 m
White-faced Starling *Sturnus albofrontatus*	⊛	VU	Lowland and hill forests in the wet zone at intermediate altitudes between 460 and 1,220 m
Sri Lanka Magpie *Urocissa ornata*	⊛	VU	Tall trees in undisturbed lowland and hill forests in the wet zone up to above 2,135 m

⊛ = breeds only in this forest region; ✈ = non-breeding visitor from another region; s = also occurs in another EBA in this region

Evergreen forests are still extensive in parts of the Western Ghats.

PHOTO: AMEEN AHMED

stature dry deciduous forest was originally extensive over the Deccan plateau of central India, but this has mostly been cleared for agriculture or replaced by plantations, leaving scattered patches of habitat suitable for Forest Owlet. In the Western Ghats, large areas of forest have been cleared or disturbed in the north (Maharashtra) as a result of shifting cultivation, collection of timber and fuelwood, cattle grazing, and the construction of roads and dams. In the south, large areas of forest and montane grassland have been replaced by agriculture and plantations (of tea, coffee, etc.), but substantial areas of natural habitat still remain. Sri Lanka has suffered large-scale forest loss and degradation as a result of logging, clearance of forest for agriculture and plantations (particularly of tea), fuelwood-gathering, gem mining and urbanisation. The wet zone in the south and west, the most important part of the island for threatened birds, was recently estimated to retain only 9% forest cover. Protected areas include most remaining wet zone forests, but some have been subject to illegal logging and encroachment, despite a moratorium passed in 1990 to protect all wet zone forests from logging.

Red-faced Malkoha requires tall primary forest, and has declined because of forest clearance and degradation.

PHOTO: TIM LOSEBY

CONSERVATION ISSUES AND STRATEGIC SOLUTIONS (summarised in Table 3)

Forest loss and degradation

FORESTRY AND ILLEGAL LOGGING

Large-scale commercial logging is no longer taking place in this region, although all known Forest Owlet sites in central India are in reserve forests which are managed for timber extraction rather than conservation, leading to a general scarcity of trees with suitable nest holes. A moratorium on logging is currently in place on Sri Lanka, and only very limited logging is allowed in the Western Ghats (following a ban in 1991), on some private lands and in plantations for personal use and 'shade regulation'. However, small-scale illegal logging operations are a serious problem in the Western Ghats and Sri Lanka, mainly affecting reserve forests and private lands. The logging moratorium on Sri Lanka and the logging ban in the Western Ghats should be maintained. Measures should be taken to prevent illegal logging, backed up by community-based forest conservation initiatives, focusing on sites of high biodiversity importance. In central India, forestry department policies and practices should be adjusted to ensure that old growth forests are maintained in the areas that support the Forest Owlet. Nest boxes should be provided for Forest Owlet, on an experimental basis, in areas which lack suitable nest holes.

CONVERSION TO AGRICULTURE AND PLANTATIONS

Large areas of forest in central India, the Western Ghats and Sri Lanka have been converted into croplands or plantations, and a tendency to remove peripheral or riverine forest patches within plantations often reduces their value for wildlife. All Forest Owlet sites in central India are under pressure from agricultural encroachment: an area of 50 km^2 near Toranmal Reserve Forest, the site where the species was rediscovered, was recently cleared for people displaced by a dam. In the Western Ghats, large areas in Maharashtra are affected by shifting cultivation, and in the south plantations of tea, coffee, cardamom, *Eucalyptus* and *Acacia* (wattle) have been spreading rapidly and replacing high-altitude shola forests and montane grasslands. Many cardamom and coffee plantations have native shade trees, and are of considerable value to birds; however, these are being converted to tea, or their natural shade trees are being replaced by exotics, mainly silver oak *Grevillea robusta*, which is also widely used as a shade tree in tea estates. In Sri Lanka, encroachment into forest patches continues to threaten many sites and increase habitat fragmentation. The development of numerous vegetable farms in upland areas (such as Nuwara Eliya and Horton Plains) is removing habitat, and producing insecticide run-off which almost certainly affects stream-dependent birds, such as Sri Lanka Whistling-thrush.

There is a need to control agricultural encroachment at key sites for threatened birds. Plantation managers should enact best environmental practices, including the retention of forest patches along streams and in peripheral areas, and minimise the use of agrochemicals. In the range of Forest Owlet in central India (Madhya Pradesh, Maharashtra and Orissa), a moratorium should be imposed on clearance of forest, with plantations only expanded in areas which have already been cleared. In the Western Ghats, the expansion of plantations needs to be halted (through government policy changes), and wattle (which has spread like an invasive weed) needs to be controlled. The practice of replacing native shade trees in plantations with silver oak

Large areas of forest and montane grassland have been replaced by tea and other plantations in the Western Ghats.

PHOTO: ASAD RAHMANI

Table 3. Conservation issues and strategic solutions for birds of the Indian peninsula and Sri Lankan forests.

Conservation issues	Strategic solutions
Forest loss and degradation	
■ *FORESTRY AND ILLEGAL LOGGING* ■ *CONVERSION TO AGRICULTURE AND PLANTATIONS* ■ *EXPLOITATION OF FOREST PRODUCTS* ■ *DEVELOPMENT (URBAN, INDUSTRIAL, ETC.)*	➤ Continue the logging moratorium on Sri Lanka and logging ban in the Western Ghats ➤ Prevent illegal logging, particularly at key sites for threatened birds ➤ Maintain old growth forest at Forest Owlet sites in central India, and experiment with the provision of nest boxes ➤ Promote best environmental practices in plantations, including the retention of natural forest patches, and the use of native rather than exotic shade trees ➤ Impose a moratorium on further clearance of natural forest for plantations in central India ➤ Halt the expansion of plantations in the Western Ghats, and control the spread of wattle ➤ Control exploitation of forest products at key sites, including through the development of community forestry plantations ➤ Assess the environmental impact of development projects in forest and grassland areas, and minimise development at key sites
Protected areas coverage and management	
■ *GAPS IN PROTECTED AREAS SYSTEM* ■ *WEAKNESSES IN RESERVE MANAGEMENT*	➤ Establish new protected areas for Forest Owlet in Maharashtra and Madhya Pradesh ➤ Improve protection of key Western Ghats habitats by establishing new protected areas and extending existing reserves ➤ Increase protected areas coverage in the wet zone of Sri Lanka ➤ Strengthen the technical and management capacity of government agencies responsible for protected area management
Gaps in knowledge	
■ *INADEQUATE DATA ON THREATENED BIRDS*	➤ Continue studies of Forest Owlet in central India ➤ Study the distribution and ecology of threatened birds in the Western Ghats and Sri Lanka, to locate new sites and improve management ➤ Investigate the causes of forest die-back on Sri Lanka

and other exotic trees should be stopped, and reversed by replanting native trees in place of these exotics. Conservation awareness work is needed, stressing the ecological services that forests provide (e.g. maintenance of water supplies) as well as their value for biodiversity.

■ *EXPLOITATION OF FOREST PRODUCTS*

In this densely populated region, natural forests are intensively exploited for fuelwood and building materials, as well as minor forest products (e.g. cinnamon, bamboo). This causes severe forest degradation as well as disturbance to wildlife, and needs to be controlled particularly in protected areas and other important sites. Efforts should also be made to reduce pressure on natural forests through the development of community forestry plantations.

■ *DEVELOPMENT (URBAN, INDUSTRIAL, ETC.)*

An underlying cause of much forest loss has been the construction of new roads, which allow access by loggers and settlers to remote regions. Mining of minerals and gems also affects habitat in the Western Ghats and Sri Lanka. Several reservoirs in the Western Ghats have inundated large areas of forest, and the construction of the Sardar Sarovar Dam in central India led to the clearance of a large area of forest at Toranmal Reserve Forest (the best site known for Forest Owlet) to rehabilitate the many people displaced by the dam (see *Conversion to agriculture and plantations* above); the proposed Upper Tapi Irrigation Project Stage II seeks to submerge 2.44 km^2 of Forest Owlet habitat, in an area that was removed from the Melghat Sanctuary by the government of Maharashtra in 1994. Recent forest die-back in the high-altitude forests in the wet zone of Sri Lanka has been blamed on acidic clouds, rain and mist caused by industrial air pollution (presumably from the adjacent lowlands). Environmental impact assessments should be conducted for development projects that have the potential to damage forested areas, and new developments should be avoided near protected areas and other sites of high biodiversity value.

Protected areas coverage and management

■ *GAPS IN PROTECTED AREAS SYSTEM*

None of the known sites for Forest Owlet is in a protected area, and there is an urgent need to establish reserves where its habitat can be protected and managed: the largest known population is in Toranmal Reserve Forest in Maharashtra, and other current sites are Taloda Reserve Forest and Melghat Wildlife Sanctuary (in an area that was deleted from the reserve in 1994) in Maharashtra, and Khaknaar Forest Range in Madhya Pradesh. There are about 50 protected areas in the Western Ghats (covering c.10% of the geographical area), which represent all of the key habitats for threatened birds; however, some of the reserves are small, and large areas of wet evergreen forest and other habitats remain outside reserves. Recent proposals by the Wildlife Institute of India to improve protection of key Western Ghats habitats by establishing several new protected areas and extending existing reserves should be implemented. Sri Lanka has an extensive system of protected areas, including over 14% of its total land area, but coverage is least extensive in the wet zone, where most of the threatened birds are found; the protected areas system in this part of the island needs to be revised, following the recommendations in the recent National Conservation Review.

■ *WEAKNESSES IN RESERVE MANAGEMENT*

While management of reserves in India and Sri Lanka is relatively well funded and organised, there are still many problems affecting protected areas in both countries (some of which are outlined above). Many of these could be addressed through improved financial, technical and management capacity in the relevant government agencies, leading to better patrolling, boundary demarcation, staff training and equipment at the reserves.

Gaps in knowledge

■ *INADEQUATE DATA ON THREATENED BIRDS*

Since its rediscovery in 1997, there have been several studies of Forest Owlet, which have helped formulate a conservation strategy for the species. However, more surveys are required (possibly using satellite images) to locate additional sites, and to clarify the sizes of the remaining populations, and their conservation requirements. Some of the threatened birds of the Western Ghats and Sri Lanka have been studied, but there is scope for surveys to locate further sites for conservation action, particularly in less studied areas such as the central and northern sections of the Western Ghats, and ecological research to help improve habitat management at known sites. These could focus on the least known species, such as Broad-tailed Grassbird and Kashmir Flycatcher in the Western Ghats and Green-billed Coucal and Sri Lanka Whistling-thrush on Sri Lanka. A study is needed to investigate the causes of the forest die-back that has recently been observed in the high-altitude wet zone forests on Sri Lanka.

The Forest Owlet was rediscovered in central India in 1997, after a gap of over 100 years, but all known sites are under threat.

PHOTO: FARAH ISHTIAQ

INDO-BURMESE FORESTS

THE Indo-Burmese region includes the moist tropical and subtropical forests which extend from north-east India and southern China across the lowlands and isolated mountains of South-East Asia, as well as the Andaman and Nicobar islands. The forests in this region support 24 threatened bird species, 18 of which breed nowhere else; the other six include four which also occur in the Sundaland forests (F07) and two from the South-east Chinese forests (F03). Seven threatened species are relatively widespread within this region (Table 2), and 17 are confined to one of the region's six Endemic Bird Areas and two Secondary Areas. Nine species of this region are Endangered, including three low-density species which inhabit forested wetlands, and six restricted-range species affected by deforestation.

- **Key habitats** Lowland evergreen and semi-evergreen tropical rainforest, moist deciduous forest and hill evergreen forest, and associated wetlands.
- **Altitude** 0–2,400 m.
- **Countries and territories** **China** (Yunnan, Hainan); **India** (Arunachal Pradesh, Assam, Meghalaya, Nagaland, Manipur, Andaman and Nicobar islands); **Bhutan**; **Bangladesh**; **Myanmar**; **Thailand**; **Laos**; **Cambodia**; **Vietnam**.

Threatened species

	CR	EN	VU	Total
⊙	—	7	11	18
◖	—	2	3	5
✈	—	—	1	1
Total	—	9	15	24

Key: ⊙ = breeds only in this forest region.
◖ = also breeds in other region(s).
✈ = non-breeding visitor from another region.

The Indo-Burmese forests region corresponds closely to Conservation International's Indo-Burma Hotspot (see pp.20–21).

Much of the original forest in Thailand has been cleared, and most of the remaining forests are inside protected areas such as Nam Nao National Park. PHOTO: MIKE CROSBY/BIRDLIFE

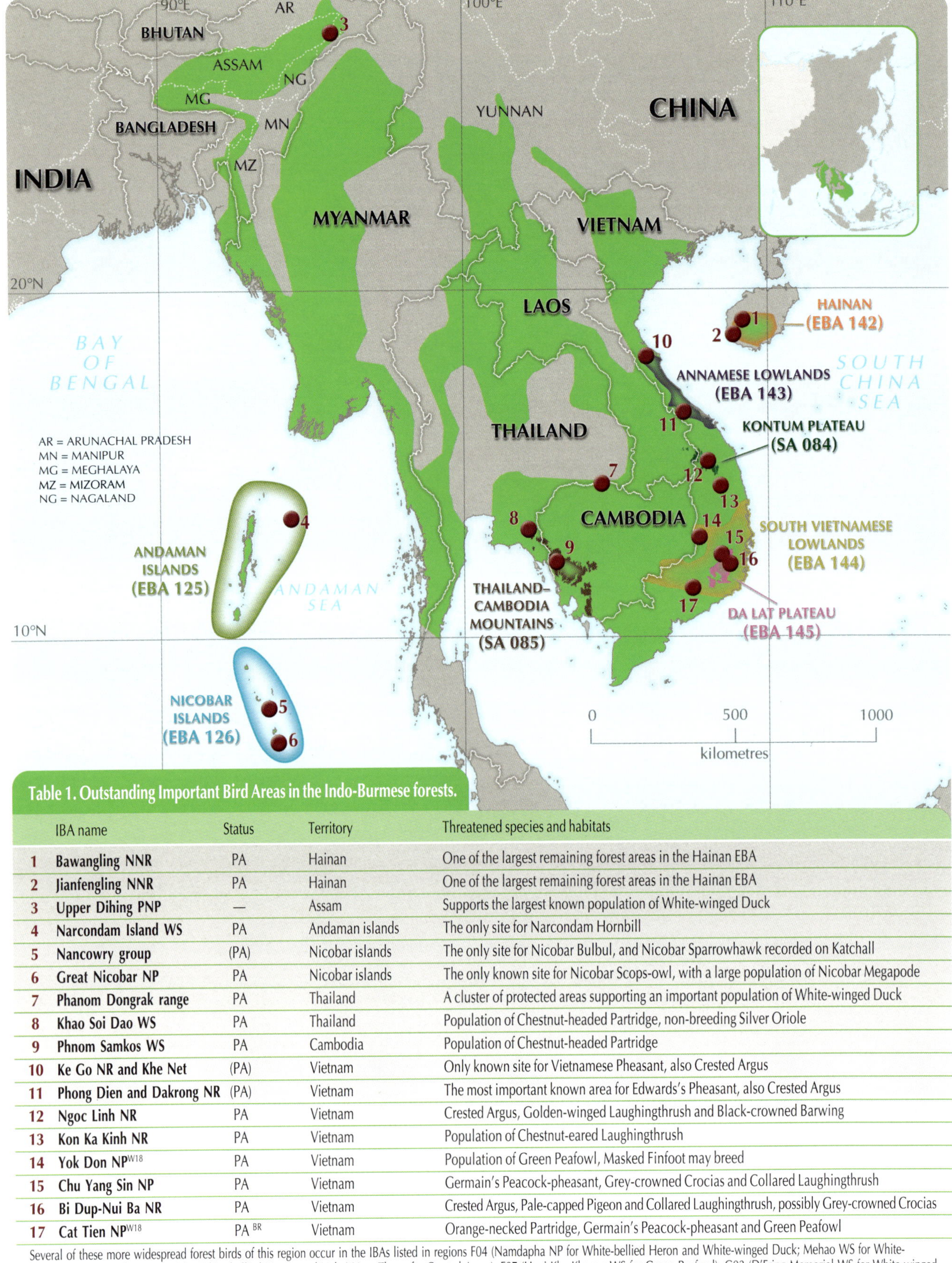

Table 1. Outstanding Important Bird Areas in the Indo-Burmese forests.

	IBA name	Status	Territory	Threatened species and habitats
1	**Bawangling NNR**	PA	Hainan	One of the largest remaining forest areas in the Hainan EBA
2	**Jianfengling NNR**	PA	Hainan	One of the largest remaining forest areas in the Hainan EBA
3	**Upper Dihing PNP**	—	Assam	Supports the largest known population of White-winged Duck
4	**Narcondam Island WS**	PA	Andaman islands	The only site for Narcondam Hornbill
5	**Nancowry group**	(PA)	Nicobar islands	The only site for Nicobar Bulbul, and Nicobar Sparrowhawk recorded on Katchall
6	**Great Nicobar NP**	PA	Nicobar islands	The only known site for Nicobar Scops-owl, with a large population of Nicobar Megapode
7	**Phanom Dongrak range**	PA	Thailand	A cluster of protected areas supporting an important population of White-winged Duck
8	**Khao Soi Dao WS**	PA	Thailand	Population of Chestnut-headed Partridge, non-breeding Silver Oriole
9	**Phnom Samkos WS**	PA	Cambodia	Population of Chestnut-headed Partridge
10	**Ke Go NR and Khe Net**	(PA)	Vietnam	Only known site for Vietnamese Pheasant, also Crested Argus
11	**Phong Dien and Dakrong NR**	(PA)	Vietnam	The most important known area for Edwards's Pheasant, also Crested Argus
12	**Ngoc Linh NR**	PA	Vietnam	Crested Argus, Golden-winged Laughingthrush and Black-crowned Barwing
13	**Kon Ka Kinh NR**	PA	Vietnam	Population of Chestnut-eared Laughingthrush
14	**Yok Don NP**^{W18}	PA	Vietnam	Population of Green Peafowl, Masked Finfoot may breed
15	**Chu Yang Sin NP**	PA	Vietnam	Germain's Peacock-pheasant, Grey-crowned Crocias and Collared Laughingthrush
16	**Bi Dup-Nui Ba NR**	PA	Vietnam	Crested Argus, Pale-capped Pigeon and Collared Laughingthrush, possibly Grey-crowned Crocias
17	**Cat Tien NP**^{W18}	PA^{BR}	Vietnam	Orange-necked Partridge, Germain's Peacock-pheasant and Green Peafowl

Several of these more widespread forest birds of this region occur in the IBAs listed in regions F04 (Namdapha NP for White-bellied Heron and White-winged Duck; Mehao WS for White-winged Duck; Hponkanrazi WS for White-bellied Heron; and Nakai-Nam Theun for Crested Argus), F07 (Huai Kha Khaeng WS for Green Peafowl), G02 (D'Ering Memorial WS for White-winged Duck; Manas NP for White-bellied Heron; Dibru-Saikhowa NP for White-bellied Heron, White-winged Duck and Pale-capped Pigeon; and Kaziranga NP for White-bellied Heron and Pale-capped Pigeon), W15 (Sundarbans for Masked Finfoot) and W18 (Xe Kong plains and Dong Khanthung for White-winged Duck, Green Peafowl and Masked Finfoot; Prek Toal and Boeung Chhmar / Moat Khla for Masked Finfoot, Western Siem Pang and Upper Stung Sen catchment for Green Peafowl, Lomphat WS for White-winged Duck and Masked Finfoot, and Chhep for White-winged Duck and Green Peafowl).

Note that more IBAs in this region will be included in the *Important Bird Areas in Asia*, due to be published in early 2004.

Key *IBA name*: NP = National Park; NR = Nature Reserve; NNR = National Nature Reserve; PNP = proposed national park; WS = Wildlife Sanctuary.
Status: PA = IBA is a protected area; (PA) = IBA partially protected; — = unprotected; BR = IBA is wholly or partially a Biosphere Reserve (see pp.34–35); W18 = also supports threatened waterbirds of region W18.

Green Peafowl is widespread in mainland South-East Asia, but it is now very thinly distributed because of deforestation and other pressures.

PHOTO: SMITH SUTIBUT

Table 2. Threatened birds of the Indo-Burmese forests.

Species			Distribution and habitat
White-bellied Heron *Ardea insignis*	breeds only in this forest region	EN	Forested rivers and swamps at up to 1,500 m in India (Arunachal Pradesh, Assam, Nagaland), Bhutan and Myanmar, non-breeding birds sometimes in non-forest habitats
White-winged Duck *Cairina scutulata*	also breeds in other region(s)	EN	In this region, widespread but very local in forested wetlands at up to 1,400 m from north-east India to mainland South-East Asia
Crested Argus *Rheinardia ocellata*	also breeds in other region(s)	VU	In this region, found in lowland and hill forests from the level lowlands to 1,900 m in Vietnam and adjacent parts of Laos
Green Peafowl *Pavo muticus*	also breeds in other region(s)	VU	In this region, widely but thinly distributed in open forest, mainly in lowlands but sometimes above 2,000 m, from north-east India and Yunnan to mainland South-East Asia
Masked Finfoot *Heliopais personata*	also breeds in other region(s)	VU	In this region, widely but thinly distributed in forested wetlands in the lowlands, from north-east India to mainland South-East Asia
Pale-capped Pigeon *Columba punicea*	breeds only in this forest region	VU	Widespread but very local and irregular in forest at up to c.1,600 m, from north-east India and southern China to mainland South-East Asia
Silver Oriole *Oriolus mellianus*	non-breeding visitor from another region	VU	Non-breeding visitor to evergreen forest at c.450–1,300 m in Thailand and Cambodia
			ANDAMAN ISLANDS (EBA 125)
Narcondam Hornbill *Aceros narcondami*	breeds only in this forest region	VU	Confined to the forests on tiny Narcondam island (6.82 km^2)
The Data Deficient Andaman Crake *Rallina canningi* is known only from forested wetlands on Middle and South Andaman islands			
			NICOBAR ISLANDS (EBA 126)
Nicobar Sparrowhawk *Accipiter butleri*	breeds only in this forest region	VU	Forest, only definitely known from Katchall (in the Nancowry group) and Car Nicobar
Nicobar Megapode *Megapodius nicobariensis*	breeds only in this forest region	VU	Occurs throughout the forests of the Nicobars (other than Car Nicobar), but concentrated near the coast
Nicobar Bulbul *Hypsipetes nicobariensis*	breeds only in this forest region	VU	Found in forest and man-modified habitats only in Nancowry group of islands
The Data Deficient Nicobar Scops-owl *Otus alius* is known only from coastal forest at a single locality on Great Nicobar			
			HAINAN (EBA 142)
White-eared Night-heron *Gorsachius magnificus*	also breeds in other region(s)	EN	Forest in the hills in the centre of the island
Hainan Partridge *Arborophila ardens*	breeds only in this forest region	VU	Forest in the hills in the centre and south of the island
Hainan Leaf-warbler *Phylloscopus hainanus*	breeds only in this forest region	VU	Forest in the hills in the centre and south of the island
			ANNAMESE LOWLANDS (EBA 143)
Edwards's Pheasant *Lophura edwardsi*	breeds only in this forest region	EN	Tropical evergreen forest up to c.600 m, found to the south of the core range of Vietnamese Pheasant
Vietnamese Pheasant *Lophura hatinhensis*	breeds only in this forest region	EN	Tropical evergreen forest up to c.300 m, in the northern part of the EBA
Imperial Pheasant *Lophura imperialis*, which is known only from the Annamese lowlands, is currently listed as Data Deficient, but new research has shown it to be of hybrid origin, and it will therefore be removed from the 2004 Red List.			

(symbol) = breeds only in this forest region; (symbol) = also breeds in other region(s); (symbol) = non-breeding visitor from another region.

... continued

Table 2 ... continued. Threatened birds of the Indo-Burmese forests.

Species			Distribution and habitat
			KON TUM PLATEAU (SA 084[1])
Golden-winged Laughingthrush *Garrulax ngoclinhensis*	◐	VU	Recorded in primary evergreen forest at 2,000–2,200 m on Ngoc Linh massif and Mt Ngoc Boc
Chestnut-eared Laughingthrush *Garrulax konkakinhensis*[2]	◐	VU	Recently described species recorded in evergreen forest above c.1,000 m on Mt Kon Ka Kin and Mt Ngoc Boc
Black-crowned Barwing *Actinodura sodangorum*	◐	VU	Evergreen forest at c.1,000–2,400 m, known from Ngoc Linh massif in Vietnam and the Dakchung plateau in Laos
			SOUTH VIETNAMESE LOWLANDS (EBA 144)
Orange-necked Partridge *Arborophila davidi*	◐	EN	Forest and secondary habitats on low hills (of up to 400 m), very small known distribution in Vietnam and eastern Cambodia
Germain's Peacock-pheasant *Polyplectron germaini*	◐	VU	Tropical evergreen and semi-evergreen forest at 0–1,200 m or higher, widespread in southern Vietnam and recorded in eastern Cambodia
			DA LAT PLATEAU (EBA 145)
Collared Laughingthrush *Garrulax yersini*	◐	EN	Dense undergrowth in evergreen forest at 1,500–2,300 m
Grey-crowned Crocias *Crocias langbianis*	◐	EN	Closed-canopy evergreen forest at 900–1,700 m
			THAILAND-CAMBODIA MOUNTAINS (SA 085)
Chestnut-headed Partridge *Arborophila cambodiana*	◐	EN	Evergreen forest above c.400 m

[1] The description of the three restricted-range species listed in the table means that the Kon Tum plateau now qualifies as an EBA, according to the criteria used by Stattersfield *et al.* (1998).
[2] This species was described by Eames and Eames (2001) after the publication of the *Threatened birds of Asia*; its status was subsequently evaluated, and it was added to the 2002 IUCN Red List as Vulnerable.

◐ = breeds only in this forest region; ◖ = also breeds in other region(s); ✈ = non-breeding visitor from another region

OUTSTANDING IBAs FOR THREATENED BIRDS (see Table 1)

Seventeen IBAs have been selected, which together support populations of all of the region's threatened forest birds, and include some of the largest and richest forests that remain. Several of these IBAs are unique, as they support the only (or by far the largest) known populations of one or more threatened species, for example Narcondam Island Wildlife Sanctuary for Narcondam Hornbill and Ke Go Nature Reserve (together with Khe Ket forest) for Vietnamese Pheasant. Several of the more widespread species in this region occur at low densities and are migratory or nomadic, such as Masked Finfoot and Pale-capped Pigeon, and require extensive networks of sites to ensure their survival. Many other forest sites with significant populations of these and other threatened forest birds will be documented during BirdLife's ongoing IBA Project.

CURRENT STATUS OF HABITATS AND THREATENED SPECIES

Vast expanses of lowland and hill rainforest once extended from north-east India and southern China to most of mainland South-East Asia. However, extensive forest loss and degradation has occurred in this densely populated region, particularly in recent decades, through logging, conversion to agriculture, development and overexploitation of forest products. The remaining forests (and their populations of threatened birds) are highly fragmented, and under great pressure from conversion and exploitation, as well as hunting. Many protected areas have been established, including some that are large and relatively well protected, but also many that are small, isolated and gradually being eroded by illegal encroachment. The following sections give an overview of the status of lowland and hill forests throughout the region; although the condition of forests is generally poor, there are still some large unprotected areas that remain intact and available for conservation.

China: The area of natural tropical forest has decreased considerably on Hainan. Estimates of the extent of this decline are from 16,920 km² of forest in 1943 to 3,000 km² in 1994, or from 8,630 km² (25.7% of the island) in 1949 to about 2,420 km² (7.2%) in 1991. Much of the remaining forest is disturbed and unlikely to support many threatened forest birds, but some blocks of higher-quality forest are protected in nature reserves. In Yunnan, forest cover declined from c.55% to c.30% between the early 1950s and 1975; subsequent loss of habitat in China has, if anything, been even more rapid, but logging was banned in 1998 (see F03: South-east Chinese forests). However, some tracts of lowland and hill forest are protected in nature reserves.

India: The plains and foothills of north-east India originally had extensive tracks of rainforest, but most plains forest was lost during the nineteenth century, and in recent years the foothill forests have been rapidly cleared, at a rate of at least 1,000 km² annually during the 1970s and 1980s. However, some relatively large areas survive with important populations of several threatened birds, and logging and timber export was recently banned in Arunachal Pradesh and other north-east Indian states. Forests on the Andaman and Nicobar islands have been widely logged, but remain relatively intact, although a few islands have suffered significant forest loss and degradation through unsustainable commercial logging and conversion for agriculture and plantations.

Bangladesh: The forests of Bangladesh are now reduced to a few small, mostly highly disturbed fragments, as a result of logging and conversion to agriculture and settlements. By the 1960s, only c.16 km² of primary forest remained in Sylhet, previously one of the strongholds of forest in the country, and virtually no primary forest now remains in the Chittagong Hill Tracts.

Myanmar: In the 1980s the rate of forest clearance in Myanmar was estimated at up to 6,000 km^2 per annum, one of the highest deforestation rates in the world. By the early 1990s, the area of forest remaining was estimated to cover 47.4% of the country, 39.1% being intact forest and 8.3% degraded forest. Much of the remaining forest is montane, although some extensive areas of lowland and hill forest suitable for the birds of this region may remain.

Thailand: Forest cover apparently fell from an estimated 70–80% of total land area in the 1940s to well under 30% in the 1980s. Logging was officially banned in 1989, but forest has continued to be rapidly cleared and degraded through conversion to plantations and illegal logging.

Laos: Although level lowland forest has been cleared in most parts of the country, relatively large areas of foothill and hill forest remain, notably in Xe Pian NBCA. However, the remaining forests are under intense pressure from commercial logging operations, and some large areas of forest are scheduled to be submerged by hydropower projects.

Cambodia: Large areas of forest in Cambodia have been degraded or cleared in recent decades because of unsustainable logging and conversion to agriculture. Some vast tracts of lowland dry-deciduous forest remain in the north of the country, and support populations of White-winged Duck and Green Peafowl, as well as a selection of threatened waterbirds (see W18).

Vietnam: More than 80% of the original area of closed-canopy forest in Vietnam has been cleared. There is some evidence that the rate of loss of natural forest has slowed in the past decade, but the quality of that which remains continues to decline. Absolute forest cover may be increasing because statistics include plantations of exotics for pulp and commodity tree-crops like coffee and cashew as well as natural forests.

CONSERVATION ISSUES AND STRATEGIC SOLUTIONS (summarised in Table 3)

Forest loss and degradation

FORESTRY AND ILLEGAL LOGGING

Commercial clear felling and selective logging have been major factors in the loss of large areas of lowland forest in this region, but logging has now ceased in many areas. Excessive timber extraction was a major problem on Hainan until, in 1994, logging of primary forest was officially banned, although some illegal logging is reported to continue. A national logging ban was enacted in China in 1998. In north-east India, logging has depleted forests over much of lowland Assam; however, logging and timber export was recently banned in north-east Indian states, although some illegal logging takes place and it is still legal to cut timber for local plywood and veneer factories. Large-scale logging operations continue in Myanmar, to produce timber for export to Thailand and China (where logging is now banned). In Cambodia, despite a total ban on the export of sawn logs and timber in 1995, there are major logging operations (including in national parks), illegal trade continues and most forest remains under concession. Commercial timber extraction is also a threat at lower altitudes elsewhere in Indochina.

Given the massive reduction in lowland forest in this region, existing bans on logging should be enforced (and similar bans promoted elsewhere), and reforestation programmes (using native species of trees) adopted. Where legal logging continues, sustainable forestry practices need to be promoted, including allowing natural regeneration, lower timber quotas and preservation of old-growth patches, and low-impact logging practices to reduce structural damage to the forest and allow more rapid regeneration. Measures are needed to minimise illegal logging, particularly in protected areas and other sites of high biodiversity value, and to control illegal trading of timber.

Together with nearby Ke Go Nature Reserve, the forests at Khe Net are vital for the survival of the endemic Vietnamese Pheasant.

PHOTO: J. C. EAMES

CONVERSION TO AGRICULTURE AND PLANTATIONS

The clearance of forest for cultivation and plantations is a continuing threat, linked to human population growth and the movement of people into formerly remote regions. In north-east India areas of forest are being converted to tea gardens, paddyfields and teak plantations, and on the Nicobar islands forests have been converted into rice paddy and plantations (including cashew nut, coconut and rubber). In Vietnam, forest is being lost to plantations of cash crops, such as coffee and cashew, even inside Cat Tien National Park (where conversion to fishponds has posed an additional threat), which stems in part from politically motivated settlement programmes. The large areas of open dry deciduous forest habitat that remain in Dak Lak province are likely to suffer from agricultural development, as it has the highest immigration rate of any province in Vietnam.

There is a general need to control encroachment into forest for agriculture and plantations in protected areas and at other key sites for threatened birds and other biodiversity. This means investigating the underlying causes behind the movements of people into forest areas and forest conversion to cash crops, including the effects of global commodity prices; the conservation community needs to influence these policies, to minimise the pressures to convert natural forests. The establishment of new nature reserves may be required at some key sites, with effective legislation and patrolling to prevent encroachment. The development of more efficient agriculture (through the introduction of improved and

Local people use forests as a source of timber and other products, but management is required to prevent forest degradation.

PHOTO: GERRY GOMEZ/BIRDLIFE

appropriate techniques) to help alleviate poverty, if carefully implemented, also has the potential to reduce the pressure on remaining areas of natural habitat. Plantation managers should retain forest patches along streams and in peripheral areas, and minimise their use of agrochemicals. These measures should be promoted through conservation awareness work, stressing the ecological services that forests provide (e.g. maintenance of water supplies) as well as their value for biodiversity.

SHIFTING AGRICULTURE

The habitats of some of the region's higher-altitude species, such as the species endemic to the mountains of Vietnam, are affected by shifting agriculture (see F04 for further details). This is increasingly causing forest degradation, as the numbers of shifting cultivators increase and the area of forest decreases. Measures to control unsustainable shifting agriculture could include: provision of sustainable economic alternatives to shifting cultivation, including improved agricultural practices and agro-forestry, enabling farmers to settle in established clearings; rehabilitation of abandoned land, including through the development of community forestry plantations; and projects at selected key sites that integrate conservation and land-use development, and work in collaboration and participation with the local people.

EXPLOITATION OF FOREST PRODUCTS

The forests of this region face increasing pressure from exploitation as human populations grow and natural habitat diminishes. The areas around many reserves are densely populated; for example, over 100,000 people live in the buffer zone of Cat Tien National Park in Vietnam, and a large town separates Cat Loc (stronghold of Orange-necked Partridge) from the other sectors. Local people use forests as a source of timber, fuelwood (including charcoal) and rattan; these activities cause severe degradation, and disturbance to wildlife. Human access to and exploitation of forests need to be controlled in protected areas and other important sites for threatened birds. Some reserves may need to be redesigned, with broad buffer zones to absorb human needs, and rigorously patrolled core areas to prevent forest exploitation. The establishment of community forests should be promoted, to provide a sustainable source of forest products for local people, and their management improved through training in forestry techniques. At some sites, alternative livelihoods could be developed (e.g. fish-farming, ecotourism, floriculture, cultivation of medicinal plants, etc.) to reduce local dependence on forest products. These should be promoted through conservation awareness work, stressing the ecological services that forests provide, for example maintenance of water supplies.

DEVELOPMENT (URBAN, INDUSTRIAL, ETC.)

Habitat is threatened by the region's expanding network of new roads, which allow access by loggers and settlers to once-remote regions, and often result in near-total forest loss. For example, the ongoing construction of the Ho Chi Minh highway in Vietnam, and the associated human settlement, are causing habitat loss and placing increased pressure on forest products at numerous key sites for threatened birds, including Khe Net and Phong Dien / Dakrong. Numerous hydropower developments are planned for South-East Asia, for example in the Mekong catchment, especially in Yunnan and Laos; these dams threaten enormous stretches of riverine forest, especially in the Mekong basin (see W18), and cause a suite of problems by improving access for loggers and hunters, and displacing people from inundated areas. The Andaman and Nicobar islands are also threatened by development. Forest is being lost through the establishment and expansion of settlements with associated roads, airstrips and defence installations, as well as a hydroelectric project. The most alarming threat lies in a proposal to develop Great Nicobar as a free-trade port and to create a dry dock and refuelling base for international shipping at the mouth of the Galathea river. Environmental impact assessments should be conducted for development projects that have the potential to damage forested areas, with appropriate mitigation plans. Wherever possible, new developments should be avoided near protected areas and other sites of high biodiversity value. The proposed developments on the Andaman and Nicobar islands need to be assessed with particular care, because of the vulnerably of the endemic biodiversity on these often small oceanic islands.

DISTURBANCE

This region has a large and rapidly growing human population, with a high proportion of these people live in rural areas, meaning that most natural habitats are affected by human disturbance. Lowland forests and their accessible valleys and waterways are particularly susceptible. Species such as White-winged Duck, Green Peafowl and Masked Finfoot are strongly tied to pools and rivers within or near forest, and it is precisely these areas that are cleared and colonised first, or visited most frequently by people; they are seriously threatened with extinction in many areas as lowland wilderness disappears. Networks of protected areas need to be developed and managed, through the modification of existing reserves and the establishment of new areas, to protect extensive river systems and networks of forest pools from human disturbance, habitat loss and other threats. Some potentially suitable large areas of relatively intact forest wetlands still exist, for example in Assam and Cambodia.

Protected areas coverage and habitat management

GAPS IN PROTECTED AREAS SYSTEM

Many protected areas have already been established, but there are some important gaps in their coverage of threatened birds; new reserves and the modification of existing ones, are urged by country below. It should be noted that some species are wide-ranging and generally occur at low densities, including ones which make migratory or nomadic movements such as Masked Finfoot and Pale-capped Pigeon; protected areas alone cannot guarantee the survival of these species, which also require habitat protection and management at the landscape level. *China*: In Yunnan, the existing protected areas support only small numbers of Green Peafowl, and new reserves are required to protect the largest populations, for example in the area where Chuxiong city meets Shuangbai and Lufeng counties. *India*: The protection of the important White-winged Duck populations in eastern Assam needs to be improved through a network of specially protected and managed sites, including gazettement of Upper Dihing (west block) and the adjacent forests of Joypur and Dirok as a new protected area. In the Nicobar islands, protected areas should be established on Camorta and/or Katchall islands. There are two national parks on Great Nicobar, but their boundaries need to be revised to address the following three weaknesses: (1) the coastal forests important for Nicobar Megapode are relatively unprotected and thus most vulnerable; (2) the central road across Great Nicobar and the gap between national parks leaves the habitat open to fragmentation; (3) the current buffer zone covers large tracts of land that are uninhabited, and as such should be fully protected. *Myanmar*: Effective new protected areas need to be established for major areas of lowland forest (and associated wetlands), following surveys to locate suitable sites. *Vietnam*: Proposed changes to the protected areas system include: in the Annamese lowlands, support the establishment of the proposed Phong Dien and Dakrong Nature Reserves, and create a new nature reserve at Khe Net (165 km^2) in Quang Binh province; in the South Vietnamese lowlands, establish and consolidate Lo Go Xa Mat National Park and Bu Gia Map Nature Reserve, and expand Cat Tien National Park to encompass the surrounding old forestry enterprises; on the Kon Tum plateau, upgrade three contiguous sites—Ngoc Linh (Kon Tum) Nature Reserve and Ngoc Linh (Quang Nam) and Song Thanh proposed nature reserves—to a national park, ensure that management regimes in the forest areas between Kon Ka Kinh and Kon Cha Rang Nature Reserves are consistent with the maintenance of habitat corridors, and maintain forest cover between Ngoc Linh, Ngoc Boc and Kon Ka Kinh.

Current protected areas networks do not adequately cover forested rivers and wetlands, with riverine species particularly poorly represented because of their linear distributions. For species such as White-winged Duck and Masked Finfoot, improved reserve design and management is necessary, involving: (1) use of ridges rather than rivers to delimit protected areas; (2) conservation programmes targeting entire river systems and networks of pools, which aim to protect and manage forest wetlands both inside and outside the protected areas. Education campaigns are needed to promote these proposals, particularly to highlight the importance of natural river systems, for example as spawning grounds for fish.

■ *INADEQUATE LEGISLATION*

In several countries the laws designed to protect habitats are weak or ambiguous, or newly introduced and unfamiliar to the officials who must enforce them. National Protected Areas in Laos, for example, had their legal status clarified

Table 3. Conservation issues and strategic solutions for birds of the Indo-Burmese forests.

Conservation issues	Strategic solutions
Forest loss and degradation	
■ *FORESTRY AND ILLEGAL LOGGING* ■ *CONVERSION TO AGRICULTURE AND PLANTATIONS* ■ *SHIFTING AGRICULTURE* ■ *EXPLOITATION OF FOREST PRODUCTS* ■ *DEVELOPMENT (URBAN, INDUSTRIAL, ETC.)* ■ *DISTURBANCE*	➤ Maintain logging bans in Hainan and north-east India, and support similar bans if they are required elsewhere ➤ Promote sustainable forestry, including low-impact logging practices and replanting with native tree species ➤ Develop measures to control illegal logging and trading of timber ➤ Investigate the underlying causes of forest conversion to cash crops, and seek to reduce this pressure by influencing relevant government policies ➤ Promote sustainable forms of upland agriculture in the mountains of Vietnam that do not result in net loss of natural forest ➤ Reduce local dependence on forest products by establishing sustainable community forests and developing alternative livelihoods ➤ Assess the environmental impact of development projects in forested areas, and minimise development at key sites for threatened birds ➤ Reconsider proposed development projects on the Andaman and Nicobar Islands
Protected areas coverage and management	
■ *GAPS IN PROTECTED AREAS SYSTEM* ■ *INADEQUATE LEGISLATION* ■ *WEAKNESSES IN RESERVE MANAGEMENT*	➤ Establish new reserves, and modify existing reserves, to fill gaps in coverage of threatened species in China, India, Myanmar and Vietnam ➤ Develop conservation programmes for selected river catchments ➤ Review and improve protected areas legislation in Laos, Myanmar and Cambodia ➤ Strengthen reserve management through improved funding, infrastructure and staff training
Exploitation of birds	
■ *HUNTING AND TRAPPING*	➤ Improve enforcement of hunting laws, including through education programmes, control of gun ownership and more effective patrolling of protected areas ➤ Survey wildlife markets to monitor hunting pressure on threatened species
Gaps in knowledge	
■ *INADEQUATE DATA ON THREATENED BIRDS*	➤ Locate and survey lowland forests in Myanmar, to identify priority sites for conservation action ➤ Survey poorly known species in the Andaman and Nicobar islands ➤ Continue surveys in the Kon Tum plateau of Vietnam and adjacent parts of Laos ➤ Study the movements of Masked Finfoot and Pale-capped Pigeon
Other conservation issues	
■ *INTRODUCED SPECIES*	➤ Investigate the impact on Nicobar Bulbul of competition with Red-whiskered Bulbul, to help develop a strategy for its conservation ➤ Completely eradicate feral goats from Narcondam Island

only in June 2003. Protected areas in both Myanmar and Cambodia lack strict protective legislation. For reserve networks to be successful they need to be supported by well-defined laws and the manpower (with high-level political support and motivation) to enforce them.

■ *WEAKNESSES IN RESERVE MANAGEMENT*

Many of the region's protected areas receive little or no active management, and increased resources are required to ensure their long-term security, through improved management planning and training of reserve staff, and hence patrolling, boundary demarcation, etc. Special management is required for the threatened species in some protected areas, for example planting of native fig trees and provision of nest-boxes for Narcondam Hornbill in Narcondam Island Wildlife Sanctuary, and the retention of large old trees and vegetation along the banks of rivers in reserves with nesting White-winged Duck. Governmental departments responsible for habitat conservation in this region often suffer from a shortage of skills, funding and motivation. These issues need to be addressed through exchange programmes and training, supported by appropriate injections of funding.

Exploitation of birds

■ *HUNTING AND TRAPPING*

Hunting with both snares and firearms is extremely common in the Indo-Burmese forests. In China, Thailand, Laos and Vietnam, some larger-bodied birds have almost been hunted out, and snaring has reduced many forest-floor species to very low densities. The collection of eggs and chicks of birds for food is also frequent. The widespread threatened species include several which are a quarry of hunters because of their size, such as White-winged Duck, Crested Argus and Green Peafowl, and Pale-capped Pigeon is susceptible to hunting when it congregates at fruiting trees. Green Peafowl is often sold in markets for food, and live or dead birds are even traded internationally for their meat and feathers. Nicobar Megapode is subject to heavy hunting and egg collection.

The impact of hunting on forest birds is generally poorly understood, but this appears to be the principal threat to several threatened birds in this region. Protected areas need to be patrolled more effectively, to intercept hunters and remove snares, backed up by firm enforcement of existing hunting laws. Surveys of wildlife markets should monitor (and provide the basis for control of) hunting pressure on threatened species. The governments of China, Laos and Vietnam are controlling gun ownership, a measure that in Vietnam has apparently already benefited populations of larger-bodied bird species, and this measure could be applied in other countries, particularly near important protected areas. These efforts to reduce hunting need to be promoted through public education programmes concerned with forest conservation, threatened species and the hunting laws.

In some areas, larger-bodied birds have almost been hunted out, and snaring has reduced many forest-floor species to very low numbers.

PHOTO: MARK EDWARDS/BIRDLIFE

Gaps in knowledge

■ *INADEQUATE DATA ON THREATENED BIRDS*

In recent decades intensive survey work has improved knowledge of the status and distribution of habitat and threatened birds in many parts of the Indo-Burmese region. The most important exception is Myanmar, for which there is little current information, as this country could be found to support significant populations of several of the more widespread threatened species, such as White-bellied Heron, White-winged Duck, Green Peafowl and Masked Finfoot. It is therefore a priority to pinpoint any extensive stands of forest in lowland Myanmar, and to determine which of these deserve conservation attention. As an example, the wooded swamps of the Myitmaka river north of Yangon formerly held a large breeding population of Masked Finfoot, but recent information is not available. On the Andaman and Nicobar islands, there is a need to improve understanding of the status and conservation requirements of the poorly known Nicobar Sparrowhawk, and the Data Deficient Andaman Crake and Nicobar Scops-owl. In Vietnam, recent surveys in the Kon Tum plateau have led to the discovery of three bird species, all of which are considered to be threatened, and further surveys are required there and in the adjacent mountains in Laos to improve understanding of their status and to look for any more undescribed species. A study is underway to investigate whether Edwards's Pheasant and Vietnamese Pheasant, both of which are confined to the Annamese lowlands EBA, are conspecific. Two widespread species, Masked Finfoot and Pale-capped Pigeon, are migratory or nomadic and research is required (perhaps using satellite-tracking) to improve understanding of their status and movements, and their tolerance of habitat change.

Other conservation issues

■ *INTRODUCED SPECIES*

Alien species in the Andaman and Nicobar archipelagos might be affecting some of the threatened species. The Red-whiskered Bulbul *Pycnonotus jocosus* was introduced to the Nicobar islands in the late nineteenth century and has subsequently flourished in the Nancowry group (i.e. throughout the range of Nicobar Bulbul). The population of Nicobar Bulbul appears to have declined substantially on most islands, and it is possible that competitive exclusion is occurring. This needs study, and a strategy to minimise the effects of any competition on the numbers of Nicobar Bulbul. On Narcondam Island, police staff introduced a small population of goats in the 1970s, growing by 1998 to 130–150 domestic animals in the police camp and over 250 feral animals in the forest. As a result, there was very little natural woodland regeneration, posing a serious long-term threat to the Narcondam Hornbill. Most of the goats were recently removed from the island, but this programme should be continued until they are completely eradicated.

SUNDALAND FORESTS

THE Sundaland (or Sundaic) region includes the moist tropical lowland and montane forests of the Thai-Malay peninsula and the Greater Sunda islands. It supports 47 threatened bird species, 38 of which breed nowhere else. Twenty-eight threatened species are particularly associated with Sundaland's (once) extensive lowland forests (Table 2); 22 of these are endemic to the region, including four unique to the Thai-Malay peninsula (including Gurney's Pitta), one known only from Sumatra (Rueck's Blue-flycatcher), five confined to Borneo and one small island specialist (Silvery Wood-pigeon). Nineteen (mainly montane) threatened species are found in the region's three Endemic Bird Areas, including one confined to the Bornean mountains, six to the Sumatra mountains, two to the Peninsular Malaysia mountains, and seven to the Java and Bali forests. This region also supports a remarkable total of 106 Near Threatened species, including 79 lowland specialists.

- **Key habitats** Lowland evergreen and semi-evergreen rain forest, peat swamp forest, heath forest, moist deciduous forest, lower montane and upper montane rain forest, savanna and cultivated areas.
- **Altitude** 0–3,350 m.
- **Countries and territories** **Myanmar**; **Thailand**; **Malaysia** (Peninsular, Sabah, Sarawak); **Singapore**; **Brunei**; **Indonesia** (Sumatra, Kalimantan, Java and Bali).

Threatened species

	CR	EN	VU	Total
(breeds only in this forest region)	5	4	29	38
(also breeds in other region(s))	1	1	3	5
(non-breeding visitor from another region)	—	—	4	4
Total	6	5	36	47

Key: = breeds only in this forest region.
= also breeds in other region(s).
= non-breeding visitor from another region.

The Sundaland forests region corresponds closely to Conservation International's Sundaland Hotspot (see pp.20–21).

Khao Nor Chuchi is now the only significant area of level lowland forest remaining in peninsular Thailand, and it is under constant pressure from encroachment. PHOTO: P. ROUND/BIRDLIFE

F07

OUTSTANDING IBAs FOR THREATENED BIRDS (see Table 1)

Twenty-eight IBAs have been selected, which together support populations of all threatened forest birds of the Sundaland forests, apart perhaps from the poorly known Silvery Wood-pigeon, Black-browed Babbler and Rueck's Blue-flycatcher. Several of these sites are unique, as they support the only (or by far the largest) known populations of one or more threatened species; for example Bali Barat National Park in Indonesia holds the only known population of Bali Starling. However, many other significant areas of forest with populations of threatened birds survive in this huge region, which need to be protected through the habitat management approaches detailed below under *Forest loss and degradation*.

CURRENT STATUS OF HABITATS AND THREATENED SPECIES

The Sundaland region was originally almost entirely forested, with what are widely acknowledged to be amongst the most biologically diverse tropical forests on earth. However, they are being cleared or degraded at an alarming rate, and lowland forests, the richest habitat of all, are under serious threat of almost complete clearance. The main causes of this deforestation are clear-felling and logging (both legal and illegal) for timber and pulp fibre, conversion to

plantations (e.g. rubber, oil palm, and for pulp and paper) and conversion for agriculture. They are transforming once-continuous forests into a scattered archipelago of relatively small blocks; the vulnerability of true primary forest species is thereby greatly increased, since they have reduced or are entirely denied opportunities for dispersal to other areas, and their populations are exposed to heightened risks from fires, climate change, edge effects and inbreeding.

In Tenasserim, southern Myanmar, extensive lowland forests remain in Tanintharyi division, but in accessible level lowland areas near to main roads they are rapidly being cleared for oil palm plantations. In peninsular Thailand, lowland forests were still extensive at the start of the twentieth century, but almost entirely cleared by the 1980s; forest below 100 m covered only 20–50 km² in 1987, and most of this was cleared soon after. Pressure on habitat in Peninsular Malaysia is slightly less intense: total forest cover declined from 90% at the beginning of the twentieth century to 43% by 1990, but level lowland forest will soon be restricted to a few patches inside protected areas. In Sabah and Sarawak, forest cover was estimated at 39% and 64% respectively in 1990, but declined greatly in the following decade. There has been little forest loss in the small Sultanate of Brunei, which retains c.81% forest cover, of which c.59% is primary. Over the period 1985 to 1997, about 90,000 km² of forest was cleared in Kalimantan, a decrease in forest cover from 75% to 59%, and about 67,000 km² of forest was cleared in Sumatra, a decrease in

Table 1. Outstanding Important Bird Areas in the Sundaland forests

	IBA name	Status	Territory	Threatened species and habitats
1	**Southern Tanintharyi Division**	—	Myanmar	Extensive lowland forests with Gurney's Pitta and Plain-pouched Hornbill
2	**Huai Kha Khaeng WS**[F04,F06]	PA [WH]	Thailand	Large reserve with populations of White-fronted Scops-owl and Plain-pouched Hornbill
3	**Kaeng Krachan NP**	PA	Thailand	Large reserve with populations of White-fronted Scops-owl and Plain-pouched Hornbill
4	**Khao Nor Chuchi**	(PA)	Thailand	Only known site for Gurney's Pitta
5	**Belum**	—	Peninsular Malaysia	Extensive lowland forests with several threatened birds, notably very large numbers of Plain-pouched Hornbill
6	**Taman Negara NP**	PA	Peninsular Malaysia	Large reserve with populations of almost all threatened lowland and montane forest birds of Peninsular Malaysia
7	**Cameron Highlands WR**	PA	Peninsular Malaysia	Montane forests with populations of Mountain Peacock-pheasant and Malayan Whistling-thrush
8	**Endau-Rompin SP**	PA	Peninsular Malaysia	Extensive lowland forests with populations of several threatened birds
9	**Gunung Leuser NP**	PA [BR]	Sumatra	Very large reserve with populations of most threatened lowland and montane forest birds of Sumatra, notably Aceh Pheasant
10	**Bukit Tigapuluh-Teso Nilo complex**	(PA)	Sumatra	Extensive lowland forests with populations of several threatened birds
11	**Berbak NP**[W20]	PA [R]	Sumatra	Extensive swamp forests supporting several threatened birds; a key site for White-winged Duck
12	**Kerinci Seblat NP**	PA	Sumatra	Very large reserve with populations of most threatened lowland and montane forest birds of Sumatra, notably Salvadori's Pheasant
13	**Bukit Barisan Selatan NP**	PA	Sumatra	Large reserve with several threatened lowland and montane forest birds, notably the only recent records of Sumatran Ground-cuckoo
14	**Danau Sentarum NP**[W20]	PA [R]	Kalimantan	Large reserve with populations of several threatened lowland forest birds
15	**Gunung Palung NP**	PA	Kalimantan	Large reserve with populations of several threatened lowland forest birds, notably Bornean Peacock-pheasant
16	**Tanjung Puting NP**	PA [BR]	Kalimantan	Large reserve with populations of several threatened lowland forest birds
17	**Barito Ulu**	—	Kalimantan	Extensive lowland forests with populations of several threatened birds
18	**Upper Mahakam River**	—	Kalimantan	Extensive lowland forests, appears to be the Bornean stronghold of White-shouldered Ibis
19	**Kayan Mentarang NP**	PA	Kalimantan	Very large reserve with extensive lowland and montane forests and populations of several threatened birds
20	**Ulu Temburong NP**	PA	Brunei	Large reserve with several threatened lowland and montane forest birds, including Storm's Stork and Mountain Serpent-eagle
21	**Kinabalu NP**	PA [WH]	Sabah	Large reserve with extensive montane forests and populations of several threatened birds, notably Mountain Serpent-eagle
22	**Lower Kinabatangan WS**	PA	Sabah	Large reserve with several threatened lowland forest birds, notably a large population of Storm's Stork
23	**Tabin WS**	PA	Sabah	Large reserve with populations of several threatened lowland forest birds
24	**Danum Valley CA**	PA	Sabah	Large reserve with populations of several threatened lowland forest birds
25	**Gunung Mulu NP**	PA [WH]	Sarawak	Large reserve with populations of several threatened lowland and montane forest birds
26	**Gunung Halimun NP**	PA	Java	Large reserve with populations of several threatened forest birds, including all those confined to western Java
27	**Gunung Gede-Pangrango NP**	PA [BR]	Java	Large reserve with populations of several threatened forest birds, including all those confined to western Java
28	**Gunung Raung**	—	Java	Extensive montane forests with populations of several threatened birds, notably White-faced Hill-partridge
29	**Bali Barat NP**	PA	Bali	Only site for Bali Starling

Several of the forest birds of this region occur in two IBAs listed in region W20 (Sembilang and Way Kambas NP).
Note that more IBAs in this region will be included in the *Important Bird Areas in Asia*, due to be published in early 2004.

Key *IBA name*: CA = Conservation Area; NP = National Park; SP = State Park; WR = Wildlife Reserve; WS = Wildlife Sanctuary.
Status: PA = IBA is a protected area; (PA) = IBA partially protected; — = unprotected; BR = IBA is wholly or partially a Biosphere Reserve (see pp.34–35); R = IBA is wholly or partially a Ramsar Site (see pp.31–32); WH = IBA is wholly or partially a World Heritage Site (see p.34); F04/F06 = supports threatened forest birds of regions F04/F06; W20 = supports threatened waterbirds of region W20.

F07

Table 2. Threatened birds of the Sundaland forests.

Species			Distribution and habitat
			SUNDALAND LOWLANDS
Storm's Stork *Ciconia stormi*	◉	EN	PSB: forest in the level lowlands and on lower hill slopes, particularly swamp forest
White-shouldered Ibis *Pseudibis davisoni*	◆	CR	B: forested rivers in a restricted area of lowland Borneo
White-winged Duck *Cairina scutulata*	◆	EN	PSJ: wetlands in lowland forest, extinct on Java and near extinction on the Thai–Malay peninsula
Wallace's Hawk-eagle *Spizaetus nanus*	◉	VU	PSB: forest in the level lowlands and on lower hill slopes
Black Partridge *Melanoperdix nigra*	◉	VU	PSB: forest in the level lowlands and on lower hill slopes
Crestless Fireback *Lophura erythrophthalma*	◉	VU	PSB: forest in the level lowlands and on lower hill slopes
Wattled Pheasant *Lobiophasis bulweri*	◉	VU	B: forest on hill slopes at low and mid-elevations
Malaysian Peacock-pheasant *Polyplectron malacense*	◉	VU	P: forest in the level lowlands
Bornean Peacock-pheasant *Polyplectron schleiermacheri*	◉	EN	B: lowland forest below c.1,000 m, apparently favouring forest on black alluvial soils
Masked Finfoot *Heliopais personata*	✦	VU	PSJ: wetlands in lowland forest, including mangroves; may breed locally in the region
Silvery Wood-pigeon *Columba argentina*	◉	CR	I: mangroves, woodland and coconut groves on small islands; no confirmed records since 1931
Large Green-pigeon *Treron capellei*	◉	VU	PSBJ: forest in the level lowlands and on lower hill slopes
Grey Imperial-pigeon *Ducula pickeringii*	◆	VU	I: forest and cultivation on small islands off northern Borneo
Short-toed Coucal *Centropus rectunguis*	◉	VU	PSB: forest in the level lowlands and on lower hill slopes
White-fronted Scops-owl *Otus sagittatus*	◉	VU	P: forest in the level lowlands and on lower hill slopes
Sunda Nightjar *Caprimulgus concretus*	◉	VU	SB: lowland forest usually below 500 m, often near water
Blue-banded Kingfisher *Alcedo euryzona*	◉	VU	PSBJ: streams in lowland forest, mainly below 850 m (but up to 1,400 m on Borneo)
Plain-pouched Hornbill *Aceros subruficollis*	◉	VU	P: lowland forest below c.1,000 m
Gurney's Pitta *Pitta gurneyi*	◉	CR	P: forest in the level lowlands, known from a handful of sites in southern Myanmar and Thailand
Blue-headed Pitta *Pitta baudii*	◉	VU	B: lowland forest below c.500 m
Fairy Pitta *Pitta nympha*	✦	VU	B: recorded in forest below 1,070 m
Straw-headed Bulbul *Pycnonotus zeylanicus*	◉	VU	PSBJ: secondary forest and scrub, often near rivers, usually below 1,000 m
Hook-billed Bulbul *Setornis criniger*	◉	VU	SB: peat swamp and heath forest, mainly in the lowlands
Black-browed Babbler *Malacocincla perspicillata*	◉	VU	B: known by a single specimen collected in the 1840s, apparently in lowland forest
Bornean Wren-babbler *Ptilocichla leucogrammica*	◉	VU	B: lowland forest below c.600 m
Brown-chested Jungle-flycatcher *Rhinomyias brunneata*	✦	VU	PB: non-breeding birds are apparently confined to mature forest in the level lowlands
Rueck's Blue-flycatcher *Cyornis ruckii*	◉	CR	S: known by two specimens collected in the lowlands of northern Sumatra in 1917–1918
Large-billed Blue-flycatcher *Cyornis caerulatus*	◉	VU	SB: lowland forest below c.500 m

◉ = breeds only in this forest region; ◆ = also breeds in other region(s); ✦ = non-breeding visitor from another region

Distribution: P = Thai–Malay peninsula (and northward along the Thailand–Myanmar border); S = Sumatra; B = Borneo; J = Java; I = small islands.

... continued

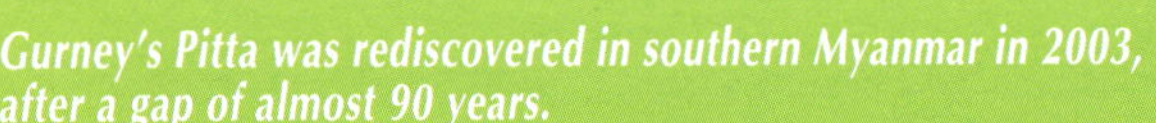

Gurney's Pitta was rediscovered in southern Myanmar in 2003, after a gap of almost 90 years.

PHOTO: KANIT KHANIKUL/GURNEY'S PITTA CONSERVATION GROUP

White-winged Duck inhabits forested wetlands in lowland Sumatra, but its habitat there is being rapidly lost and fragmented.

PHOTO: JACOB WIJPKEMA

F07

Javan Hawk-eagle, Indonesia's national bird, is confined to the pockets of forest that remain on Java.

PHOTO: BAS VAN BALEN

forest cover from 49% to 35%. It is mainly the lowland forests which have been converted and degraded, and without immediate and fundamental changes in policies and management (including improved law enforcement), the World Bank has predicted that virtually all lowland forest will have been cleared in Sumatra by the year 2005 and in Kalimantan by 2010. Much of Java was deforested centuries ago, and by the 1980s only 2,590 km^2 of lowland forest and 2,640 km^2 of montane forest remained there, all in isolated fragments of varying size.

CONSERVATION ISSUES AND STRATEGIC SOLUTIONS (summarised in Table 3)

Forest loss and degradation

FORESTRY AND ILLEGAL LOGGING

Forest ecosystems have been under intense pressure from commercial harvest by legal and illegal enterprises for several decades. Massive areas of natural habitat have been and continue to be selectively logged or clear-felled. The forests in the level lowlands have been most seriously affected, as the topography favours accessibility, and timber extraction is most profitable there. However, forest in the 1,000–2,000 m altitudinal zone is coming under increasing pressure from logging as timber resources are exhausted in the lowlands. Peat swamp forest is also increasingly targeted for extraction: large-scale exploitation of coastal forests in south Sumatra began in the mid-1970s and between 1982 and 1995 two-thirds of the remaining primary swamp forest was logged; on Borneo only scattered remnants of virgin peat swamp forest remain, most of which are scheduled to

Table 2 ... continued. Threatened birds of the Sundaland forests.

Species			Distribution and habitat
			BORNEAN MOUNTAINS (EBA 157)
Mountain Serpent-eagle *Spilornis kinabaluensis*	◉	VU	Montane forest above 800 m in northern and central Borneo
			SUMATRA MOUNTAINS (EBA 158)
Aceh Pheasant *Lophura hoogerwerfi*	◉	VU	Montane forest at c.1,200–2,000 m
Salvadori's Pheasant *Lophura inornata*	◉	VU	Montane forest at c.800–2,200 m
Sumatran Ground-cuckoo *Carpococcyx viridis*	◉	CR	Hill and lower montane forest at c.300–1,400 m
Schneider's Pitta *Pitta schneideri*	◉	VU	Montane forest at c.900–2,400 m
Graceful Pitta *Pitta venusta*	◉	VU	Hill and lower montane forest at c.400–1,400 m
Sumatran Cochoa *Cochoa beccarii*	◉	VU	Montane forest at c.800–2,400 m
			PENINSULAR MALAYSIA MOUNTAINS (EBA 158)
Mountain Peacock-pheasant *Polyplectron inopinatum*	◉	VU	Montane forest at c.820–1,800 m
Crested Argus *Rheinardia ocellata*	◐	VU	Occupies the transition between hill and lower montane forest at c.650–1,150 m
Malayan Whistling-thrush *Myophonus robinsoni*	◉	VU	Montane forest at c.750–1,750 m
Rufous-headed Robin *Luscinia ruficeps*	✈	VU	Single record of a presumed migrant mist-netted in ericaceous scrub at 2,030 m
			JAVA AND BALI FORESTS (EBA 160)
Javan Hawk-eagle *Spizaetus bartelsi*	◉	EN	Lowland and montane forest up to 2,500 m on Java, with an optimum altitude of c.500–1,000 m
White-faced Hill-partridge *Arborophila orientalis*	◉	VU	Known from montane forest at a handful of localities in East Java
Green Peafowl *Pavo muticus*	◐	VU	Semi-deciduous forest and partially open habitats on Java, both in the lowlands and mountains
Javan Scops-owl *Otus angelinae*	◉	VU	Montane forest on Java at c.900–2,500 m
Javan Cochoa *Cochoa azurea*	◉	VU	Montane forests in West and Central Java at c.900–3,000 m
Java Sparrow *Padda oryzivora*	◉	VU	Woodland, savanna and cultivated areas on Java and Bali, mainly in the lowlands
Black-winged Starling *Sturnus melanopterus*	◉	EN	Savanna, forest and cultivated areas on Java and Bali, mainly in the lowlands
Bali Starling *Leucopsar rothschildi*	◉	CR	Confined to a small area of open woodland and savanna in the lowlands of western Bali

◉ = breeds only in this forest region; ◐ = also breeds in other region(s); ✈ = non-breeding visitor from another region

Table 3. Conservation issues and strategic solutions for birds of the Sundaland forests.

Conservation issues	Strategic solutions
Forest loss and degradation	
■ FORESTRY AND ILLEGAL LOGGING ■ PULP AND PAPER INDUSTRY ■ CONVERSION TO PLANTATIONS (ESPECIALLY OIL PALM) ■ CONVERSION TO AGRICULTURE ■ FOREST FIRES ■ DEVELOPMENT (URBAN, INDUSTRIAL, ETC.) ■ MINING ■ TRANSMIGRATION	➤ Advance sustainable forest management and certification, including improved implementation of existing logging policies and regulations that are beneficial to forest biodiversity ➤ Encourage timber traders, retailers and users to take greater corporate responsibility for natural forest management and conservation ➤ Protect natural forests in active or logged concessions from conversion to other land-uses ➤ Secure substantial long-term funding to provide incentives for government and civil society to protect and restore natural forests ➤ Increase inter-governmental cooperation and action to stop illegal logging ➤ Develop a public register of Indonesian government regulations and decisions relating to forest and land-use ➤ Support national and local environmental NGOs in promoting regional concern about forest conservation ➤ Reject new proposals to clear natural forests for pulp fibre, and only establish new plantations on land that is already cleared or degraded ➤ Encourage plantation companies to take corporate responsibility for forest conservation, by developing 'best practice' in the establishment and management of industrial timber, oil palm and other plantation crops ➤ Conduct environmental audits of the pulp and paper sector, to ensure against illegal or unsustainable practices ➤ Promote increased corporate responsibility amongst investors, traders, and corporate uses of pulp and paper, oil palm and coffee, to help prevent these sectors impacting further on natural forests ➤ Establish a permanent forest fire monitoring and prevention network, and manage and protect commercial forests to prevent the outbreak and spread of fires ➤ Enforce the ASEAN no burning policy in the plantation sector ➤ Assess the environmental impact of road building, mining, and new transmigration schemes in forested areas, and minimise development near sites of high biodiversity value
Protected areas coverage and management	
■ GAPS IN PROTECTED AREAS SYSTEM ■ WEAKNESSES IN RESERVE MANAGEMENT	➤ Improve protected areas systems to fill gaps in coverage of threatened bird species and their habitats ➤ Rapidly advance the protection of the largest remaining blocks of lowland forest in Sumatra and Kalimantan, and southern Myanmar ➤ Strengthen reserve management (particularly in Sumatra and Kalimantan) through improved funding, infrastructure and staff training
Exploitation of birds	
■ HUNTING AND TRAPPING ■ WILD BIRD TRADE	➤ Improve enforcement of existing hunting laws, including by increased patrolling of protected areas ➤ Control the wild bird trade in Indonesia by strengthening law enforcement and awareness campaigns
Gaps in knowledge	
■ OUTDATED INFORMATION ON FOREST COVER ■ INADEQUATE DATA ON THREATENED BIRDS	➤ Locate remaining primary lowland forest patches in Sumatra and Kalimantan, and assess their long-term conservation prospects ➤ Survey poorly known threatened species, notably Silvery Wood-pigeon, Black-browed Babbler and Rueck's Blue-flycatcher, and Gurney's Pitta in southern Myanmar

The World Bank has predicted that virtually all lowland forest will have been cleared in Sumatra by the year 2005 and in Kalimantan by 2010.

PHOTO: MARCO LAMBERTINI/BIRDLIFE

Illegal logging is a major problem, including inside protected areas, and needs to be addressed through increased international cooperation and action.

PHOTO: JEREMY HOLDEN

be cleared or converted to other land-uses. Recent political and economic upheavals in Indonesia have resulted in an escalation of illegal logging and land conversion, with deliberate targeting of all remaining stands of valuable timber, including those inside protected areas. For example, industrial-scale logging is currently taking place within the boundaries of Gunung Leuser National Park, Sumatra, and Tanjung Puting National Park, Kalimantan, with regional authorities and outside agencies unable to prevent it.

The most extensive remaining forests are in Sumatra and Kalimantan, but are being very rapidly exploited, with logging concessions covering 67,000 km^2 in Sumatra and 118,000 km^2 in Kalimantan. Concessions are currently managed unsustainably, and increasingly subject to illegal logging, either through over-cutting (by the concessionaires) or by third parties. Over-capacity in the timber industry as a whole is fuelling massive illegal logging both inside and outside concessions. Once logged, there is considerable pressure to convert forestland to other uses (e.g. plantations). This pressure is expected to become progressively stronger, especially given the over-capacity in the pulp industry and oil palm expansion plans.

It is vital that sustainable forest management is rapidly achieved throughout the region, supported by the development of credible national and international forest certification schemes. Many existing policies and regulations relating to logging would benefit threatened birds and other biodiversity, and should be fully implemented, including: reduced-impact logging; logging restrictions on slopes, along watercourses and adjacent to protected areas; and preservation of representative pieces of primary habitat. Timber traders, retailers and users must take greater corporate responsibility for natural forest management and conservation. The area of natural forest within logging concessions, particularly in the lowlands, far exceeds that covered by protected areas; conservation efforts should focus on natural areas that have been logged or are scheduled for logging, particularly as many concessions are coming to the end of their cutting plans and are likely to come under considerable pressure from conversion proposals. The funding for these projects needs to be sourced from the global community, through initiatives such as carbon offset agreements and 'ecological taxes' on commodities such as paper and palm oil, with the aim of providing long-term incentives to governments and civil society in the region to protect and restore natural forests; however, a major challenge is to find ways to pay for forest conservation without fuelling corruption. The problem of illegal logging needs to be addressed through increased international inter-governmental cooperation and action. Conservation in Indonesia would benefit from complete transparency of the Ministry of Forestry's regulations and decisions, particularly those relating to concessions and land-use status (and proposed changes in land-use), through the development of an on-line, publicly accessible register. The capacity of national and local environmental NGOs in Indonesia needs strengthening, so that they can promote awareness about forest conservation and help monitor and protect key forest areas. Local NGOs should be supported by the conservation community when they challenge in court illegal activities and proposals by local government to release natural forest areas for other forms of land use.

PULP AND PAPER INDUSTRY

The Indonesian pulp and paper industry has expanded rapidly since the late 1980s, and the country is now one of the world's top 10 producers; the Indonesian Ministry of Industry and Trade recently proposed to make pulp and paper one of four key 'engines of macro economic recovery', and similar expansion is planned in East Malaysia. The corresponding increase in demand for wood fibre is causing large-scale forest clearance and degradation, particularly in Sumatra, through clear-felling of forest for pulp wood and for the establishment of pulp wood plantations. There are six large pulp mills in Sumatra and one in East Kalimantan, and three new mills are being considered, in Sabah, Sarawak and South Kalimantan (with associated industrial timber plantations, e.g. of 3,000 km^2 in place of logging concessions in Sabah). The pressure that the industry is placing on the forests of this region is set to increase substantially over the next 10 years as the new mills come online, and as capacity expansions are carried out at existing mills.

In line with a moratorium declared by the Government of Indonesia, all plans to clear new areas of natural forest (even degraded secondary forest) for pulp-fibre or to establish new timber plantations should be rejected. Any expansion of plantations in Sumatra and Kalimantan should be on already cleared and degraded land which now lacks significant forest cover. The plantation industry should take greater corporate responsibility for forest conservation, and, with advice from the conservation community, it should develop 'best practice' in the

The pulp and paper industry is putting huge pressure on Sundaland's forests, and must take greater corporate responsibility for forest conservation.

PHOTO: MARCO LAMBERTINI/BIRDLIFE

F07

establishment and management of plantations, including: a commitment not to clear natural forest for plantations; protection and management of natural forest patches within existing concessions; adherence to ASEAN's policy not to use fire to clear land; and prohibition of hunting within concessions. Plantations adjacent to protected areas or other important areas of natural forest should assist with and respect boundary demarcation, and could also assist in providing a buffer in any strategy to protect these areas. To ensure the sustainability of the pulp and paper industry in Indonesia, environmental audits should be conducted to fully assess the financial and environmental risks involved with the pulp and paper sector, and to ensure against illegal or unsustainable practices. In particular, investors, traders and buyers should agree a strategy with the pulp and paper companies to end the current use by mills of timber from natural forests, involving reduced processing capacity and using only sustainably harvested pulpwood from independently certified plantations.

CONVERSION TO PLANTATIONS (ESPECIALLY OIL PALM)

Extensive areas of natural forest are being cleared and replaced with plantations of cash-crops, principally oil palm, but also rubber, coffee and fruit. Palm oil is Indonesia's most important agricultural commodity in terms of foreign exchange, and between 1984 and 1997 an estimated 17,000 km^2 of forest on Sumatra was replaced by oil palm estates, with a further 21,000 km^2 cleared (mainly for oil palm) but not planted; the corresponding figures for Kalimantan are lower, but are expected to increase. Large areas of forest in Peninsular Malaysia have been cleared for oil palm and rubber plantations, and the area of oil palm is now rapidly expanding in East Malaysia (Sabah and Sarawak); for example, the Sarawak State Government plans to plant 15,000 km^2 of oil palm. The accessible lowland forests in southern Myanmar are also being rapidly cleared for oil palm plantations. Indonesia's oil palm sector is dominated by corporate groups that are also active in the timber and pulp industry; when concessions have been selectively logged, these groups apply (usually successfully)

Palm oil is Indonesia's most important agricultural commodity in terms of foreign exchange, and large areas of forest in Sumatra and Kalimantan have been converted to plantations.

PHOTO: MARCO LAMBERTINI/BIRDLIFE

for the land-use to be changed to conversion forest, for plantation development, with the residual timber being used as pulp-fibre. Palm oil is used primarily for food stuffs (e.g. cooking oil, ice cream), as well as soaps, lubricants and cosmetics; it is exported throughout the world and many European and US multinationals are major users.

The oil palm industry should take greater corporate responsibility for forest conservation, and, with advice from the conservation community, should develop 'best practice' in the establishment and management of plantations, with any expansion restricted to cleared and degraded land which lacks significant forest cover. It is in the vegetable oil industry's self-interest to develop long term strategies towards ecologically and socially sustainable land use and production methods, given that consumers will expect the industry to take care of these issues. Manufacturers should identify the suppliers of the oil that they use, and their plantations, and ensure that their consumption is not leading to the clearance of natural tropical forests.

CONVERSION TO AGRICULTURE

In some areas, the forests were long ago cleared for cultivation, notably on Java, where the rich volcanic soils support intensive agriculture and a large human population. Small-scale investors and small-holders continue to have a significant involvement in deforestation, particularly in Sumatra and Kalimantan, although the proportion of forest land they clear is very hard to determine; they have often been marginalised by concessions or plantations, or are sponsored or spontaneous inter-provincial transmigrants. Typically they clear and manage small plots for a variety of tree crops (rubber, coffee, cocoa, cinnamon, etc.) or subsistence crops. Often land is cleared, with backing from local investors, for land speculation and land claim purposes. The clearance of large areas of forest at higher altitudes is of particular concern, e.g. for cinnamon in Kerinci Seblat National Park and for coffee in Bukit Barisan Selatan National Park. Fire is used to clear or prepare land for cultivation, adding to the risk of forest fires. Encroachment into forest needs to be prevented in protected areas and at other key sites for biodiversity. Improved agricultural efficiency and techniques should be used to retain soil fertility and reduce the need to clear more forest, with cleared land rehabilitated so it can be used for cultivation.

FOREST FIRES

The Sundaland region is affected by the El Niño–Southern Oscillation (ENSO) cycle, which causes periodic droughts and renders the region's logged-over areas intensely vulnerable to fire, especially in the lowlands. Many swamp forests grow on a peat soil that easily burns when dry and they are then much more vulnerable to severe fire damage. Drought conditions in 1997 and 1998 led to very extensive fires, which were estimated to have damaged 3,000 km^2 of lowland forest and 3,000 km^2 of swamp forest in Sumatra, and 23,000 km^2 of lowland forest and 7,500 km^2 of swamp forest in Kalimantan. Nearly all burning was related to human activities, particularly commercial land clearance for oil palm and timber estates (estimated by some to account for 80% of fires in Sumatra and Kalimantan). Future ENSO events could lead to a resurgence of forest burning for plantation development; a permanent forest fire monitoring and prevention network is needed, to help with efforts to reduce the incidence and impact of forest fires. The ASEAN no-burning policy should be enforced, with appropriate penalties for plantation companies which transgress

Fires are used to clear forest for agriculture and plantations, but often burn out of control during the periodic droughts associated with the El Niño–Southern Oscillation cycle.

PHOTO: MARCO LAMBERTINI/BIRDLIFE

(although it may be necessary to allow some strictly controlled burning within existing plantations to prevent the build-up of combustible materials). Commercial forests should be managed to protect against the outbreak and spread of fires, and laws that protect forests from arson should be strongly enforced.

■ *DEVELOPMENT (URBAN, INDUSTRIAL, ETC.)*
Some important forest areas have been affected or are threatened by development. The Chew Larn dam in Thailand, for example, submerged the country's only Storm's Stork locality. In Peninsular Malaysia, the proposed Highland Resorts Road (now apparently on hold) would damage the montane habitats of Mountain Peacock-pheasant and Malayan Whistling-thrush, by causing large-scale deforestation and erosion, development (urban and tourism), encroachment for highland agriculture (e.g. vegetables), and increased incursion by loggers and hunters. Kerinci Seblat and Gunung Leuser National Parks on Sumatra and Kayan Mentarang National Park in Kalimantan are threatened by road development projects, which would facilitate access to forest areas. Environmental impact assessments should be conducted for projects that could damage forested areas, with appropriate mitigation plans. New developments should be avoided near protected areas and other sites of high biodiversity value. Plans for the Highland Resort Road in Malaysia should be dropped.

■ *MINING*
In Kalimantan, large mining operations have caused considerable damage to forests, pollution and the disruption of local communities. Under Indonesian law, new mining projects are subject to environmental impact assessments; these should fully take into account the potential damage to forest biodiversity, with mitigation for any negative effects. No further open-cast coal mining should be allowed in forest areas.

■ *TRANSMIGRATION*
It has long been policy in Indonesia to re-settle people mainly from Java to develop the country's less populated regions, with Kalimantan a major destination. The 'Million hectare peat swamp project', which aimed to clear large areas of swamp forest in Central Kalimantan for rice cultivation, has recently been cancelled, but only after huge areas of forests had been clear-felled or drained. The environmental impact of any new resettlement schemes needs to be carefully assessed, following the existing legal process (the AMDAL regulations); in general, they should be sited away from protected areas and other key forest sites.

Protected areas coverage and management

■ *GAPS IN PROTECTED AREAS SYSTEM*
There are extensive networks of protected areas in most parts of this region, but there are some important gaps in coverage of threatened birds and their habitats. Although forest designated for conservation covers c.48,000 km^2 in Sumatra and c.44,000 km^2 in Kalimantan, the most extensive and impressive national parks are in hill and montane (rather than lowland) areas, and only one has been fully gazetted; measures are urgently needed to advance the protection of the largest remaining blocks of lowland forest. The following sites and habitats need official protection or alternative measures to ensure that their habitat and biodiversity are conserved, recognising that reserves (together with any adjacent forested areas) need to be large enough to support functional forest ecosystems and bird communities: (1) all forest inhabited by Gurney's Pitta at Khao Nor Chuchi in Thailand, and at selected sites in southern Tanintharyi Division, southern Myanmar; (2) a large protected area in the Main Range in Peninsular Malaysia, for Mountain Peacock-pheasant and Malayan Whistling-thrush; (3) enlargement of Gunung Leuser National Park, Sumatra, to cover all areas of high biodiversity value, especially in the lowlands; (4) management of the Teso Nilo forests (including Bukit Tigapuluh National Park, Kerumutan, Bukit Rimbang and Bukit Baling), Sumatra, for elephant and tiger; (5) a protected area for Silvery Wood-pigeon, once surveys have located a suitable site, (6) inclusion of lowland logging concessions into the Kerinci Seblat National Park, Sumatra, especially the Sipurak Hook area; (7) a major protected area in the Sebuku Sembakung region of East Kalimantan (which is adjacent to extensive areas of managed natural forest in Sabah).

■ *WEAKNESSES IN RESERVE MANAGEMENT*
Some protected areas in this region are well planned and managed, but most lack the financial, technical and management capacity required for their protection. Given the rapid rate of deforestation in this part of Indonesia, there is an urgent need to preserve the integrity of the existing protected areas systems, and especially to prevent illegal logging and land clearance in lowland areas within reserves. National and international efforts are needed to support and monitor government efforts to secure these areas; this should develop and advance a reform agenda for protected areas as an integral part of the emerging National Forest Programme. Funding is needed to strengthen patrolling and enforcement, to finance participatory boundary demarcation in critical areas, and to promote protected areas at local government level and among the Indonesian public. The donor community should place primary importance on preserving the integrity of the protected areas system in its lending and grant-making programmes.

In parts of the region, e.g. Sarawak, there is a trend towards privatised reserve management, with the aim of generating revenue to cover management costs; at protected areas with high landscape and wilderness value it may be possible to raise sufficient income from tourism, but reserves with lower tourism potential could be starved of resources. In Indonesia, recent autonomy legislation to decentralise power from national to local governments is causing problems in some protected areas, e.g. at Kayan Mentarang National Park in Kalimantan the local government has approved the construction of a new road, apparently without reference to central government; it is vital that this new local legislation (and its implementation) is harmonised with existing national laws (and the mechanisms used to enforce them), to prevent conflicts of this type. Many reserves in the region lack adequate management plans, and these should be prepared, fully taking into account all relevant national and local legislation and the needs of local communities. Other measures for site conservation could include local land-use planning backed by local government legislation and the establishment of site conservation partnerships with local stakeholders.

Exploitation of birds

HUNTING AND TRAPPING

Hunting is a problem, especially of larger birds such as pheasants, storks, raptors, pigeons and hornbills. Many people carry firearms and shoot opportunistically in forest areas. Snaring of terrestrial birds (particularly pheasants and partridges) to supply food to villages and logging camps is very common and must pose a serious threat to some species. Existing hunting laws need to be more strictly enforced, particularly in protected areas, through increased patrolling and removal of snares. In logged forest restrictions on hunting and trapping will have long-term economic benefits through the pollination, seed dispersal and other ecological services provided by wildlife, many of which speed the recovery of degraded habitat.

WILD BIRD TRADE

Cage birds play an important role in Indonesian culture, and the wild bird trade is a serious problem, particularly on Java and Bali where Javan Hawk-eagle, Black-winged Starling and, especially, Bali Starling are severely threatened largely because of this pressure. The widespread decline in the Straw-headed Bulbul, a popular song bird that once occurred throughout this region, has been one of the most dramatic of any species in Asia. Improved roads, including the opening up of forest by logging concessions and plantations, have greatly increased access for hunters and trappers. Designing conservation strategies to counter the threat of trade is especially difficult. In the case of Bali Starling and Javan Hawk-eagle (Indonesia's 'national bird'), raising awareness through the national media runs the risk of increasing demand. The financial value of these birds means that people will go to great lengths to capture them, and attempts to control trade may be undermined by corruption. However, efforts to control the trade must be continued, by strengthening law enforcement and awareness campaigns.

Gaps in knowledge

OUTDATED INFORMATION ON FOREST COVER

Given the rapid rate of deforestation, forest-cover maps are now outdated. A thorough assessment (using satellite images, aerial surveys and ground surveys) is urgently needed to locate all significant remaining patches of non-swampy primary forest in the lowlands of Sumatra and Kalimantan; the prospects for their long-term conservation should be assessed, and they should become the focus of national forest monitoring initiatives. The potential impacts of the pulp and paper and the oil palm industries on natural forests need to be fully taken into account in this assessment.

INADEQUATE DATA ON THREATENED BIRDS

Ornithological coverage of this region is very incomplete; a few areas have been reasonably well studied (e.g. Peninsular Malaysia), but large parts of Sumatra and Kalimantan, for example, have never been surveyed. Research is therefore needed to assess the distribution, population and habitat needs of many Sundaland species, with a particular focus on protected areas and other key sites for biodiversity, especially the remaining patches of primary lowland forest. Three species have not been recorded in recent decades, and need to be searched for: Silvery Wood-pigeon (offshore islands), Black-browed Babbler (lowland Borneo) and Rueck's Blue-flycatcher (lowland Sumatra). Gurney's Pitta was known from a single site in peninsular Thailand until surveys in 2003 located it in southern Myanmar, where research should be continued to identify the largest remaining areas of suitable lowland forest habitat and to develop an appropriate conservation strategy. Sumatran Ground-cuckoo was rediscovered in Bukit Barisan Selatan National Park in 1997, but needs further study. Other poorly known species which require surveys include: White-faced Hill-partridge (East Java), Bornean Peacock-pheasant (lowland Borneo), Mountain Serpent-eagle (Bornean montains), White-fronted Scops-owl (Thailand and Peninsular Malaysia) and Javan Scops-owl (Javan mountains). The taxonomic position of Aceh Pheasant, which may be conspecific with Salvadori's Pheasant, should be assessed.

Straw-headed Bulbul is a popular cagebird because of its beautiful song, but excessive capture for trade has caused a dramatic decline in its range.

PHOTO: CHRISTIAN ARTUSO

WALLACEA

WALLACEA includes the Indonesian regions of Nusa Tenggara (almost equivalent to the Lesser Sundas), Sulawesi and Maluku (almost equivalent to the Moluccas). The entire region is remarkable for the high degree of localised endemism, and has been subdivided into 10 Endemic Bird Areas (four in Nusa Tenggara, three in Sulawesi and three in Maluku) and one Secondary Area (in Sulawesi). In several of these EBAs (e.g. Timor and Wetar; Sulawesi; Buru) the threatened species include both lowland and montane forest specialists, and some threatened species are highly localised (e.g. Black-chinned Monarch is confined to the tiny island of Boano); conservation measures are therefore required to protect both lowland and montane forests in these EBAs, and in the areas which support highly localised species. The remarkable total of 27 highly threatened species mainly comprises birds affected by habitat loss within their small ranges, but also several species under pressure from exploitation for the wild bird trade (e.g. Chattering Lory and Yellow-crested Cockatoo) or for their eggs (e.g. Maleo).

- **Key habitats** Tropical lowland and montane rainforest, moist and dry deciduous forest, mangrove forest and sago swamps, and associated grassland scrub and cultivation.
- **Altitude** 0–3,000 m.
- **Countries and territories** **Indonesia** (Nusa Tenggara, Sulawesi, Maluku); **Timor-Leste (East Timor)**.

Threatened species	CR	EN	VU	Total
[breeds only in this forest region]	7	20	22	49
[also breeds in other region(s)]	—	—	1	1
[non-breeding visitor from another region]	—	1	—	1
Total	7	21	23	51

Key: = breeds only in this forest region.
= also breeds in other region(s).
= non-breeding visitor from another region.

The Wallacea region corresponds to Conservation International's Wallacea Hotspot (see pp.20–21).

The Gunung Sahendaruman IBA on Sangihe supports five highly threatened bird species, three of which are unique to this site. PHOTO: PHIL BENSTEAD

F08

Table 1. Outstanding Important Bird Areas in Wallacea.

	IBA name	Status	Island	Threatened species and habitats
1	**Komodo NP**	PA BR, WH	Komodo	Important population of Yellow-crested Cockatoo
2	**Mbeliling**	—	Flores	Several Northern Nusa Tenggara EBA species, notably Flores Monarch and Flores Hanging-parrot
3	**Ruteng NRP**	PA	Flores	Several Northern Nusa Tenggara EBA species, notably Flores Scops-owl
4	**Wolo Tadho NR**	(PA)	Flores	Extensive forests with several Northern Nusa Tenggara EBA species
5	**Manupeu-Tanadaru**	PA	Sumba	Large protected area, with populations of all of Sumba's threatened forest birds
6	**Laiwangi-Wanggameti NP**	PA	Sumba	Large protected area, with populations of all of Sumba's threatened forest birds
7	**Gunung Mutis**	PA	Timor	Populations of several Timor and Wetar EBA species, notably Timor Imperial-pigeon
8	**Paitchau-Iralalora**	—	Timor-Leste	Large area of forest, with populations of several Timor and Wetar EBA species
9	**Arnau**	—	Wetar	Populations of several Timor and Wetar EBA species
10	**Pulau Damar**	—	Damar	The only site for Damar Flycatcher
11	**Karakelang HR**	PA	Talaud	Populations of Talaud Rail, Grey Imperial-pigeon and Red-and-blue Lory
12	**Gunung Sahendaruman**	—	Sangihe	The only site for Sangihe Shrike-thrush, Cerulean Paradise-flycatcher and Sangihe White-eye, also Sangihe Hanging-parrot and Elegant Sunbird
13	**Pulau Siau**	—	Siau	The only site for Siau Scops-owl
14	**Bogani Nani Wartabone NP**	PA	Sulawesi	Extensive forests supporting several Sulawesi EBA species, notably Cinnabar Hawk-owl, Matinan Flycatcher and Maleo nesting grounds
15	**Lore Lindu NP**	PA BR	Sulawesi	Extensive forests supporting several Sulawesi EBA species, notably Sulawesi Eared-nightjar and Maleo nesting grounds
16	**Lompobatang PF**	PA	Sulawesi	The only site for Lompobatang Flycatcher
17	**Tanahjampea**	—	Tanahjampea	The only site for White-tipped Monarch
18	**Taliabu PNR**	—	Taliabu	The only site for Taliabu Masked-owl
19	**Wayabula**	—	Morotai	Supports several North Maluku EBA species, notably Dusky Friarbird
20	**Lalobata**	—	Halmahera	Extensive forests supporting several North Maluku EBA species
21	**Pulau Obi**	—	Obi	Moluccan Woodcock on the main peak and Carunculated Fruit-dove in the lowlands
22	**Kapalat Mada**	—	Buru	The only site known for Black-lored Parrot and Rufous-throated White-eye, probably also Blue-fronted Lorikeet
23	**Waibula**	—	Seram	Extensive forests supporting several Seram EBA species
24	**Manusela NP**	PA	Seram	Extensive forests supporting several Seram EBA species
25	**Pulau Boano**	—	Boano	The only site for Black-chinned Monarch

Note that more IBAs in this region will be included in the *Important Bird Areas in Asia*, due to be published in early 2004.

Key *IBA name*: HR = Hunting Reserve; NP = National Park; NR = Nature Reserve; NRP = Nature Recreation Park; PF = Protection Forest; PNR = proposed nature reserve.
Status: PA = IBA is a protected area; (PA) = IBA partially protected areas; — = unprotected; BR = IBA is wholly or partially a Biosphere Reserve (see pp.34–35); WH = IBA is wholly or partially a World Heritage Site (see p.34).

Table 2. Threatened birds of Wallacea.

Species			Distribution and habitat
			NORTHERN NUSA TENGGARA (EBA 162)
Flores Green-pigeon *Treron floris*	◉	VU	Throughout EBA in lowland forest
Yellow-crested Cockatoo *Cacatua sulphurea*	◉ s	CR	Occurs sparsely throughout EBA in lowland forest
Flores Hanging-parrot *Loriculus flosculus*	◉	EN	Endemic to Flores, where very local in mid-elevation semi-evergreen rainforest
Flores Scops-owl *Otus alfredi*	◉	EN	Endemic to Flores, known from two localities in montane forest above 1,000 m
Flores Monarch *Monarcha sacerdotum*	◉	EN	Endemic to Flores, where very local in mid-elevation semi-evergreen rainforest
Flores Crow *Corvus florensis*	◉	EN	Endemic to Flores, inhabits lowland forest below 950 m
			SUMBA (EBA 163)
Sumba Buttonquail *Turnix everetti*	◉	VU	Sparse dry grassland with patches of bushes in the lowlands
Red-naped Fruit-dove *Ptilinopus dohertyi*	◉	VU	Mostly confined to montane forest
Yellow-crested Cockatoo *Cacatua sulphurea*	◉ s	CR	Depends on closed-canopy primary lowland forest with tall trees
Sumba Hornbill *Aceros everetti*	◉	VU	Mostly confined to large patches of undisturbed lowland forest
			TIMOR AND WETAR (EBA 164)
Slaty Cuckoo-dove *Turacoena modesta*	◉	VU	Primary and tall secondary lowland and hill forests on Timor and Wetar
Wetar Ground-dove *Gallicolumba hoedtii*	◉	EN	Lowland and hill forests on Timor and Wetar
Timor Green-pigeon *Treron psittacea*	◉	EN	Primary or tall secondary forest on Timor, chiefly in the extreme lowlands
Timor Imperial-pigeon *Ducula cineracea*	◉	EN	Forest at 500–2,200 m in the mountains of Timor and Wetar
Yellow-crested Cockatoo *Cacatua sulphurea*	◉ s	CR	Lowland and hill forests on Timor
Timor Sparrow *Padda fuscata*	◉	VU	Savanna, scrub and cultivation in the lowlands of Timor
			BANDA SEA ISLANDS (EBA 165)
Damar Flycatcher *Ficedula henrici*	◉	VU	Common in forest on the tiny island of Damar

The Data Deficient Lesser Masked-owl *Tyto sororcula* has been recorded in forest on two islands of the Tanimbar group

◉ = breeds only in this forest region; (bird symbol) = also breeds in other region(s); (flying bird symbol) = non-breeding visitor from another region; s = also occurs in other EBA(s) and/or SA(s) in Wallacea

... continued

Flores Green-pigeon is widespread but localised in the fragmented lowland forests of northern Nusa Tenggara.

PHOTO: COLIN TRAINOR/BIRDLIFE

Yellow-crested Cockatoo has declined rapidly in many parts of Wallacea because of capture for the wild bird trade.

PHOTO: BIRDLIFE

F08

Table 2 continued. Threatened birds of Wallacea.

Species			Distribution and habitat
			SULAWESI (EBA 166)
Japanese Night-heron *Gorsachius goisagi*	✈ s	EN	Several non-breeding records from Sulawesi
Maleo *Macrocephalon maleo*	●	EN	Widespread in lowland and mid-elevation forests, nests colonially in warm sand or soil
Snoring Rail *Aramidopsis plateni*	●	VU	Widespread in lowland and mid-elevation forests
Blue-faced Rail *Gymnocrex rosenbergii*	● s	VU	Widespread in lowland and mid-elevation forests
Yellow-crested Cockatoo *Cacatua sulphurea*	● s	CR	Widespread but extremely scarce in lowland and mid-elevation forests and open habitats
Sulawesi Golden Owl *Tyto inexspectata*	●	VU	Recorded from northern and central Sulawesi in lowland and mid-elevation forests
Cinnabar Hawk-owl *Ninox ios*	●	VU	Known only from montane forest on the Minahassa peninsula
Sulawesi Eared-nightjar *Eurostopodus diabolicus*	●	VU	Recorded from northern and central Sulawesi in lowland and montane forests
Lompobatang Flycatcher *Ficedula bonthaina*	●	EN	Confined to montane forest on Gunung Lompobatang in South Sulawesi
Matinan Flycatcher *Cyornis sanfordi*	●	EN	Known only from montane forest on the Minahassa peninsula
			SANGIHE AND TALAUD (EBA 167)
Talaud Rail *Gymnocrex talaudensis*	●	EN	Known only from Karakelang in the Talaud islands, in grassland and rank vegetation near forest
Grey Imperial-pigeon *Ducula pickeringii*	◐	VU	Small island specialist that occurs on the Talaud islands
Red-and-blue Lory *Eos histrio*	●	EN	Extinct on Sangihe and declining rapidly on Talaud, in forest and plantations
Sangihe Hanging-parrot *Loriculus catamene*	●	EN	Endemic to Sangihe, where widespread in forest, coconut plantations and scrub
Siau Scops-owl *Otus siaoensis*	●	CR	Known by a single specimen collected on the tiny island of Siau in 1866
Caerulean Paradise-flycatcher *Eutrichomyias rowleyi*	●	CR	Endemic to Sangihe, confined to mid-altitude forest on Gunung Sahendaruman
Sangihe Shrike-thrush *Colluricincla sanghirensis*	●	CR	Endemic to Sangihe, confined to mid-altitude forest on Gunung Sahendaruman
Elegant Sunbird *Aethopyga duyvenbodei*	●	EN	Possibly now confined to Sangihe, where widespread in forest, coconut plantations and scrub
Sangihe White-eye *Zosterops nehrkorni*	●	CR	Endemic to Sangihe, confined to mid-altitude forest on Gunung Sahendaruman
			BANGGAI AND SULA ISLANDS (EBA 168)
Blue-faced Rail *Gymnocrex rosenbergii*	● s	VU	In this EBA, known by a single record on Peleng in the 1930s
Taliabu Masked-owl *Tyto nigrobrunnea*	●	EN	Known only by two records on Taliabu, in lowland forest
Banggai Crow *Corvus unicolor*	●	EN	Known by two specimens from the Banggai group in the 1880s, possibly seen in mossy forest on Peleng in 1991
			TANAHJAMPEA (SA 110)
Yellow-crested Cockatoo *Cacatua sulphurea*	● s	CR	Recorded, but now almost extirpated
White-tipped Monarch *Monarcha everetti*	●	EN	Fairly common in evergreen forest, scrub and mangroves
			BURU (EBA 169)
Moluccan Megapode *Eulipoa wallacei*	● s	VU	Breeds colonially on Buru's beaches and occurs in montane forest
Blue-fronted Lorikeet *Charmosyna toxopei*	●	CR	Historically found in mid-altitude forest, but no confirmed records for many years
Black-lored Parrot *Tanygnathus gramineus*	●	VU	Inhabits montane forest, seldom recorded and thought to be nocturnal
Rufous-throated White-eye *Madanga ruficollis*	●	EN	Known from montane forest in western Buru
The Data Deficient Lesser Masked-owl *Tyto sororcula* is historically known from the lowlands of Buru, but there are no recent records			
			SERAM (EBA 170)
Moluccan Megapode *Eulipoa wallacei*	● s	VU	Nesting colonies on Boano, Seram, Ambon and Haruku, and occurs in montane forest
Purple-naped Lory *Lorius domicella*	●	VU	Inhabits mid-elevation evergreen forests on Seram, records on other islands possibly involved escaped birds
Salmon-crested Cockatoo *Cacatua moluccensis*	●	VU	Inhabits lowland evergreen forest on Seram, records on other islands possibly involved escaped birds
Black-chinned Monarch *Monarcha boanensis*	●	CR	Endemic to the tiny island of Boano, in semi-evergreen forest in the foothills
			NORTHERN MALUKU (EBA 171)
Japanese Night-heron *Gorsachius goisagi*	✈ s	EN	Two non-breeding records from Halmahera
Moluccan Megapode *Eulipoa wallacei*	● s	VU	Nesting colonies on Halmahera, and occurs in montane forest
Invisible Rail *Habroptila wallacii*	●	VU	Flightless rail known only from Halmahera, in lowland sago swamps
Moluccan Woodcock *Scolopax rochussenii*	●	EN	Known only from Obi and Bacan, believed to inhabit montane forest
Carunculated Fruit-dove *Ptilinopus granulifrons*	●	VU	Endemic to Obi, in lowland forest and wooded cultivation
Chattering Lory *Lorius garrulus*	●	EN	Widespread in the EBA, in lowland and montane forests
White Cockatoo *Cacatua alba*	●	VU	Found in lowland forest on Halmahera, Bacan and associated smaller islands
Sombre Kingfisher *Todiramphus funebris*	●	VU	Known only from Halmahera, in forest, sago swamps and mangroves
Purple Dollarbird *Eurystomus azureus*	●	VU	Found in lowland forest on Halmahera, Bacan and associated smaller islands
Dusky Friarbird *Philemon fuscicapillus*	●	VU	Recorded from Morotai, Bacan and Halmahera, in forest and plantations

● = breeds only in this forest region; ◐ = also breeds in other region(s); ✈ = non-breeding visitor from another region; s = also occurs in other EBA(s) and/or SA(s) in Wallacea

The Cerulean Paradise-flycatcher was feared extinct, but was rediscovered during recent surveys of the few forest remnants on Sangihe.

PHOTO: JON RILEY

OUTSTANDING IBAs FOR THREATENED BIRDS (see Table 1)

Twenty-five IBAs have been selected in Wallacea, which represent all EBAs and SAs and together support populations of almost all threatened forest birds of this region. Several sites are unique, as they support the only (or by far the largest) known populations of one or more threatened species, notably Gunung Sahendaruman which is the only site for the Critically Endangered Sangihe Shrike-thrush, Cerulean Paradise-flycatcher and Sangihe White-eye. The other sites include some of the largest and richest forests remaining in this region; however, many other forest sites with significant populations of threatened forest birds will be documented during BirdLife's ongoing IBA Programme.

CURRENT STATUS OF HABITATS AND THREATENED SPECIES

This region is made up of many thousands of islands. The extent of habitat loss on these islands varies widely, and is related to their accessibility and history of human occupation, climate and soil fertility, and whether their forests have been exploited. Some small islands remain pristine, but many other islands have been largely converted to coconut and other plantations. On the larger islands, the coastal lowlands are typically cleared and cultivated, with forest remaining in the interior, although on many of them cultivation and pastureland extend well inland. Commercially valuable forests are mainly found in the lowlands, and most are covered by logging concessions (except for Nusa Tenggara, where the forests are largely protected from commercial forestry); large areas on some islands have already been commercially logged. In comparison to western Indonesia, only small areas have been converted for plantations, but this is likely to become an increasing pressure. Some extensive areas of forest have been cleared for mining, for example on Halmahera.

Nusa Tenggara: This part of Indonesia has a relatively dry, seasonal climate, and the natural forest types are dry deciduous 'monsoon' forest, with semi-evergreen forests in the moister areas, mostly on mountain slopes. The drier forests are relatively easy to clear, because they can be burned, and large areas have been converted to pasture. The semi-evergreen and montane forests, which are naturally more restricted in extent, are often relatively intact but have been affected by cutting for timber and fuelwood and conversion for agriculture. On Flores, the lowlands have been extensively deforested, with only relatively small patches of intact habitat remaining; at higher altitudes habitat is more extensive but suffers continuous erosion at its lower fringes. Sumbawa has a lower population density and more extensive forests, including the best monsoon

The monsoon forests of Nusa Tenggara are relatively easy to clear, because they can be burned, and large areas on Sumba and other islands have been converted to pastureland.

PHOTO: PHIL BENSTEAD

forests remaining in Indonesia. On Sumba, around 60% of the island's forest cover was cleared between 1927 and 1990, leaving 34 fragments varying in size from 0.16 to 425 km² (15.5% of the total land area). Tropical monsoon forest now occupies less than 4% of West Timor, and is distributed among seven or so remnant and isolated patches (none protected), the largest being only 90 km². In Timor-Leste (East Timor) forest cover declined from 37% in the early 1980s to 15% in the mid-1990s, although forests there are still much more extensive than those in West Timor.

Sulawesi: The lowlands of Sulawesi have recently suffered rapid forest loss. Around 1975, the island retained from its original cover 53% of wet, 26% moist, 24% dry lowland forest on alluvium, 4%, 33% and 7% respectively on limestone, and 6%, 10% and 3% respectively on volcanic soils. During the subsequent two decades it lost over 67% of remaining wet lowland forest to timber production and agriculture. Extensive forests remain in the mountains, but large areas have been cleared or degraded by encroachment for agriculture. North of Sulawesi, the island of Sangihe has lost almost all of its original forest, with the only remnant on the Gunung Sahendaruman massif, which retains around 8 km² of habitat, of which the largest continuous tract of primary forest is only 2.25–3.40 km² in size. The three Critically Endangered species confined to this IBA all have specialised habitat requirements and occupy only part of the forest. Clearance of forest on the tiny island of Siau has been near-total, with as little as 0.5 km² of logged woodland remaining. The Talaud islands retain extensive forest cover, but this is now under great pressure from logging and agriculture. East of Sulawesi, the Banggai and Sula Islands have suffered extensive forest loss and degradation, particularly in the lowlands; on Taliabu, the largest of the Sula islands, most forest below 800 m has been commercially logged, and on Banggai the last areas of intact rainforest were being selectively logged in the 1990s and primary habitat was restricted to montane areas. South of Sulawesi, up to half of Tanahjampea remained under forest in 1993, but much of it had been selectively logged.

Maluku: The forests in the coastal lowlands of most larger islands in Maluku have been converted for agriculture, but the interior forests are relatively intact; however, most remaining lowland forests are covered by logging concessions, and large-scale commercial logging operations are now underway on several islands. In the mid-1990s nearly 75% of Buru retained some forest cover (totalling 6,250 km²), but no fewer than 3,866 km²—62%, almost two-thirds—have been designated for conversion to other land uses such as agriculture. On Seram, huge areas of forest remain in the mountainous interior, but the coastal fringe is largely cleared or degraded. Logging concessions cover 48% of Seram's forests, leaving less than 5,096 km² of lowland forest outside concessions, and logging became increasingly intense during the 1990s, even occurring inside Manusela National Park. In northern Maluku, 88.5% of the total area of Morotai, Halmahera and Bacan was reported to remain forested in the early 1990s, but these islands are rich in economically valuable timber and intensive commercial logging is now underway in the lowlands. In southern Maluku, the remote island of Damar remains heavily forested other than the coastal fringe, and in the Tanimbar Islands natural habitat on Larat has long been seriously degraded, but Yamdena retains extensive forest cover. The relatively remote and unpopulated island of Wetar retained over 90% forest cover in the mid-1980s, but logging of the larger, more valuable timber trees has commenced.

Most of the remaining lowland forests in Maluku are covered by logging concessions, and large-scale commercial operations are now underway on several islands.

PHOTO: P. JEPSON/BIRDLIFE

CONSERVATION ISSUES AND STRATEGIC SOLUTIONS (summarised in Table 3)

Forest loss and degradation

■ *FORESTRY AND ILLEGAL LOGGING*

The forestry sector has an important (although declining) role in the Indonesian economy. In many areas most commercially valuable forest has already been logged, and timber extraction is no longer the major threat, although illegal or small-scale logging operations often continue to degrade the remaining forests. Major commercial logging enterprises are now focused on islands in Maluku, especially Halmahera and Seram, where most remaining forest is covered by concessions. In Nusa Tenggara, large-scale logging is really only an issue in Sumbawa, there being no production forests in Flores, Sumba and Timor; but small-scale extraction, usually illegal, is a widespread threat to remaining forest areas in these islands.

Selective logging is the primary extractive technique, but cutting plans and regulations are rarely followed, and so forests are badly damaged by extraction techniques, and areas that should be left uncut (e.g. along watercourses and on hill slopes) are frequently logged. Selective logging also targets mature trees with a (probably) disproportionate impact on parrots, hornbills, owls and other hole-nesters as fewer nest sites remain. Moreover, commercial logging necessitates networks of roads, which can lead to secondary problems such as increased hunting and clearance for agriculture and settlements. In several areas, logging has recently occurred inside protected areas, for example Bogani Nani Wartabone National Park on Sulawesi and Manusela National Park on Seram. Once selective logging has been concluded there are frequently pressures to change the landuse from production forestry to agriculture or plantations.

The Indonesian government is committed to sustainable forest management, and has commendable laws and regulations in place. However, these laws are frequently flouted, in part due to widespread corruption in the forest sector. Given that extensive commercially valuable forests remain, sustainable forest management is a major priority. Certification schemes need to be rapidly advanced, based on best practices in forest management and independent monitoring. Concessions should not be granted inside gazetted or proposed protected areas, i.e. on land officially allocated or designated for nature conservation (with the national conservation plan of 1981 remaining the principal reference). Concessions should be cancelled, and concessionaires penalised, whenever logging operations encroach on protected areas.

■ *EXPLOITATION OF FOREST PRODUCTS*

In some areas, such as the remaining lowland forests on Flores, habitat is being heavily degraded by exploitation for firewood and building material. In some moist forest areas throughout the region, timber and rattan collection is intense, especially near settlements and wherever habitat has been fragmented into small patches. On Sulawesi, some villages are so close to Maleo nesting grounds that opportunistic felling of trees and collection of rattan by villagers is likely to deter breeding birds. Protected areas need to be zoned, with ample buffers to absorb human pressures, and core areas free of exploitative activities. Greater community participation in forest management and conservation should be developed, respecting traditional lifestyles; this should be promoted through conservation awareness work (stressing the ecological services that forests provide, including maintenance of water supplies), particularly around nature reserves and IBAs, to help reduce damage and disturbance in the most important areas of habitat for threatened birds.

■ *CONVERSION TO AGRICULTURE AND PLANTATIONS*

Large areas of forest continue to be converted for agriculture and plantations, especially in Maluku. Near the coast, forest has been replaced with coconut, banana, cacao and oil palm plantations. Inland, forest on rich alluvial soil is liable to be converted to agricultural fields, e.g. at Ruteng on Flores. Many important protected areas are threatened by agricultural encroachment, including Karakelang Hunting Reserve on Talaud and Gunung Sibela Strict Nature Reserve on Bacan. This threat is twofold because it leads to influxes of people and puts greater pressure on forest resources, so the impact on a reserve extends well beyond the area encroached upon. For example, the southern boundary of Bogani Nani Wartabone National Park on Sulawesi is almost entirely degraded by coconut plantations and other cultivation, and people venture within the park for a variety of extractive purposes, including timber, rattan and Maleo eggs. Sago swamp on Halmahera (apparently vital to Invisible Rail) has been extensively cleared, and the threat to the remaining tracts is high, involving commercial sago extraction, irrigation schemes, conversion for wet rice and, potentially, fishpond development. Grassland on Sumba is being burnt, overgrazed and converted to agriculture, all to the likely detriment of Sumba Buttonquail, which appears to avoid man-made grasslands. The lower edges of the vital last patches of forest on Sangihe are gradually being cleared for shifting cultivation.

The development of more efficient agriculture (through improved and appropriate techniques) to help alleviate

Table 3. Conservation issues and strategic solutions for birds of Wallacea.

Conservation issues	Strategic solutions
Forest loss and degradation	
■ *FORESTRY AND ILLEGAL LOGGING* ■ *EXPLOITATION OF FOREST PRODUCTS* ■ *CONVERSION TO AGRICULTURE AND PLANTATIONS* ■ *LIVESTOCK GRAZING AND FIRE* ■ *TRANSMIGRATION* ■ *DEVELOPMENT (URBAN, INDUSTRIAL, ETC.)* ■ *PESTICIDES*	➤ Manage forests sustainably, with certification schemes based on best practices in forest management and independent monitoring ➤ Cancel logging concessions within gazetted and proposed protected areas ➤ Develop greater community participation in forest management and conservation ➤ Promote the development of more efficient agriculture, to help reduce the pressure on the remaining areas of natural habitat ➤ Maintain forest patches on Sangihe, and restore the forest at Gunung Sahendaruman ➤ Introduce fire management programmes, and measures to reduce grazing in forests, in Nusa Tenggara ➤ Develop new transmigration schemes, roads and mines following the existing legal process, including environmental impact assessment, to integrate biodiversity conservation with regional development ➤ Eliminate use of illegal insecticides on plantations
Protected areas coverage and management	
■ *GAPS IN PROTECTED AREAS SYSTEM* ■ *WEAKNESSES IN RESERVE MANAGEMENT*	➤ Establish new protected areas to fill gaps in coverage of threatened birds and their habitats ➤ Strengthen the PKA through training, improved terms and conditions, and equipment for staff ➤ Improve reserve management through more intensive patrolling, boundary demarcation and stricter law enforcement
Exploitation of birds	
■ *HUNTING AND TRAPPING* ■ *EGG-COLLECTION* ■ *WILD BIRD TRADE*	➤ Improve enforcement of existing hunting laws, especially in protected areas ➤ Promote sustainable community management of Maleo and Moluccan Megapode nesting grounds ➤ Implement the Yellow-crested Cockatoo Recovery Plan, and adapt it for other threatened parrots
Gaps in knowledge	
■ *INADEQUATE DATA ON THREATENED BIRDS*	➤ Survey poorly known threatened species, islands and sites, including proposed protected areas ➤ Monitor populations of species that are exploited by man, including megapodes and parrots

poverty is part of the Indonesian government's long-term planning strategy, and if carefully implemented has the potential to reduce the pressure on the remaining areas of natural habitat. These improvements should be promoted away from protected areas (and particularly away from their core areas) and IBAs, and efforts should be made to prevent further encroachment into existing and proposed reserves. Special conservation awareness efforts should be made in the most critical areas (stressing the uniqueness, rarity and endangerment of the birds and their habitats), for example on Sangihe local people should be encouraged not to clear remnant forest patches and native trees, and to plant native tree species in agricultural areas: at Gunung Sahendaruman, the forest should be restored, to increase the area of habitat available to the three Critically Endangered species unique to the site.

■ *LIVESTOCK GRAZING AND FIRE*

Intensive livestock grazing is a problem in some areas, for example on Timor where excessive grazing pressure has severely inhibited regeneration in most lowland and mid-altitude forest. In areas with a relatively dry climate and seasonal rainfall, mainly in Nusa Tenggara, dry-season fires are used to clear land and to encourage new growth. These often burn out of control and damage or destroy forests, especially where these are already fragmented. Although periodic fires are a natural phenomenon, they now occur so frequently that vegetation has little chance to recover. Fire management programmes, and measures to reduce grazing pressure in forested areas, are now urgently needed.

■ *TRANSMIGRATION*

It has long been policy in Indonesia to re-settle people mainly from Java to develop the less populated regions of the country, with Sulawesi and Maluku being major destinations. Whilst the scale of transmigration has been reduced over the past decade, the recent unrest in Maluku and East Timor has led to large-scale movement of people. In some areas transmigration schemes have had serious negative effects on the environment, involving forest clearance for agriculture, hunting and unsustainable slash-and-burn farming. A planned transmigration project could damage the 60 km^2 Kao sago swamp on Halmahera (a potential site for Invisible Rail), and on Sulawesi the arrival of transmigrants in the Dumoga valley may result in the loss of lowland forests and important montane areas within Bogani Nani Wartabone National Park. Transmigration has also led to the breakdown of traditionally controlled egg-collecting systems at nesting grounds of Maleo and Moluccan Megapode, causing local extinctions. New resettlement schemes need to be carefully developed, following the existing legal process including environmental impact assessment. In general, schemes should be sited away from protected areas and IBAs, and not on islands with high biodiversity value. Awareness programmesshould be instituted in transmigration areas to educate people about the ecology and traditions of their new home, in particular the ecological services provided by forests.

■ *DEVELOPMENT (URBAN, INDUSTRIAL, ETC.)*

Improvement of the transportation infrastructure is an essential part of regional development, but new roads can cause serious environmental problems, notably uncontrolled forest clearance by impoverished slash-and-burn farmers. Roads on mountainous islands tend to follow the coast, and often lead to loss and degradation of lowland habitats. Mining and prospecting is degrading forest and polluting watercourses in some areas, including important areas such as Karakelang Hunting Reserve on Talaud, Gunung Sahendaruman on Sangihe and Gunung Sibela Strict Nature Reserve on Bacan. Much forest has been cleared for mining in Halmahera and Wetar. Halmahera has deposits of nickel and gold, and further mining projects are proposed in Central Halmahera. Previous proposals for a large cement factory, with quarry and dam, close to Manusela NP, Seram appear to have been abandoned. Exploitation of sand for local road construction has affected some Moluccan Megapode nesting grounds on Seram. New roads and mines need to be carefully developed, following the existing legal process including environmental impact assessment, with the aim of integrating biodiversity conservation with regional development. They should be sited away from protected areas and IBAs.

■ *PESTICIDES*

The use of pesticides has increased since the 1980s, and poisons have been applied to reduce pig numbers. The current (at least occasional) use of technically illegal insecticides (azodrin and other 'monokrotofos' pest-control chemicals) to eliminate locusts (*Sexava*) from coconut plantations reportedly causes deaths in lories a few days after application, and could be affecting other threatened parrots. Existing regulations should be enforced to prevent the use of these chemicals in plantations.

Protected areas coverage and management

■ *GAPS IN PROTECTED AREAS SYSTEM*

There are some important gaps in the protected areas system in Wallacea, particularly in Maluku where Manusela

The small area of forest remaining at Gunung Sahendaruman on Sangihe is the most important site for bird conservation in Asia.

PHOTO: JIM WARDILL

National Park is the only reserve large enough to ensure biodiversity conservation. Several important new protected areas are planned, awaiting full gazettement, and others are needed to fill the remaining gaps. The following sites and habitats need official protection or alternative measures to ensure that their habitat and biodiversity are conserved: (1) Mbeliling (Kerita Mese) on Flores, to protect semi-evergreen forest below 1,000 m; (2) natural grassland on Sumba, for Sumba Buttonquail; (3) a protected area enclosing Gunung Timau and Gunung Mutis on Timor; (4) Gunung Arnau proposed nature reserve on Wetar; (5) Yamdena proposed nature reserve in the Tanimbar islands; (6) forest on the island of Damar; (7) selected Maleo nesting grounds on Sulawesi; (8) Gunung Lompobatang proposed nature reserve in South Sulawesi; (9) locate and protect an area on Sulawesi that combines mangrove, beach forest, swamp forest and lowland forest connecting to hill forest (e.g. the north-west corner of the Minahassa peninsula); (10) Gunung Sahendaruman proposed wildlife sanctuary on Sangihe; (11) Taliabu proposed nature reserve in the Sula islands; (12) Gunung Kapalatmada proposed wildlife sanctuary on Buru; (13) eastern Seram, ideally in the Wae Fufa and adjacent catchments; (14) Lalobata and Ake Tajawe proposed national parks on Halmahera; (15) Wayabula proposed wildlife sanctuary on Morotai; (16) Pulau Obi proposed nature reserve.

In Timor-Leste, 15 Protected Wild Areas were listed during a brief period of administration by the United Nations (UNTEAT) in 2000, but these reserves have no management arrangements on the ground. The government is in need of support to establish these areas, including the designation of the Monte Paitchau–Iralalora area as the country's first national park.

WEAKNESSES IN RESERVE MANAGEMENT

Responsibility for reserve management in Indonesia lies with the Directorate General of Forest Protection and Nature Conservation (PKA), but effective management is constrained by shortage of staff, expertise and budget, compounded by the vastness and remoteness of the areas which need to be protected. The situation was made worse by the recent civil unrest, which forced the cut-back of conservation programmes managed from Ambon and Ternate. Some protected areas are virtually unsupervised, allowing local communities to hunt and clear forest, and logging or mining corporations to ignore protected area legislation. Land disputes have caused problems in some reserves, for example at Ruteng Nature Recreation Park on Flores, where local communities dispute the current boundaries, and in 1998 it was reported that a land-ownership dispute on Sulawesi had led to occupation of the southern section of Rawa Aopa Watumohai National Park. On Sulawesi, in almost all cases protected status has failed to deter exploitation of Maleo eggs.

It is necessary to strengthen the PKA in the region through better training, pay and equipment for reserve staff, and to improve reserve management through (a) more intensive patrolling to intercept hunters and loggers; (b) clear demarcation of reserve boundaries; and (c) stricter enforcement of environmental laws. Close collaboration between conservation NGOs and the forest department, and effective management of forest officers, would also help. Community management of resources may be the most successful strategy for Maleo (and Moluccan Megapode) nesting grounds, but a deep regard for local politics is needed if it is to succeed (see below).

Exploitation of birds

HUNTING AND TRAPPING

Shooting of wildlife with shotguns, air-rifles and slingshots is widespread, affecting many threatened birds. For example, pigeons are commonly shot on Timor, while hunting, even of small passerines, is intensive on Sangihe. Villagers on Sumba trap the Sumba Buttonquail for food during the dry season, especially at the maize harvest. Trapping with snares threatens forest rails and ground-doves, and it also occurs at or adjacent to some Maleo nesting grounds, adding to the pressures of habitat loss and egg-collection. Protected areas need to be more intensively patrolled to intercept hunters and clear snares, with conservation awareness campaigns to make hunters more aware of the plight of threatened species, and of existing hunting laws.

EGG-COLLECTION

The region's two threatened megapodes, Maleo and Moluccan Megapode, nest colonially and lay large, commercially valuable eggs. The eggs sell for several times the price of a chicken egg, and even the remotest nesting grounds are exploited. Of the 120 Maleo nesting grounds whose present conservation status is known, 42 (35%) have already been abandoned largely owing to over-collection of eggs and habitat destruction. At many colonies current levels of exploitation are so unsustainable that total protection has become necessary. This is by no means straightforward, however, as guardposts, fences and hatcheries installed at protected nesting grounds have been repeatedly destroyed. Eggs are also collected from the nests of parrots, hornbills and other large species, which are raised for food or as pets for trade.

The following general management programme has been proposed for Maleo: (1) build up populations by policing all

Maleos lay large, commercially valuable eggs, and even the most remote nesting grounds are known to local people and exploited.

PHOTO: STUART BUTCHART/BIRDLIFE

nesting grounds, and enforcing a temporary ban on egg-collecting (with appropriate incentives); (2) once populations have reached an acceptable size, strictly supervise exploitation of eggs, ensuring a certain percentage is transferred to hatcheries so that the harvest is sustainable and provides a continuous income to the harvesters and protecting authorities; (3) encourage tourist interest in viewing Maleo grounds through these practices, adding to the economic incentives to implement them; (4) improve nesting ground suitability by clearing, burning or trimming vegetation (particularly the vigorous invasive *Lantana camara*) to increase insolation, and by raking sand over burrows to make detection of eggs by poachers more difficult. Similar measures are required at Moluccan Megapode nesting grounds, perhaps with 10% of each colony sealed off as no-go areas.

■ *WILD BIRD TRADE*

Cage birds play an important role in Indonesian culture. Parrots are particularly popular, and the massive demand for cockatoos and lories, both domestically and internationally, fuels a vigorous trade. Capture for trade is the most important factor in the decline of several threatened parrots in this region, notably Yellow-crested Cockatoo: even after trade was banned, hundreds of birds were traded openly in Jakarta and thousands were annually smuggled abroad. White and Salmon-crested Cockatoos are also declining for the same reason. At least 10,000 Salmon-crested Cockatoos were being trapped annually in the 1980s and the number probably still exceeds 4,000 per annum. As cockatoos tend to raid plantations and agricultural fields, they are persecuted as crop pests wherever they cause damage, with farmers using lime to capture the birds, and then selling them as pets. Three species of lory are also under serious threat from trapping. The Red-and-blue Lory is now confined to the Talaud Islands, where as many as 1,000–2,000 birds still leave Karakelang annually, 80% to the Philippines. In North Maluku, trappers apparently remove around 10% of the world population of Chattering Lory annually, while in Seram large numbers of Purple-naped Lories are trapped to meet domestic demand.

A recovery plan for Yellow-crested Cockatoo should be implemented, and adapted for other threatened parrots, with the aim of reducing trapping of all species to sustainable levels. National legislation needs to be strengthened, followed by collaboration with wildlife traders' associations, national carriers, airport authorities, CITES management authorities, local NGOs and Asian NGOs, in order to impose the law and monitor the situation. The measures required include awareness campaigning, improved protection for key habitats and sites, dockside control in relevant ports, and fair but firm deterrents imposed on offenders, from local trappers (light fines) to middlemen and exporters (heavy fines). In North Maluku, zero quotas should remain in place for all threatened parrots at least until a reliable system of trade management is in place. Population monitoring is required to detect trends.

Parrots are particularly popular cagebirds, and the massive demand for attractive species such as Red-and-blue Lory fuels a vigorous trade.

PHOTO: A. COMPOST/BIRDLIFE

Gaps in knowledge

■ *INADEQUATE DATA ON THREATENED BIRDS*

The vast number of islands in this region, many of which are remote and inaccessible, means that much basic biological survey and inventory work is still required, together with investigations of the ecology of the threatened birds. Studies are needed of the most poorly known threatened species, to clarify their distribution and habitats, notably: Sumba Buttonquail (grasslands on Sumba); Lesser Masked-owl, Taliabu Masked-owl, Sulawesi Golden-owl, Flores Scops-owl, Siau Scops-owl and Cinnabar Hawk-owl (using tape-recording, spotlighting and interviews with local people); Banggai Crow (Banggai and Sula islands); Blue-fronted Lorikeet and Black-lored Parrot (Buru); and Invisible Rail, Moluccan Woodcock and Dusky Friarbird (northern Maluku).

Threatened birds and their habitats need to be surveyed in poorly known islands and sites, including: all proposed protected areas listed above; unsurveyed lowland forests on Flores, notably in the east; Timor-Leste and Wetar, to help develop a comprehensive conservation strategy; Gunung Sahendaruman on Sangihe. The populations of species that are exploited by man need to be monitored, to determine the impact of this threat and the effects of ongoing conservation measures, including: numbers and breeding success at Maleo and Moluccan Megapode nesting grounds; and wild populations of threatened parrots and numbers captured for trade. The threat posed by introduced predators (e.g. cats) to ground-dwelling birds, especially to flightless species such as Invisible Rail, needs study.

PHILIPPINE FORESTS

F09

THE Philippine archipelago is remarkable for its high degree of endemism, with c.44% of the (c.395) breeding bird species, and 54 of the 58 threatened forest species which occur in the country, unique to the islands. There are also high levels of localised endemism within the archipelago, which is subdivided here into seven Endemic Bird Areas and two Secondary Areas. In several EBAs (Mindoro; Luzon; Negros and Panay; Mindanao and the Eastern Visayas) the threatened species include both lowland and montane forest specialists, and some of them are highly localised (e.g. Negros Striped-babbler is virtually confined to Cuernos de Negros in southern Negros); conservation measures are therefore required to protect both lowland and montane forests in these EBAs, and in the areas which support localised species. The remarkable total of 22 highly threatened species mainly comprises birds affected by habitat loss within their small ranges, but also includes Philippine Eagle, found at low densities in the remaining forests of Luzon, Samar, Leyte and Mindanao, and Philippine Cockatoo, which is under huge pressure from capture for the wild bird trade. The conservation of waterbirds in the Philippines is covered in W19.

- **Key habitats** Tropical lowland and montane forest.
- **Altitude** 0–2,700 m.
- **Countries and territories** **Philippines**.

Threatened species

	CR	EN	VU	Total
breeds only in this forest region	11	10	33	54
also breeds in other region(s)	—	—	1	1
non-breeding visitor from another region	—	1	2	3
Total	11	11	36	58

Key: = breeds only in this forest region. = also breeds in other region(s). = non-breeding visitor from another region.

The Philippine forests region corresponds to Conservation International's Philippines Hotspot (see pp.20–21).

The northern Sierra Madre is the only place in the Philippines where unbroken forests extend from the coast to the mountain peaks, but the proposed construction of new roads into the area could lead to rapid deforestation. PHOTO: MICHAEL POULSEN/BIRDLIFE

120°E
20°N
10°N
SOUTH CHINA SEA
PACIFIC OCEAN
SULU SEA
PHILIPPINES
LUZON (EBA 151)
LUZON
CORDILLERA CENTRAL
NORTHERN SIERRA MADRE
SOUTHERN SIERRA MADRE
MINDORO (EBA 150)
TABLAS
TABLAS, ROMBLON AND SIBUYAN (SA 095)
PALAWAN (EBA 156)
PALAWAN
PANAY
NEGROS
NEGROS AND PANAY (EBA 152)
CEBU (EBA 153)
SIQUIJOR (SA 096)
SAMAR
LEYTE
MINDANAO AND THE EASTERN VISAYAS (EBA 154)
MINDANAO
SULU ARCHIPELAGO (EBA 155)
TAWITAWI
1
2
3
4
5
6
7
8
9
10
11
12
13
14
15
16
17
18
19
20
21
22
23
0
250
500
kilometres

Table 1. Outstanding Important Bird Areas in the Phillipine forests.

	IBA name	Status	Island	Habitats
1	Mt Halcon	—	Mindoro	Largest area of montane forest on Mindoro
2	Siburan	—	Mindoro	Largest remaining lowland forest on Mindoro
3	Balbalasang-Balbalan NP	PA	Luzon	Large area of montane forest
4	Mt Cetaceo	—	Luzon	Large area of montane forest, some lowland forest
5	Northern Sierra Madre Nature Park	PA	Luzon	Large area of lowland and montane forest, extending unbroken from the coast to the highest peaks
6	Central Sierra Madre mountains	—	Luzon	Large area of forest, both lowland and montane
7	Balogo watershed	—	Tablas	Largest area of forest remaining on Tablas
8	North-west Panay (Pandan peninsula)	—	Panay	Probably includes the best quality lowland forests remaining on Negros and Panay
9	Central Panay Mountains	—	Panay	Large area of forest, both lowland and montane
10	Mt Canlaon NP	PA	Negros	Large area of montane forest, some lowland forest; only known locality for Negros Fruit-dove
11	Cuernos de Negros	—	Negros	Large area of forest, both lowland and montane; main locality for Negros Striped-babbler
12	Tabunan	PA	Cebu	One of the few significant areas of forest remaining on Cebu
13	Nug-As and Mt Lantoy	—	Cebu	One of the few significant areas of forest remaining on Cebu
14	Mt Bandila-an	—	Siquijor	Largest area of forest remaining on Siquijor
15	Mt Cabalantian-Mt Capoto-an complex	—	Samar	Large area of lowland forest
16	Mt Hilong-hilong	—	Mindanao	Large area of forest, both lowland and montane
17	Mt Putting Bato-Kampalili-Mayo complex	—	Mindanao	Large area of montane forest, some lowland forest
18	Mt Kaluayan-Mt Kinabalian complex	—	Mindanao	Large area of forest, both lowland and montane
19	Mt Kitanglad	PA	Mindanao	Large area of montane forest
20	Mt Busa-Kiamba	—	Mindanao	Large area of montane forest, some lowland forest
21	Tawitawi island	—	Sulu islands	The largest area of forest known to remain in the Sulu archipelago
22	St Paul's Subterranean River NP	PA BR,WH	Palawan	Large area of lowland forest
23	Mt Mantalingajan	— BR	Palawan	Large area of forest, both lowland and montane

Some forest birds of this region breed in two of the IBAs listed in region W19.
Note that more IBAs in this region will be included in the *Important Bird Areas in Asia*, due to be published in early 2004.

Key *IBA name*: NP = National Park.
Status: PA = IBA is a protected area; (PA) = IBA partially protected; — = unprotected; BR = IBA is wholly or partially a Biosphere Reserve (see pp.34–35); WH = IBA is wholly or partially a World Heritage Site (see p.34). Note that the legal status of protected areas in the Philippines is currently being determined under the NIPAS process, so the protection status of many of these IBAs is liable to change.

OUTSTANDING IBAs FOR THREATENED BIRDS (see Table 1)

The Haribon Foundation has identified and documented 117 IBAs in the Philippines, of which 114 are considered to be important for the conservation of globally threatened bird species. Twenty-three of these have been selected as outstanding IBAs, because they include the best remaining examples of the key habitats for threatened birds in all nine subregions (seven EBAs and two SAs) of the Philippines.

CURRENT STATUS OF HABITATS AND THREATENED SPECIES

The Philippines were originally almost entirely forested, but by the end of the nineteenth century large areas had been cleared for agriculture, notably in the Visayas, where Negros, Bohol and Cebu had already lost much of their forest cover. Agricultural expansion continued throughout the twentieth century, but the most extensive and rapid deforestation was caused by commercial logging in the latter half of the century. This had a particular impact on primary lowland dipterocarp forests, the most valuable commercially, which shrunk from an estimated 10 million hectares in the 1950s to only one million by the late 1980s. New logging roads allowed access to farmers and timber collectors, who cleared more forest and prevented the regeneration of logged forest. Civil strife has affected many parts of Mindanao and the Sulu archipelago. Throughout the country insurgents may have prevented logging and

The western Visayas are highly deforested, most notably the island of Cebu, where the endemic Black Shama and Cebu Flowerpecker survive here at Tabunan and in a few other small forest patches.

PHOTO: GUY DUTSON/BIRDLIFE

F09

Table 2. Threatened birds of the Philippine forests.

Species			Distribution and habitat
			MINDORO (EBA 150)
Japanese Night-heron *Gorsachius goisagi*	✈ s	EN	Forest
Philippine Hawk-eagle *Spizaetus philippensis*	● s	VU	Forest
Mindoro Bleeding-heart *Gallicolumba platenae*	●	CR	Lowland forest
Mindoro Imperial-pigeon *Ducula mindorensis*	●	VU	Montane forest
Spotted Imperial-pigeon *Ducula carola*	● s	VU	Forest
Philippine Cockatoo *Cacatua haematuropygia*	● s	CR	Lowland forest
Black-hooded Coucal *Centropus steerii*	●	CR	Lowland forest
Mindoro Tarictic *Penelopides mindorensis*	●	EN	Lowland forest
Ashy Thrush *Zoothera cinerea*	● s	VU	Forest
Luzon Water-redstart *Rhyacornis bicolor*	● s	VU	Streams near montane forest
Scarlet-collared Flowerpecker *Dicaeum retrocinctum*	● s	VU	Lowland forest
Yellow Bunting *Emberiza sulphurata*	✈ s	VU	Scrub, cultivation
			LUZON (EBA 151)
Japanese Night-heron *Gorsachius goisagi*	✈ s	EN	Forest
Philippine Eagle *Pithecophaga jefferyi*	● s	CR	Forest
Philippine Hawk-eagle *Spizaetus philippensis*	● s	VU	Forest
Flame-breasted Fruit-dove *Ptilinopus marchei*	●	VU	Montane forest
Spotted Imperial-pigeon *Ducula carola*	● s	VU	Forest
Philippine Cockatoo *Cacatua haematuropygia*	● s	CR	Lowland forest
Green Racquet-tail *Prioniturus luconensis*	●	VU	Lowland forest
Philippine Eagle-owl *Bubo philippensis*	● s	VU	Lowland forest
Philippine Dwarf Kingfisher *Ceyx melanurus*	● s	VU	Lowland forest
Whiskered Pitta *Pitta kochi*	●	VU	Forest
Ashy Thrush *Zoothera cinerea*	● s	VU	Forest
Luzon Water-redstart *Rhyacornis bicolor*	● s	VU	Streams near montane forest
Izu Leaf-warbler *Phylloscopus ijimae*	✈	VU	Forest
White-browed Jungle-flycatcher *Rhinomyias insignis*	●	VU	Montane forest
Ashy-breasted Flycatcher *Muscicapa randi*	● s	VU	Lowland forest
Celestial Monarch *Hypothymis coelestis*	● s	VU	Lowland forest
Yellow Bunting *Emberiza sulphurata*	✈ s	VU	Scrub, cultivation
Green-faced Parrotfinch *Erythrura viridifacies*	● s	VU	Forest, bamboo
Isabela Oriole *Oriolus isabellae*	●	EN	Lowland forest

Three Data Deficient species occur on Luzon, Luzon Buttonquail *Turnix worcesteri* and Brown-banded Rail *Lewinia mirificus*, which are assumed to inhabit grasslands, and the montane Whitehead's Swiftlet *Collocalia whiteheadi*

Species			Distribution and habitat
			NEGROS AND PANAY (EBA 152)
Japanese Night-heron *Gorsachius goisagi*	✈ s	EN	Forest
Philippine Hawk-eagle *Spizaetus philippensis*	● s	VU	Forest
Negros Bleeding-heart *Gallicolumba keayi*	●	CR	Lowland forest
Negros Fruit-dove *Ptilinopus arcanus*	●	CR	Forest
Spotted Imperial-pigeon *Ducula carola*	● s	VU	Forest
Philippine Cockatoo *Cacatua haematuropygia*	● s	CR	Lowland forest
Rufous-lored Kingfisher *Todiramphus winchelli*	● s	VU	Lowland forest
Visayan Tarictic *Penelopides panini*	●	EN	Lowland forest
Visayan Wrinkled Hornbill *Aceros waldeni*	●	CR	Lowland forest
White-winged Cuckoo-shrike *Coracina ostenta*	●	VU	Forest
Flame-templed Babbler *Dasycrotapha speciosa*	●	EN	Lowland forest
Negros Striped-babbler *Stachyris nigrorum*	●	EN	Montane forest
White-throated Jungle-flycatcher *Rhinomyias albigularis*	●	EN	Lowland forest
Ashy-breasted Flycatcher *Muscicapa randi*	● s	VU	Lowland forest
Celestial Monarch *Hypothymis coelestis*	● s	VU	Lowland forest
Visayan Flowerpecker *Dicaeum haematostictum*	●	VU	Lowland forest
Scarlet-collared Flowerpecker *Dicaeum retrocinctum*	● s	VU	Lowland forest
Green-faced Parrotfinch *Erythrura viridifacies*	● s	VU	Forest, bamboo

● = breeds only in this forest region; ◗ = also breeds in other region(s); ✈ = non-breeding visitor from another region; s = also occurs in other EBA(s) and/or SA(s) in the Philippines

Table 2 ... continued. Threatened birds of the Philippine forests.

Species			Distribution and habitat
			CEBU (EBA 153)
Philippine Cockatoo *Cacatua haematuropygia*	● s	CR	Forest
Rufous-lored Kingfisher *Todiramphus winchelli*	● s	VU	Forest
Streak-breasted Bulbul *Ixos siquijorensis*	● s	EN	Forest
Philippine Leafbird *Chloropsis flavipennis*	● s	VU	Forest
Black Shama *Copsychus cebuensis*	●	EN	Forest
Cebu Flowerpecker *Dicaeum quadricolor*	●	CR	Forest
			MINDANAO AND THE EASTERN VISAYAS (EBA 154)
Japanese Night-heron *Gorsachius goisagi*	✈ s	EN	Forest
Philippine Eagle *Pithecophaga jefferyi*	● s	CR	Forest
Philippine Hawk-eagle *Spizaetus philippensis*	● s	VU	Forest
Mindanao Bleeding-heart *Gallicolumba criniger*	●	EN	Lowland forest
Mindanao Brown-dove *Phapitreron brunneiceps*	●	VU	Forest
Spotted Imperial-pigeon *Ducula carola*	● s	VU	Forest
Philippine Cockatoo *Cacatua haematuropygia*	● s	CR	Lowland forest
Giant Scops-owl *Mimizuku gurneyi*	●	VU	Lowland forest
Philippine Eagle-owl *Bubo philippensis*	● s	VU	Lowland forest
Silvery Kingfisher *Alcedo argentata*	●	VU	Lowland forest
Philippine Dwarf Kingfisher *Ceyx melanurus*	● s	VU	Lowland forest
Rufous-lored Kingfisher *Todiramphus winchelli*	● s	VU	Lowland forest
Blue-capped Kingfisher *Actenoides hombroni*	●	VU	Montane forest
Visayan Broadbill *Eurylaimus samarensis*	●	VU	Lowland forest
Mindanao Broadbill *Eurylaimus steerii*	●	VU	Lowland forest
Azure-breasted Pitta *Pitta steerii*	●	VU	Lowland forest
Philippine Leafbird *Chloropsis flavipennis*	● s	VU	Lowland forest
Ashy-breasted Flycatcher *Muscicapa randi*	● s	VU	Lowland forest
Little Slaty Flycatcher *Ficedula basilanica*	●	VU	Lowland forest
Celestial Monarch *Hypothymis coelestis*	● s	VU	Lowland forest

Three Data Deficient species occur on Mindanao and the Eastern Visayas, Brown-banded Rail *Lewinia mirificus* (Samar only), which is assumed to inhabit grassland, the montane Whitehead's Swiftlet *Collocalia whiteheadi* and Miniature Tit-babbler *Micromacronus leytensis*

● = breeds only in this forest region; ◐ = also breeds in other region(s); ✈ = non-breeding visitor from another region; s = also occurs in other EBA(s) and/or SA(s) in the Philippines

... continued

Negros Bleeding-heart is one of four threatened species in the genus Gallicolumba that are endemic to the Philippines.

PHOTO: EBERHARD CURLO/PESCP

Further deforestation must be halted on Luzon, Mindanao and the eastern Visayas if the huge Philippine Eagle is to survive.

PHOTO: TIM LAMAN

F09

Philippine Cockatoo used to be abundant, but its population has crashed because of unsustainable collection of nestlings for commercial trade.

PHOTOS: PH. GARGUIL & O. MORVAN/PYGARGUE PRODUCTIONS

Table 2 ... continued. Threatened birds of the Philippine forests.

Species			Distribution and habitat
			SULU ARCHIPELAGO (EBA 155)
Sulu Bleeding-heart *Gallicolumba menagei*	●	CR	Forest
Tawitawi Brown-dove *Phapitreron cinereiceps*	●	CR	Forest
Grey Imperial-pigeon *Ducula pickeringii*	◆ s	VU	Forest
Philippine Cockatoo *Cacatua haematuropygia*	● s	CR	Forest
Blue-winged Racquet-tail *Prioniturus verticalis*	●	EN	Forest
Rufous-lored Kingfisher *Todiramphus winchelli*	● s	VU	Forest
Sulu Hornbill *Anthracoceros montani*	●	CR	Forest
Sulu Woodpecker *Picoides ramsayi*	●	VU	Forest
Celestial Monarch *Hypothymis coelestis*	● s	VU	Forest
			PALAWAN (EBA 156)
Japanese Night-heron *Gorsachius goisagi*	✈ s	EN	Forest
Philippine Hawk-eagle *Spizaetus philippensis*	● s	VU	Forest
Palawan Peacock-pheasant *Polyplectron emphanum*	●	VU	Lowland forest
Grey Imperial-pigeon *Ducula pickeringii*	◆ s	VU	Forest on small islands
Philippine Cockatoo *Cacatua haematuropygia*	● s	CR	Lowland forest
Blue-headed Racquet-tail *Prioniturus platenae*	●	VU	Lowland forest
Palawan Hornbill *Anthracoceros marchei*	●	VU	Lowland forest
Falcated Wren-babbler *Ptilocichla falcata*	●	VU	Lowland forest
Palawan Flycatcher *Ficedula platenae*	●	VU	Lowland forest
			TABLAS, ROMBLON AND SIBUYAN (SA 095)
Philippine Dwarf Kingfisher *Ceyx melanurus*	● s	VU	Forest
Rufous-lored Kingfisher *Todiramphus winchelli*	● s	VU	Forest
Streak-breasted Bulbul *Ixos siquijorensis*	● s	EN	Forest
Celestial Monarch *Hypothymis coelestis*	● s	VU	Forest

Spotted Imperial-pigeon *Ducula carola* is known by old records from Sibuyan, and Philippine Cockatoo *Cacatua haematuropygia* from Tablas, but they have not been found on these islands during recent surveys and are assumed to be locally extinct

Species			Distribution and habitat
			SIQUIJOR (SA 096)
Japanese Night-heron *Gorsachius goisagi*	✈ s	EN	Forest
Philippine Hawk-eagle *Spizaetus philippensis*	● s	VU	Forest
Spotted Imperial-pigeon *Ducula carola*	● s	VU	Forest
Philippine Cockatoo *Cacatua haematuropygia*	● s	CR	Forest
Rufous-lored Kingfisher *Todiramphus winchelli*	● s	VU	Forest
Streak-breasted Bulbul *Ixos siquijorensis*	● s	EN	Forest

● = breeds only in this forest region; ◆ = also breeds in other region(s); ✈ = non-breeding visitor from another region; s = also occurs in other EBA(s) and/or SA(s) in the Philippines

agricultural development, but sometimes they may have promoted these activities, and the deliberate conflagration of forests on Mindanao—associated with insurgency—is a problem, particularly on the Zamboanga peninsula.

Today, forest cover varies considerably across the archipelago but is everywhere drastically reduced—according to satellite data from the late 1980s, Mindoro retained 8.5% forest cover, Luzon 24%, Mindanao 29% and Palawan 54%. In the Eastern Visayas, Samar retained 33%, Leyte 14% and Bohol 6%, and in the Western Visayas, Negros 4% and Panay 8%. These figures are, however, probably overestimates, and only a proportion of the cover estimated on each island was closed-canopy forest. Further forest loss and degradation has taken place since these estimates were made, as a result of *kaingin* farming (otherwise termed 'slash-and-burn' or shifting cultivation), fire-maintained pasture and the harvesting of non-timber forest products (such as rattans and other palms). These alarming statistics on remaining forest cover fit badly with a Philippines government report which asserted that the country needs 46% of its land area under forest for both its economic and environmental wellbeing. Deforestation has been most extensive in the lowlands, and the lowland-forest species tend to be the most highly threatened. Only a few small fragments of lowland forest remain on Mindoro, Negros and Panay, Cebu is almost completely deforested, and the only substantial forests known to remain in the Sulu archipelago are on Tawitawi (although it is possible that some forest remains on Jolo island, where the security situation precludes surveys).

CONSERVATION ISSUES AND STRATEGIC SOLUTIONS (summarised in Table 3)

Habitat loss and degradation

■ *CONVERSION TO AGRICULTURE*

By the late 1980s, almost all commercially valuable timber in the Philippines had been extracted, and since then the greatest threat to forests has been conversion to shifting (*kaingin*) and permanent agriculture practised by landless peasants: well over half of the IBAs in the Philippines are affected by *kaingin*. About 38 million people live in rural areas, half of them, the poorest in the country, in the uplands. The presence of landless people, who rapidly colonise areas when they are opened up by new roads or logging operations, poses enormous difficulties to site and habitat conservation initiatives.

Marginalised farmers should be given greater control over the land that they occupy (through appropriate tenurial instruments), to reduce further encroachment into forested areas. Their *kaingin* practices need to be replaced with agroforestry and improved farming methods, which would enable them to remain longer on established clearings by slowing the loss of soil fertility; simple socio-economic incentives, and awareness and training programmes, could be used to promote these changes. Efforts should be made to rehabilitate abandoned land, for example through community forestry plantations. At selected key sites, projects are required that integrate conservation and local land-use development. An integrated land management project is particularly urgent at Tabunan IBA on Cebu, to restore the remaining tiny area of forest (through replanting with native species to increase and link the isolated forest patches) whilst offering incentives to conserve the area and develop alternative sources of income.

■ *FORESTRY AND ILLEGAL LOGGING*

Commercial logging was the main cause of deforestation in the Philippines after 1950, but declined from the late 1980s as the harvest of commercially valuable timber was exhausted. In the 1990s logging was banned in all provinces (i.e. 64 out of 73) where forest cover is less than 40%. Commercial logging on Palawan has been suspended by presidential decree, but despite this nearly all of the island's forests remain under concession. Legal logging operations continue in some areas, for example at Bislig on Mindanao, where good primary forest is being clear-felled (under the PICOP logging concession) and replaced with exotic trees for paper production. Far more worrying is the fact that illegal logging is widespread and often unchecked, including inside protected areas.

Given the perilous state of the Philippines environment and biodiversity, it is now time for an immediate total ban on the logging (and conversion to plantations) of all remaining forest in the country (including secondary growth in the lowlands which is also a critical habitat for biodiversity). There is a need for a national forest management policy, recognising that the remaining natural forests are an important and, in part, renewable resource essential for the future welfare of both Filipino people and wildlife. This policy should make habitat management the central target, and promote reforestation using native species. To implement the policy, foresters in the Philippines, whose training is currently geared to timber extraction as an economic activity, need to be retrained in forest management for biological diversity.

Visayan Tarictic is confined to the few remaining lowland forests in the Negros and Panay EBA, where it is under pressure from continuing forest clearance and hunting for food and sport.

PHOTO: TIM LAMAN

The Philippines has suffered one of the highest rates of deforestation anywhere in the world, which has reduced many formerly forested areas to wasteland.

PHOTO: MICHAEL POULSEN/BIRDLIFE

■ *EXPLOITATION OF FOREST PRODUCTS*

The harvesting of non-timber forest products (such as rattans and other palms) is widespread, and can seriously degrade the forest understorey. Sustainable use of forest products should be promoted, including through community-based tree farms under the jurisdiction of local governments.

■ *MINING*

Mining is an important potential threat, as most remaining forests are covered by mining applications, whose acceptance would give companies the right to clear forests. Ongoing mining activities are already damaging habitats in several important areas, for example northern Dinagat, and there are plans to commence granite mining at Iwahig on Palawan, and for major operations on Samar and Leyte. Mining activities should be prohibited in all critical habitats identified under the National Wildlife Act (which include NIPAS sites and the 117 IBAs identified by the Haribon Foundation). In other areas the impact of proposed mining activities on forests and their biodiversity needs to be carefully evaluated, and mitigation measures imposed to minimise their negative effects.

■ *DEVELOPMENT (URBAN, INDUSTRIAL, ETC.)*

Some of the best remaining forests in the Philippines are in the most inaccessible areas, for example in Luzon's Sierra Madre, but new roads into these areas could lead to rapid forest loss and degradation through illegal logging and settlement by *kaingin* farmers. Other development projects could cause similar projects, such as the construction of a road system for a proposed industrial complex along the coast of the Northern Sierra Madre Natural Park. A rigorous site selection procedure should therefore be followed for new roads and other proposed development projects, to avoid new development in critical habitats identified under the National Wildlife Act (including NIPAS sites and IBAs).

■ *POLLUTION/PESTICIDES*

Several threatened birds occupy riverine habitats, including Luzon Water-redstart and several kingfisher species. These may be affected by pollution and siltation caused by mining and logging, and the increased use of inorganic or synthetic fertilisers, herbicides and pesticides by farmers. Species at the terminus of food-chains may accumulate pesticides which reduce their reproductive output, including Philippine Eagle and Philippine Hawk-eagle. The Philippine government via DENR must strictly implement laws controlling the pollution of rivers. The indiscriminate use of agrochemicals (especially those already banned) needs to be controlled, and organic agricultural practices promoted.

Protected areas coverage and management

■ *GAPS IN PROTECTED AREAS SYSTEM*

The protected areas system in the Philippines is currently being redeveloped through the National Integrated Protected Area System (NIPAS) process. This fundamentally important new law (passed in 1992) is the only environmental legislation available that can override other land-use legislation, e.g. logging and mining concessions. The old system of reserves and sanctuaries (known as the 'initial component' of NIPAS) has been thoroughly reassessed. Sites no longer suitable as protected areas have been dropped, and many new ones added. A few of the c.200 sites proposed under NIPAS have already received funding; the Conservation of Priority Protected

Areas Project (CPPAP, funded by the World Bank Global Environment Facility) and the National Integrated Protected Areas Project (NIPAP, funded by the European Union and implemented by DENR) have provided major funding for 18 sites, with others supported by the Foundation for the Philippine Environment (FPE) and other initiatives. The recent Local Government Code legislation is a vital mechanism for the conservation of many sites, as it devolves responsibility for land management to Local Government Units, so that the management of NIPAS sites and local protected areas can be closely integrated with other local land-use plans.

It is vital that the NIPAS fully represents the unique biological diversity of the Philippine archipelago. In order to achieve this for birds (and for many other animal and plant groups, given that many of these share similar patterns of endemism to birds), NIPAS sites should be selected to cover the main forest types and endemic bird species in all seven EBAs and two SAs in the Philippines. The Haribon Foundation's IBA analysis, especially the outstanding IBAs listed above, provides an ideal dataset for a 'gap analysis' of the sites selected and proposed under NIPAS; for example, the sites currently funded under CPPAP and NIPAP do not include any on Cebu or the Sulu archipelago. The proposed NIPAS network therefore needs to be analysed to identify IBAs such as Tabunan on Cebu and Tawitawi in the Sulu archipelago, which should then be proposed for establishment under NIPAS. Given limited resources, and the difficulties in managing sites in parts of the Philippines, NIPAS cannot provide adequate protection for all important areas. The Haribon Foundation and others have ongoing projects at several IBAs, to promote these sites for designation and management as local protected areas under the Local Government Code. These projects are developing a model of how this legislation can be used for effective site conservation, which should be followed at many other IBAs.

■ *WEAKNESSES IN RESERVE MANAGEMENT*

Serious management problems affected the old system of reserves and sanctuaries in the Philippines, linked to a lack of resources. Most sites being designated under NIPAS do not yet have large-scale funding, and a major challenge is to provide the resources and infrastructure necessary to ensure their permanent and complete protection. The capacity and resources of the Protected Areas and Wildlife Bureau (PAWB) need to be increased at all levels, especially to develop sufficient site management capability within local PAWB offices to protect the NIPAS sites under their jurisdiction.

Exploitation of birds

■ *HUNTING*

Hunting is a major problem, with firearms widely available, and 40% of the country's threatened birds are affected. Some species are hunted for food, notably pigeons and hornbills; many are particularly vulnerable because they congregate at fruiting trees, and some ground-dwelling birds (e.g. pigeons) are trapped using snares. However, much hunting is for sport, and even Philippine Eagles are occasionally shot for trophies. Hunting of all bird species is illegal, but enforcement is lacking, and local people in many areas are likely to resist attempts at strict control. Concerted programmes of education and awareness are needed within the communities in and around key sites for threatened species, to demonstrate the effects of hunting on the threatened birds. Efforts need to be made to regulate the large-scale netting of migratory birds (known as 'ik-ik') that occurs at several sites, notably at Dalton Pass on Luzon, through the promotion of alternative livelihoods, for example tourism.

■ *WILD BIRD TRADE*

Collection for commercial trade affects almost 20% of the Philippines' threatened birds, including several pigeons, parrots and hornbills, and Palawan Peacock-pheasant. The

Table 3. Conservation issues and strategic solutions for birds of the Philippine forests.

Conservation issues	Strategic solutions
Habitat loss and degradation	
■ CONVERSION TO AGRICULTURE ■ FORESTRY AND ILLEGAL LOGGING ■ EXPLOITATION OF FOREST PRODUCTS ■ MINING ■ DEVELOPMENT (URBAN, INDUSTRIAL, ETC.) ■ POLLUTION/PESTICIDES	➤ Reduce encroachment into forest by giving marginalised farmers greater control over the land that they occupy ➤ Develop agro-forestry and improved agricultural practices to allow more permanent settlement of land cleared for *kaingin*, and rehabilitate abandoned land ➤ Declare a national logging ban, and develop a national forest management policy, with foresters retrained to implement it ➤ Promote sustainable use of forest products, including through the development of community-based tree farms ➤ Assess the environmental impact of proposed mines, roads and other developments, and minimise development at key sites ➤ Enforce laws to prevent river pollution, and promote organic agricultural practices
Protected areas coverage and management	
■ GAPS IN PROTECTED AREAS SYSTEM ■ WEAKNESSES IN RESERVE MANAGEMENT	➤ Ensure that NIPAS fully represents the unique biodiversity of the Philippines, by covering the main forest types in all seven EBAs and two SAs ➤ Establish and manage additional IBAs as local protected areas under the Local Government Code ➤ Increase the capacity of the Protected Areas and Wildlife Bureau, especially local offices
Exploitation of birds	
■ HUNTING ■ WILD BIRD TRADE	➤ Conduct education and awareness programmes around key sites, to reduce hunting of threatened species ➤ Strengthen enforcement of CITES and other legislation to combat illegal trade in Philippine Cockatoo ➤ Establish nest protection schemes for hornbills and Philippine Cockatoo
Gaps in knowledge	
■ OUTDATED INFORMATION ON FOREST COVER ■ INADEQUATE DATA ON THREATENED BIRDS	➤ Analyse satellite images or aerial surveys to locate undetected forest sites, particularly on the most deforested islands and in lowland Mindanao ➤ Survey poorly known islands and sites, to identify new areas for conservation action ➤ Study the ecology of Philippine Eagle and other keystone species

Hunting is a major problem in the Philippines, and ground-dwelling birds are trapped using snares.

PHOTO: DES ALLEN

previously abundant Philippine Cockatoo has suffered most conspicuously: huge numbers were trapped in recent decades, and its populations have plunged towards extinction in the wild. On Palawan, chicks are taken from virtually every known accessible nest, and in the south of the island tribesmen purposely leave nest-trees in otherwise cleared land. Nestling cockatoos collected have a mortality rate of 50%, and only 20% of Palawan Peacock-pheasants survive capture. Seven threatened Philippines species (six endemic) are listed on Appendix I of CITES, meaning that their international trade is prohibited, and 14 threatened species are listed on Appendix II, which permits strictly regulated international trade; however, enforcement has been problematic, while much trade is domestic and not directly affected by CITES legislation.

Drastic measures therefore need to be considered to combat illegal trade in Philippine Cockatoo, possibly including the establishment of DENR-manned posts at airports (with appropriate training and equipment for DENR staff), major ferry terminals (e.g. Puerto Princesa) and the Rio Tuba nickel-ore port. Education and awareness campaigns are needed to improve understanding of the plight of this and other species affected by trade. These campaigns could incorporate components of the parrot conservation projects in the Caribbean, whereby local communities are encouraged to become active participants in conservation efforts. Despite this, however, illegal trade may only be countered by introducing economically viable alternatives: local income-generating activities that lessen threats to birds and their habitats should be promoted. Nest protection schemes involving local communities should be considered for some highly threatened species, including Philippine Cockatoo, Visayan Tarictic and Visayan Wrinkled Hornbill.

Gaps in knowledge

■ *OUTDATED INFORMATION ON FOREST COVER*

Forest cover maps produced from satellite images and aerial surveys in the 1980s and 1990s are now outdated. During their IBA project, the Haribon Foundation used these maps to identify ornithologically unexplored areas of forest in the more inaccessible areas, but then assessed the current condition of the habitat by consulting local officials or by visiting the sites. This process identified several IBAs that are likely to be very important for threatened birds on the basis of the habitat present. Further studies are required using satellite images and aerial surveys to locate unexplored areas of forest, followed by consultations and site visits to assess habitat quality. This would be particularly valuable on the most deforested islands (e.g. Cebu, Mindoro, Negros) where even small patches of forest are of conservation value, and on Mindanao where it is unclear which are the best sites for the conservation of lowland forest birds.

■ *INADEQUATE DATA ON THREATENED BIRDS*

The distribution and status of most threatened birds in the Philippines are incompletely known. Most islands have been explored ornithologically, and most parts of the larger islands, but the data are old and incomplete. This problem is exacerbated by the pattern of habitat loss in the islands. The areas visited by naturalists in the past are also the places where access has been easiest for logging and agriculture, so many 'old' sites for threatened birds have now lost their habitats. Surveys are required at poorly known IBAs, and other forested areas located using satellite images or aerial surveys, to assess their importance and hence to identify new areas for NIPAS or other conservation initiatives. Some threatened birds are difficult to locate and survey, and require special techniques such as tape-playback (e.g. for owls) and mist-netting (e.g. for some babblers and flycatchers). Interviews with bird trappers could provide information on species affected by hunting and trade.

The ecology of most threatened forest birds is poorly known, and studies are required to clarify their needs, and hence the most appropriate ways to manage their habitats. In the case of Philippine Eagle, work is needed on its survival rates and breeding success in different types of forest and at different elevations; prey composition, and constraints on prey abundance, on the four islands it inhabits; reasons for different densities on different islands; effects of habitat fragmentation on survival and breeding; and optimum habitat and the value of corridors between areas of such habitat. Radio-telemetry (or satellite-based) studies of Philippine Eagles would help determine territory size, home-range size and dispersal capability; this technique can also elucidate the daily and seasonal movements of birds which need to locate fruiting trees, including keystone species such as pigeons (some of which may prove to be inter-island migrants) and hornbills.

G01

EURASIAN STEPPES and DESERT

THE Eurasian steppes are a vast belt of grassland extending from eastern Europe through western and central Asia to north-east Asia. The five threatened species which occur in the east Asian steppes include four which also range westwards outside the Asian region, Imperial Eagle, Lesser Kestrel and Great Bustard to Europe, and Pale-backed Pigeon to Central Asia. They occur in a variety of habitats in addition to lowland grasslands: Imperial Eagle inhabits forest-steppe, and some forest conservation measures are relevant to the species (see F01); Pale-backed Pigeon is found in open, sparsely wooded habitats in the mountains; and White-throated Bushchat breeds in mountain grasslands. The lakes and other wetlands in the steppes are important for threatened waterbirds, and are covered in W05.

- **Key habitats** Steppe grasslands, forest steppe and mountain steppe, scrub in sandy desert.
- **Altitude** Lowlands to 3,100 m.
- **Countries and territories** **Russia** (Krasnoyarsk, Khakassia, Tuva, Irkutsk, Buryatia, Chita); **Mongolia**; **China** (Heilongjiang, Jilin, Inner Mongolia, Xinjiang, Gansu); outside the Asian region, the steppes extend through central and western Asia to eastern Europe.

Threatened species

	CR	EN	VU	Total
(breeds only in this region)[1]	—	—	1	1
(also breeds in other regions)	—	—	4	4
Total	—	—	5	5

Key: (symbol) = breeds only in this grassland region.
[1] White-throated Bushchat, which nests only in this grassland region, migrates to another grassland region (G02).
(symbol) = also breeds in other region(s).

Large areas of relatively unspoilt steppe grassland remain in East Asia, particularly in Mongolia. PHOTO: UTE BRADTER

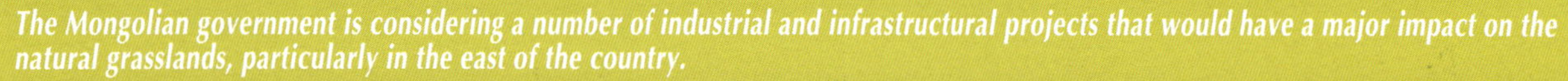

The Mongolian government is considering a number of industrial and infrastructural projects that would have a major impact on the natural grasslands, particularly in the east of the country.

PHOTO: UTE BRADTER

OUTSTANDING IBAs FOR THREATENED BIRDS

All threatened species in this region are relatively widespread and tend to occur at low densities. Their conservation is therefore best addressed at the landscape level, and no outstanding IBAs have been selected for them. However, several grassland birds of this region breed in or near to some of the wetland IBAs listed for W05, and other sites which support the threatened grassland birds will be documented during BirdLife's ongoing IBA Project.

CURRENT STATUS OF HABITATS AND THREATENED SPECIES

The steppe grasslands of eastern Russia and northern China have been greatly reduced through ploughing and conversion to agricultural land, and other threats there include overgrazing, cutting of Imperial Eagle nesting trees, too frequent steppe fires, pesticide use, human disturbance and hunting. However, all five of the threatened birds are migratory and some declines in their numbers (e.g. of Imperial Eagle in Russia in areas where there have been no

Lesser Kestrel has declined rapidly in Europe in recent decades, but it remains locally common in the East Asian steppes.

PHOTO: RAY TIPPER

Table 1. Threatened birds of the Eurasian steppes and desert.

Species			Distribution and habitat
Imperial Eagle *Aquila heliaca*	mo	VU	Recorded breeding in Krasnoyarsk, Khakassia, Irkutsk, Buryatia, Mongolia and Xinjiang, in forest-steppe and lowland grasslands
Lesser Kestrel *Falco naumanni*	mo	VU	Locally common breeding species in the steppes of Krasnoyarsk, Tuva, Mongolia, Xinjiang and Inner Mongolia, and presumably Buryatia and Gansu
Great Bustard *Otis tarda*	mo	VU	Breeds in steppe and forest-steppe in Krasnoyarsk, Khakassia, Tuva, Buryatia, Chita, Mongolia, Heilongjiang, Jilin, Inner Mongolia and Xinjiang
Pale-backed Pigeon *Columba eversmanni*	mo	VU	Breeds in open, sparsely wooded habitats in the Tien Shan mountains in China (Xinjiang, with summer records from Gansu)
White-throated Bushchat *Saxicola insignis*	m	VU	Breeds in mountain grasslands in central and western Mongolia, and (at least formerly) just across the Russian border in Altay

The Data Deficient Vaurie's Nightjar *Caprimulgus centralasicus* is known only by a single specimen collected in sandy scrub in the Taklimakan desert in Xinjiang, China

= breeds only in this grassland region; = also breeds in other region(s); o = also breeds outside the Asia region; m = migrates to other region(s) of Asia

obvious changes to their breeding habitats) have probably been caused by problems on their wintering grounds; for example, wintering Imperial Eagle and Great Bustard are hunted, and wintering White-throated Bushchat is affected by loss of natural grasslands (see G02).

In Russia, large-scale conversion of steppe to agricultural land from the 1960s to early 1980s led to substantial declines in Imperial Eagles (through reduced availability of its main prey species, the Siberian suslik *Spermophilus dauricus*) and Great Bustard; however, the ploughing of steppes has ceased in this part of Russia, and some formerly cultivated areas are reverting to steppe. Great Bustards ceased breeding on the eastern Song-nen plains in Heilongjiang following oilfield development at Daqing, mass human migration from other provinces, and a campaign to develop the 'northern wilderness' for agriculture from the 1950s to the 1970s. An estimated four million hectares of grassland was converted to farmland in Xinjiang between the 1950s and the 1990s, but much of this land was unsuitable for cultivation and is now subject to uncontrollable soil erosion, and about one million hectares has become barren ground; there were also inappropriate policies in this region to poison keystone species such as pika *Ochotona* and zokor *Myospalax*. The rate of development of the steppes for agriculture has been much lower in Mongolia, but there is growing pressure from an increasing human population (which has tripled since 1950) and an associated increase in livestock. The Mongolian government plans to develop large areas of the country, which could have a major impact on the threatened grassland species.

CONSERVATION ISSUES AND STRATEGIC SOLUTIONS (summarised in Table 2)

Habitat loss and degradation

CONVERSION FOR AGRICULTURE

Although the conversion of steppe for agriculture has now ceased in most parts of eastern Russia, this is still a threat in China, and the Mongolian government has plans for large-scale agricultural development in the steppe zone. The benefits of such conversion should be carefully considered in the light of the economic failure (and subsequent abandonment) of such schemes in several parts of the steppes, and the potential losses of biodiversity. Indeed, the evidence indicates that very little of the steppe is suitable for cultivation, and that most schemes should be halted.

Conversion of steppe for agriculture has ceased in most parts of eastern Russia, and some formerly cultivated areas are reverting to steppe, but this is still a threat in Mongolia and China.

PHOTO: OTTO PFISTER

DEVELOPMENT (URBAN, INDUSTRIAL, ETC.)

In parts of eastern Russia and northern China, steppe grasslands have been reduced and degraded by industrial, infrastructural and urban development. The steppe zone of Mongolia is currently relatively undeveloped, but the government is considering industrial and infrastructural projects that would have a major impact on natural grasslands, particularly in the east. These include the construction of a new bridge crossing from China into eastern Mongolia, passing through Nomrog Strictly Protected Area, and a 'Millennium highway' across the country. The planned Tumen River Area Development Programme (see W02) could lead to increased exploitation of natural resources in Mongolia and elsewhere. Industrial development plans in Mongolia include mining, oilfields and other exploration, and some protected areas may be degazetted as a result. These proposals should be subject to environmental impact assessments at the strategic level, to consider the potential impacts of the whole development process on the environment and threatened species, and to plan the mitigation of any negative effects.

CUTTING OF NESTING TREES

Imperial Eagles require large trees for nesting, but in several parts of Russia and China suitable trees are being lost because of wood-cutting (in Russia often for tourists' fires), as well as grazing cattle and steppe fires. Tree cutting is also a potential threat to Pale-backed Pigeon in Xinjiang. Special protection needs to be given to Imperial Eagle nesting trees, and artificial nest platforms could be erected in areas which lack nest sites.

STEPPE FIRES

Steppe fires, usually set by man in spring and early summer, are a threat to Great Bustard, Imperial Eagle (through destruction of nesting trees) and possibly other species. For example, in Chita up to 70% of steppe is burnt annually, killing many young Great Bustards. Fire is a natural phenomenon in the steppes, and improved fire management could help to maintain the grassland quality for both livestock and wildlife. The setting of grassland fires therefore needs careful planning and strict control (with measures to prevent fires being caused by negligence or deliberately by

Table 2. Conservation issues and strategic solutions for birds of the Eurasian steppes and desert.

Conservation issues	Strategic solutions
Habitat loss and degradation	
CONVERSION TO AGRICULTURE	Reconsider plans to convert steppe to agricultural land in Mongolia and China
DEVELOPMENT (URBAN, INDUSTRIAL, ETC.)	Assess the environmental impact of proposed development projects in Mongolia
CUTTING OF NESTING TREES	Specially protect Imperial Eagle nesting trees
STEPPE FIRES	Improve fire management, to avoid harm to nesting birds and their habitats
IRRIGATION	Prevent flooding of Great Bustard nests on irrigated farmland
LIVESTOCK GRAZING	Develop ecological management of grazing, encouraging nomadic herders to reduce their herds
PESTICIDES	Promote alternatives to pesticides for the control of vole outbreaks
DISTURBANCE	Control disturbance in key breeding areas of threatened birds, and provide direct protection to nest sites if required
Protected areas coverage and management	
GAPS IN PROTECTED AREAS SYSTEM	Establish new nature reserves for Imperial Eagle and Great Bustard
WEAKNESSES IN RESERVE MANAGEMENT	Strengthen the management of protected areas in eastern Mongolia
Exploitation of birds	
HUNTING	Improve enforcement of hunting laws, through training of law enforcers and conservation awareness work with hunters and the public Control gun ownership, and regularly check vehicles for guns and shot birds
Gaps in knowledge	
INADEQUATE DATA ON THREATENED BIRDS	Survey all threatened species to identify key breeding areas Study the breeding ecology of Pale-backed Pigeon and White-throated Bushchat Investigate the breeding distribution and migratory routes of Great Bustard

Livestock levels have increased since the 1950s, and in many parts of the steppes they now exceed the carrying capacity of the grasslands.

PHOTO: JACOB WIJPKEMA

miscreants), involving new fire management laws, to prevent them running out of control or affecting threatened birds during the breeding season. Current policies to try to exclude fires from some parts of the steppes should be reviewed, as this leads to a build-up of flammable material, so that when fires occur (as they inevitably will) they are more intense, destructive and difficult to control.

IRRIGATION

The sinking of boreholes to supply water to domestic herds has led to adjacent severe land erosion and generally lowered water-tables. Irrigation of agricultural land is a major threat to breeding Great Bustards in Xinjiang, as it floods nests and drowns chicks, although some farmers build small dykes to protect nests. This positive action should be promoted more widely, so that more nests are successful.

LIVESTOCK GRAZING

Livestock levels in many parts of the steppes exceed the carrying capacity of the grasslands, with the densities in some areas three to four times the levels in the 1950s, as herders tend to keep their increased wealth as livestock. This causes direct damage to the grasslands, and is linked to outbreaks in vole numbers (see *Pesticides* below). Ecological management of grazing needs to be established, particularly by preventing grazing in some areas in summer; this allows the grass to grow tall, thereby reducing the density of voles by four to five times, and providing autumn and winter grazing. Nomadic herders may need to be persuaded to reduce herd sizes, perhaps by providing alternative means to save their wealth.

PESTICIDES

The steppes suffer periodic outbreaks in the numbers of Brandts' voles *Microtus brandti*, which occur at their highest densities in short, heavily grazed grasslands; the frequency of outbreaks has increased in recent years, correlated with increased density of livestock. The voles are regarded as pests because they compete with livestock for pasture, so aerial spraying of zinc phosphide and bromadiolone, and distribution of seeds dressed with pesticides, have been used against them in Mongolia. These chemicals have caused mortality in birds, for example of Great Bustards in Russia, and of large number of birds in Mongolia in 2002. The effectiveness of pesticides as a means of controlling voles has been questioned, particularly as they kill predators that in a natural system would control vole numbers. Alternative approaches are the development (which is ongoing in China and Australia) of non-toxic chemicals which render female voles infertile, ecological management of grazing (see *Livestock grazing* above), and the reduction of hunting of mammalian predators. Pesticides are also used in eastern Russia to control insects, and this seems to be affecting the breeding success of Imperial Eagle (and other birds of prey). Awareness campaigns should be conducted to inform farmers about the dangers of pesticides and about the regulations on their use.

DISTURBANCE

Disturbance by people and domestic animals is a significant threat to nesting steppe birds. Some Imperial Eagle breeding territories in eastern Russia are subject to heavy tourist pressure, which has probably caused nest desertion. Disturbance linked to increased human population and changing agricultural practices is probably a factor in the decline of Great Bustard, which tends to abandon heavily grazed steppes. Its eggs may be destroyed by tractors; disturbance and predation by herding dogs is a problem in some areas; and when the bustards are flushed from their nests the eggs are vulnerable to crows and other predators. Disturbance should be controlled in key breeding areas of both Imperial Eagle and Great Bustard, and nest sites directly protected if necessary (see *Gaps in protected areas system* below). In areas where Great Bustards nest on farmland, studies should be conducted to help identify ways to modify agricultural practices and minimise their negative impact. Local people should be encouraged to take the initiative in Great Bustard conservation; in Russia and Mongolia this has proved successful, with farmers avoiding ploughing areas near Great Bustard nests, while in Xinjiang farmers have built dykes to prevent flooding of nests. In Russia, new legislation may be required to reduce the numbers of dogs kept by herdsmen.

Protected areas coverage and management

GAPS IN PROTECTED AREAS SYSTEM

The threatened steppe species mostly occur at low densities and range widely to feed, so site protection is of limited value for their conservation. The most practical approaches may be small, seasonally protected areas, to manage the habitat at important nesting and feeding areas during summer. New protected areas suggested for Imperial Eagle include Bratsk reservoir in Irkutsk and Barguzinskaya valley in Buryatia. New reserves proposed for key Great Bustard breeding areas in eastern Russia include: several small areas in the Khakassia steppes; Domna (Domka) valley, Yeravinski district, Buryatia; and Argunsky and Urulyunguievskaya Pad' in the Urulyunguy lowlands of Chita. Other key areas for Great Bustard that warrant official protection are Kherlen-Menen in Mongolia, Tacheng basin in Xinjiang and Baicheng Pintai in Jilin.

■ *WEAKNESSES IN RESERVE MANAGEMENT*

Imperial Eagle breeds in Pribaykal'skiy and Tunkinskiy National Parks in eastern Russia, but these reserves do not provide effective protection for the species and its habitats; many areas where it hunts are privately owned, and need to be transferred to the park (probably by paying their current owners), and its nest sites should be given special protection. Given the plans to develop the steppes of eastern Mongolia (see *Development (urban, industrial, etc.)* above), it is important to ensure the effective protection of globally and nationally important biodiversity there by strengthening the management of the nine existing and 12 proposed protected areas, and by supporting biodiversity conservation and sustainable alternative livelihoods in buffer zones.

Exploitation of birds

■ *HUNTING*

Great Bustard is protected throughout its Asian breeding range, but illegal hunting is a major problem, and Imperial Eagle is also hunted. Great Bustards are wary of people, but less so of men on horseback or in cars, so hunters often shoot birds from vehicles; the recent decline of the east Asian population appears to be closely linked to the availability of improved firearms and motorised vehicles. In China, bustards are also killed using poisoned grasshoppers, and local people sometimes collect their eggs. In the past, bustards were protected by traditional beliefs, but these appear to have broken down in many areas. Park rangers and police need to be better equipped and trained (including in the identification of protected species) to enforce wildlife protection laws. The ownership of guns should be strictly controlled, with regular checks of vehicles for guns and shot birds. Awareness campaigns are required to ensure that hunters know that killing Great Bustards is illegal, and to encourage the formation of unions that help their members to avoid violating the hunting laws.

Gaps in knowledge

■ *INADEQUATE DATA ON THREATENED BIRDS*

The distribution and breeding populations of all of the threatened species are incompletely known, and surveys are required to identify key sites for their conservation and any threats; studies are also needed of the breeding ecology of Pale-backed Pigeon and White-throated Bushchat. Protection of the east Asian population of Great Bustard is hampered by inadequate knowledge of its breeding and wintering ranges and migratory routes; surveys are required, involving analysis of vegetation maps and interviews with local people, together with satellite-tracking and possibly colour-banding (using the national colour codes already established for cranes).

INDO-GANGETIC GRASSLANDS

G02

A band of tropical grasslands (locally known as the terai and duars), now highly fragmented, extends across the plains of the Ganges and Brahmaputra rivers and into the adjacent Himalayan foothills. Eleven threatened grassland species occur there, including Bristled Grass-warbler, which is also known from scattered localities (away from the main grassland belt) in Pakistan and the Indian peninsula. Most of these birds inhabit damp lowland grasslands, but a few range into the hills, including Manipur Bush-quail and Slender-billed Babbler in the Manipur basin, and Grey-crowned Prinia, which is found to well over 1,000 m and in forest edge as well as grasslands. Manipur Bush-quail, Marsh Babbler and Black-breasted Parrotbill have restricted ranges, and are confined to the Assam plains Endemic Bird Area, and Finn's Weaver is also highly localised in distribution. The Indo-Gangetic grasslands overlap with two wetland regions, W12 and W14.

- **Key habitats** Tropical grassland along rivers and on plains, and associated scrub and woodland.
- **Altitude** Lowlands to 1,350 m.
- **Countries and territories** **Pakistan**; **India** (Haryana, Delhi, Uttaranchal, Uttar Pradesh, Bihar, West Bengal, Arunachal Pradesh, Assam, Meghalaya, Manipur); **Nepal**; **Bhutan**; **Bangladesh**.

Threatened species

	CR	EN	VU	Total
(breeds only in this grassland region)	—	—	8	8
(also breeds in other region(s))	—	1	1	2
(non-breeding visitor)	—	—	1	1
Total	—	1	10	11

Key: = breeds only in this grassland region. = also breeds in other region(s). = non-breeding visitor from another region.

The Indo-Gangetic grasslands region overlaps with part of Conservation International's Indo-Burma Hotspot (see pp.20–21).

Almost all grasslands suitable for the threatened birds are in protected areas such as Royal Chitwan National Park in Nepal. PHOTO: OTTO PFISTER

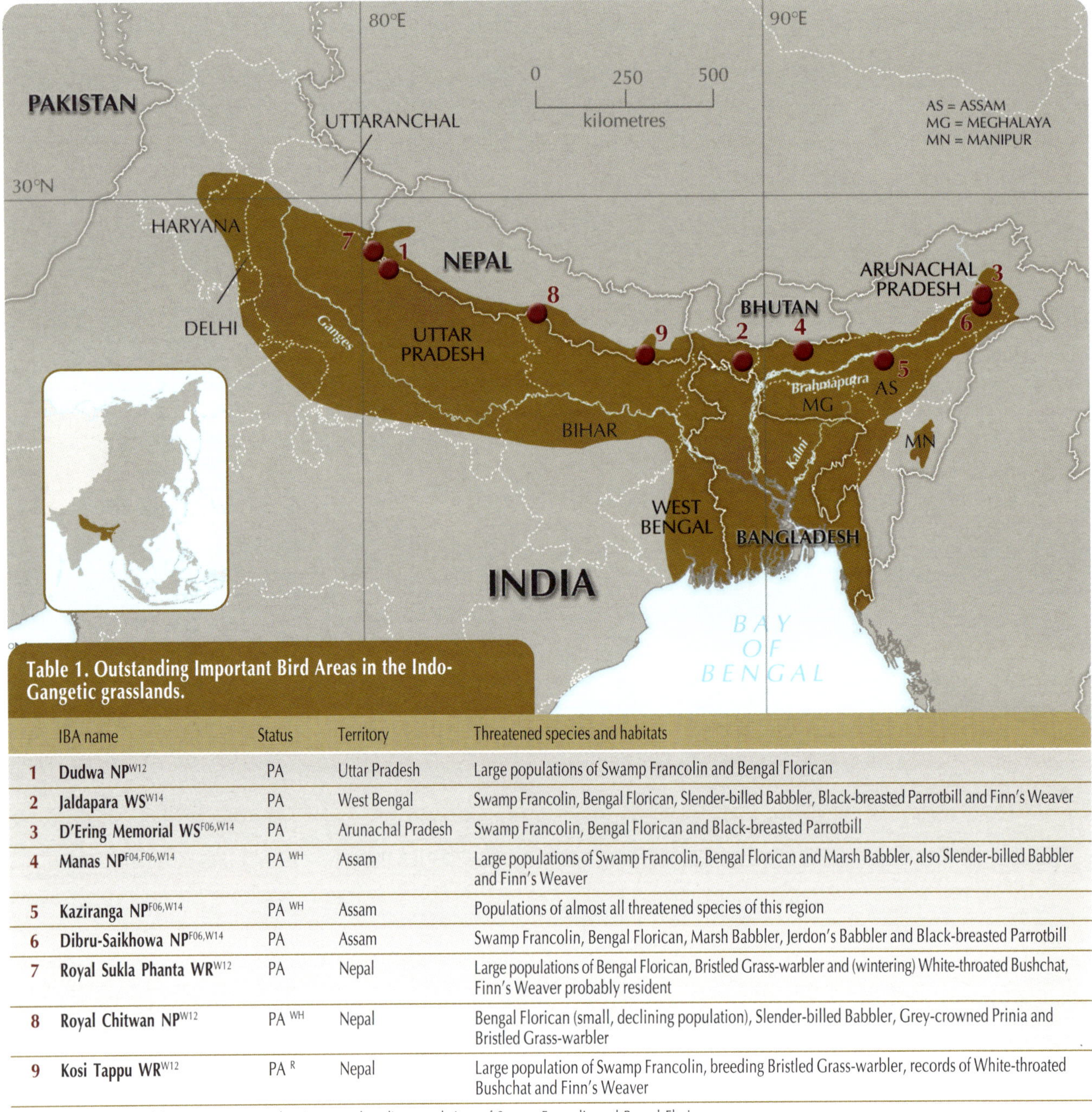

Table 1. Outstanding Important Bird Areas in the Indo-Gangetic grasslands.

	IBA name	Status	Territory	Threatened species and habitats
1	**Dudwa NP**[W12]	PA	Uttar Pradesh	Large populations of Swamp Francolin and Bengal Florican
2	**Jaldapara WS**[W14]	PA	West Bengal	Swamp Francolin, Bengal Florican, Slender-billed Babbler, Black-breasted Parrotbill and Finn's Weaver
3	**D'Ering Memorial WS**[F06,W14]	PA	Arunachal Pradesh	Swamp Francolin, Bengal Florican and Black-breasted Parrotbill
4	**Manas NP**[F04,F06,W14]	PA [WH]	Assam	Large populations of Swamp Francolin, Bengal Florican and Marsh Babbler, also Slender-billed Babbler and Finn's Weaver
5	**Kaziranga NP**[F06,W14]	PA [WH]	Assam	Populations of almost all threatened species of this region
6	**Dibru-Saikhowa NP**[F06,W14]	PA	Assam	Swamp Francolin, Bengal Florican, Marsh Babbler, Jerdon's Babbler and Black-breasted Parrotbill
7	**Royal Sukla Phanta WR**[W12]	PA	Nepal	Large populations of Bengal Florican, Bristled Grass-warbler and (wintering) White-throated Bushchat, Finn's Weaver probably resident
8	**Royal Chitwan NP**[W12]	PA [WH]	Nepal	Bengal Florican (small, declining population), Slender-billed Babbler, Grey-crowned Prinia and Bristled Grass-warbler
9	**Kosi Tappu WR**[W12]	PA [R]	Nepal	Large population of Swamp Francolin, breeding Bristled Grass-warbler, records of White-throated Bushchat and Finn's Weaver

One of the IBAs listed for W14, Orang NP, has important breeding populations of Swamp Francolin and Bengal Florican.
Note that more IBAs in this region will be included in the *Important Bird Areas in Asia*, due to be published in early 2004.

Key *IBA name*: NP = National Park; WR = Wildlife Reserve; WS = Wildlife Sanctuary.
Status: PA = IBA is a protected area; (PA) = IBA partially protected; — = unprotected; R = IBA is wholly or partially a Ramsar Site (see pp.31–32); WH = IBA is wholly or partially a World Heritage Site (see p.34); F04/F06 = also supports threatened forest birds of regions F04/F06; W12/W14 = also supports threatened waterbirds of regions W12/W14. White-rumped, Indian and/or Slender-billed Vultures of region G03 have (or had) populations in all IBAs in this region.

OUTSTANDING IBAs FOR THREATENED BIRDS (see Table 1)

The nine IBAs in Table 1 include the largest and richest grasslands remaining in this region, and together support substantial populations of all of the threatened species, apart perhaps from the poorly known Manipur Bush-quail.

CURRENT STATUS OF HABITATS AND THREATENED SPECIES

A huge expanse of grasslands and swamplands once covered much of the Ganges and Brahmaputra basins in northern India, southern Nepal, southern Bhutan and Bangladesh. Most of these grasslands have been cleared for agriculture and settlements, leaving fragments that are heavily used by people and livestock. The loss of grasslands was particularly rapid following the virtual eradication of malaria in the 1950s. Human usage can be beneficial, as healthy grasslands are usually maintained by the combined effects of fire, flood, cutting and grazing. However, many key grasslands are now threatened by conversion for agriculture and plantations, development, grass harvesting, overgrazing, excessive burning and other pressures.

Typical undeveloped terai habitat in northern India and southern Nepal (mainly *Imperata cylindrica* and *Saccharum bengalense* grasses dotted with isolated trees) has declined precipitously in area and quality. For example, continuous grassland once extended over 12 districts of Uttar Pradesh, but drastic changes in land-use patterns have left fragments in only six districts, and grasslands have virtually disappeared outside protected areas in Nepal. There are now very few, if any, extensive patches of natural grassland in Bangladesh, where the once-vast reedlands in the north-east were leased out for paper production, settled by people and apparently entirely converted to other land uses. The most extensive remaining grasslands are in Assam; this is the only part of the region where some important grasslands are not yet officially protected.

CONSERVATION ISSUES AND STRATEGIC SOLUTIONS (summarised in Table 3)

Habitat loss and degradation

■ *CONVERSION TO AGRICULTURE AND PLANTATIONS*

Large-scale conversion of natural grasslands into cropland and plantations has taken place throughout. For example, in Uttar Pradesh many grasslands have been replaced by sugarcane, and grasslands in the duars of West Bengal and Assam have been converted for a variety of agricultural uses. Plantations of commercially important timber producing trees such as *Eucalyptus*, *Dalbergia sissoo* and *Tectona grandis* (teak) have replaced large areas of grassland habitat in India, even inside protected areas (e.g. Valmikinagar Wildlife Sanctuary). Swamp Francolin may be able to survive and even breed in sugarcane fields adjacent to natural grasslands, but in general this conversion results in the almost total loss of specialised grassland birds.

In northern India, natural grasslands on Forest Department land continue to be converted to agriculture, either legally or illegally, but all further conversion needs to be prevented, and the remaining grasslands managed by the Forest Department for the benefit of the threatened birds and other grassland wildlife. The Indian Wildlife Protection Act allows exotic trees to be cut for ecological restoration, even inside protected areas, and this provision should be used to remove eucalyptus and teak plantations (developed during the 1950s and 1960s) from Dudwa National Park; new plantations should not be allowed inside protected areas. Outside protected areas, there has been a tendency to encourage forestry plantations on degraded grasslands, but these areas should be managed to encourage grassland regeneration (unlike forests, high-quality grassland can rapidly be regenerated from disturbed or overgrazed areas).

Swamp Francolin can survive in sugarcane fields, but only near to natural grasslands.

PHOTO: PERVEZ IQUBAL

■ *DAMS AND IRRIGATION*

In many areas, dams and irrigation schemes have altered or eliminated areas of wet grassland. For example, barrages across the Ganges and its tributaries control the annual floods that once maintained broad swathes of riverine grassland; this habitat is now often desiccated, and more easily cleared for agriculture (this loss is only partly offset by the development of *Phragmites* and *Typha* swamps along new irrigation canals, around reservoirs and in the seepage zones of barrage headponds). New dam and irrigation schemes which could affect wet grasslands, and the rivers which feed important grasslands, should be subject to environmental impact assessments, with plans developed to mitigate any negative effects and prevent the loss of wet grasslands. When dams and irrigation projects create new habitats, these should be managed to maximise their value for threatened grassland birds.

■ *DEFORESTATION*

Although clearance of trees is often beneficial to grassland fauna, deforestation is a threat for two threatened birds in this region, Grey-crowned Prinia and Finn's Weaver. The

Table 2. Threatened birds of the Indo-Gangetic grasslands.

Species			Distribution and habitat
Swamp Francolin *Francolinus gularis*	◐	VU	Inhabits damp grasslands and adjoining agriculture in India, Nepal and (at least formerly) Bangladesh
Manipur Bush-quail *Perdicula manipurensis*	◐	VU	Known from West Bengal, Assam and Manipur, apparently breeding in tall grasslands, but not conclusively identified since the 1930s
Bengal Florican *Houbaropsis bengalensis*	◆	EN	Inhabits grasslands in India, Nepal and (at least formerly) Bangladesh
White-throated Bushchat *Saxicola insignis*	✈	VU	Winter visitor to open short-grass plains in India, Nepal and Bhutan
Marsh Babbler *Pellorneum palustre*	◐	VU	Inhabits damp grasslands in Arunachal Pradesh, Assam, Meghalaya and Bangladesh
Jerdon's Babbler *Chrysomma altirostre*	◆	VU	Local in tall riverine grasslands from southern Nepal to eastern Assam
Slender-billed Babbler *Turdoides longirostris*	◐	VU	Local in damp grasslands from Nepal to Assam and (at least formerly) Manipur
Black-breasted Parrotbill *Paradoxornis flavirostris*	◐	VU	Dense reeds and tall damp grassland in West Bengal, Arunachal Pradesh and Assam
Grey-crowned Prinia *Prinia cinereocapilla*	◐	VU	Inhabits forest edge and shrubby grasslands from the plains up to 1,350 m in the terai and Himalayan foothills of India, Nepal and Bhutan
Bristled Grass-warbler *Chaetornis striatus*	◐	VU	Widespread in damp grasslands in Pakistan, India, Nepal and Bangladesh, but very local and erratic in occurrence
Finn's Weaver *Ploceus megarhynchus*	◐	VU	Disjunct distribution, with populations in Uttar Pradesh, Delhi (possibly escaped birds) and west Nepal, and others in Assam and West Bengal, in tall grasslands, often with scattered trees

◐ = breeds only in this grassland region; ◆ = also breeds in other region(s); ✈ = non-breeding visitor from another region

prinia prefers lightly wooded grasslands or *Shorea robusta* forest–grassland ecotones, and disappears when all bushes and trees are cleared. Similarly, Finn's Weaver requires secure trees in which to build its colonies (at least in Uttar Pradesh). Grasslands within the ranges of Grey-crowned Prinia and Finn's Weaver (which are now mainly within protected areas) need to be managed to retain light woodland and *Shorea robusta* forest for the prinia, and scattered trees for the weaver.

■ GRASS HARVESTING

Grass is a vital resource for rural people in South Asia who use it for fodder, construction (thatch for roofs and wattle for walls), and rope, mats, baskets and other items. Most surviving grasslands, even within protected areas, are heavily harvested. Annual or even biannual harvesting is often beneficial for grassland as it impedes succession to woodland, but over-harvesting damages habitat and creates disturbance. Harvesting needs to be controlled where it is degrading natural grasslands, with alternative materials promoted in communities overly dependent on grassland resources, and timed to avoid disturbance during the breeding season.

■ LIVESTOCK GRAZING

Many grasslands are being degraded by unsustainable grazing pressure, reducing the quality of habitat available to threatened birds; this is especially damaging in summer when post-burn re-growth emerges and water distribution is limited. Overgrazing is a major problem in many protected areas, including Kosi Tappu Wildlife Reserve in Nepal, where large numbers of livestock have caused severe degradation, and Katerniaghat Wildlife Sanctuary in Uttar Pradesh, which supports nearly 10 times more cattle than stipulated by its management plan. Grazing needs to be managed much more effectively, especially in protected areas. Cessation of grazing is, of course, detrimental to grassland habitat, but grazing (as well as grass harvesting and burning) should be limited and rotational, leaving areas of tall and mid-height grass. The concept of fewer but better-quality livestock should be promoted, particularly near protected grasslands, to reduce grazing pressure.

Grasslands are heavily utilised by local people, but this needs careful control to prevent habitat degradation.

PHOTO: OTTO PFISTER

Grasslands are primarily managed for large mammals in some reserves, but burning to provide fresh grasses may leave insufficient cover for grassland birds.

PHOTO: OTTO PFISTER

■ BURNING

Grasslands, even within protected areas, are frequently burnt by people to remove all cover before cultivating or settling the land, or to stimulate fresh growth of grasses for grazing or harvesting. In some protected areas, e.g. Royal Chitwan National Park in Nepal and Kaziranga National Park in Assam, grasslands are primarily managed for large mammals; burning is used to provide fresh grasses for these mammals, but at the expense of many grassland bird species. Annual fires may help prevent succession of grassland to forest, but burning several times a year may alter the composition of grassland, rendering it unsuitable for threatened species, and burning in the breeding season is extremely damaging as nests and eggs are destroyed. In addition, burning is often too comprehensive, leaving no shelter for grassland wildlife. Burning should be rotational, so that a mosaic of patches provides both tall grass for cover and short grass favoured by species such as Bengal Florican as foraging habitat. All burning should be carried out in January or early February in north India and Nepal, before the breeding season, and illegal burning prevented in protected areas. Conservation awareness programmes are required, given the need to manage grass harvesting, grazing and burning to balance the needs of grassland wildlife and sustainable exploitation by local communities. These could aim to explain the importance of properly managed grasslands for wildlife and people, and the value and legal status of threatened species, perhaps using the spectacular Bengal Florican as a flagship for the conservation of the region's grasslands.

■ DISTURBANCE

Most grasslands, even in protected areas, are subject to high levels of disturbance by people and livestock. In Bangladesh, for example, human population density is very high and most remaining grassland areas are highly fragmented, heavily used and harvested three times a year. In Nepal, people are allowed into protected areas for 7–10 days annually to cut grass, at which time the grasslands are also burned; in the case of Royal Chitwan National Park this involves an influx of 70,000 people. Off-road driving, which is common in some protected areas, causes additional disturbance. In many areas shy birds such as floricans are constantly flushed and frightened away, and during the breeding season eggs or chicks risk being trodden on by cattle or people. The usage of grasslands in protected areas by people, cattle and all-terrain vehicles needs careful control, particularly while birds are nesting. Steps should also be taken to prevent new settlements in grassland protected areas.

■ *EROSION AND FLOODING*

Heavy flooding, exacerbated by greater run-off from deforested river catchments, is an increasing problem. Seasonal flooding is a natural phenomenon which helps to maintain grassland habitat, but if too heavy or frequent it reduces grassland quality. Floods in Kaziranga and Dibru-Saikhowa National Parks, for example, have washed away large grassy islands and waterlogged other areas, resulting in grasslands becoming overgrown or excessively sandy. Erosion caused by flooding has reduced the extent of grassland in these crucial reserves. Moreover, grassland habitat away from rivers has often been destroyed, leaving no shelter for grassland fauna during floods. This problem is most extreme in parts of Bangladesh, where any grasslands that remain are inundated for two-thirds of the year with no alternative refugia. Sufficient grassland needs to be maintained outside the riverine floodplains to provide habitat for grassland birds during heavy floods, and forest conservation and reforestation are required in the upper catchments in the Eastern Himalayas (see F04).

■ *PESTICIDES*

Agricultural pesticides and herbicides may be affecting grassland birds either by direct mortality or by reduction of invertebrate prey. Finn's Weaver often nests with Black Drongo *Dicrurus macrocercus*, which protects it from predation by House Crow *Corvus splendens*; however, the drongo appears to be declining in numbers, possibly because of pesticides, which may be allowing increased predation by crows and reducing the weaver's nesting success. The use of agrochemicals should be controlled near important grasslands, with traditional organic agricultural practices encouraged wherever possible.

■ *CIVIL STRIFE AND CONFLICT OVER LAND*

In parts of the region the presence of military or rebellious factions severely reduces opportunities for effective wildlife conservation, even in protected areas such as Manas National Park. The rapidly growing human population is placing extreme pressure on remaining fragments of habitat within many protected areas, and in several reserves conservation action is complicated by multiple land rights claims. Lagga Bagga Sanctuary in Uttar Pradesh once contained high-quality grasslands, but local politicians promised land to encroachers, which generated an influx and rapidly increased deforestation and grazing. This type of land claim in protected areas need to be resolved through careful negotiation involving all stakeholders, to ensure the protection of natural grasslands and other habitats whilst maximising the economic and social benefits to the local communities.

Protected areas coverage and management

■ *GAPS IN PROTECTED AREAS SYSTEM*

There are some large grassland reserves in parts of this region, for example in northern India and Nepal, but elsewhere grasslands are generally poorly represented in

Table 3. Conservation issues and strategic solutions for birds of the Indo-Gangetic grasslands.

Conservation issues	Strategic solutions
Habitat loss and degradation	
■ *CONVERSION TO AGRICULTURE AND PLANTATIONS* ■ *DAMS AND IRRIGATION* ■ *DEFORESTATION* ■ *GRASS HARVESTING* ■ *LIVESTOCK GRAZING* ■ *BURNING* ■ *DISTURBANCE* ■ *EROSION AND FLOODING* ■ *PESTICIDES* ■ *CIVIL STRIFE AND CONFLICT OVER LAND*	➤ Prevent further conversion of Indian Forest Department grasslands to agriculture ➤ Restore former grasslands in Dudwa National Park by removing eucalyptus and teak plantations, and prohibit new plantations in protected areas ➤ Encourage the regeneration of degraded grasslands ➤ Assess the environmental impact of new dam and irrigation schemes, with mitigation measures to maintain wet grasslands ➤ Retain the woodland and scattered trees required by Grey-crowned Prinia and nesting Finn's Weaver ➤ Encourage rotational burning, grazing and grass harvesting practices, to provide a mosaic of tall and short grass for nesting and foraging ➤ Avoid grass burning and harvesting during the breeding season, excessive grass harvesting, and overgrazing ➤ Minimise disturbance in grassland protected areas by people, cattle and vehicles, particularly while birds are nesting ➤ Control the use of agrochemicals, and encourage traditional organic farming methods ➤ Resolve land rights claims in protected areas, to ensure the protection of grasslands whilst maximising the local economic and social benefits
Protected areas coverage and management	
■ *GAPS IN PROTECTED AREAS SYSTEM* ■ *WEAKNESSES IN RESERVE MANAGEMENT*	➤ Improve the coverage of grasslands in the protected areas system of Assam ➤ Develop reserve management regimes which balance the conservation of all grassland biodiversity with sustainable use by local communities ➤ Increase resources available for protected areas management, to equip and train staff better
Exploitation of birds	
■ *HUNTING* ■ *WILD BIRD TRADE*	➤ Enforce existing laws to prevent hunting in protected areas, and control gun ownership ➤ Improve the enforcement of laws to prevent trade in Finn's Weaver
Gaps in knowledge	
■ *INADEQUATE DATA ON THREATENED BIRDS*	➤ Investigate the optimal grazing, burning, cutting and flooding regimes for key grasslands ➤ Search for Manipur Bush-quail in the Manipur basin and elsewhere ➤ Survey any surviving areas of natural or semi-natural grassland in Bangladesh ➤ Study the distributions of non-breeding Bengal Florican and White-throated Bushchat, and the possible link between pesticide use and predation of Finn's Weaver by House Crow

protected-area systems, e.g. in Assam. New protected areas and extensions to existing reserves should also be considered to improve coverage of the relatively extensive grasslands of Assam. For example, in Assam the Amarpur area near Dibru-Saikhowa National Park should be brought under protection as a satellite core area of the park, and the Deobal-Jalah grassland should be protected.

■ *WEAKNESSES IN RESERVE DESIGN AND MANAGEMENT*
Many protected areas are small and/or isolated, and surrounded by densely populated and intensively cultivated farmland, and the populations of threatened species which they support are thus susceptible to local extinction. Wherever possible the boundaries of small parks should be expanded, and corridors of grassland restored to link them to other nearby reserves, to improve the chances for long-term survival of the populations of larger birds such as Bengal Florican. The zoning of protected areas is critical, to provide sufficiently large (relatively) undisturbed core areas for their wildlife, as well as ample grassland buffer zones to cater for local people.

The grassland reserves need direct habitat management to maximise their value to wildlife, typically aimed to provide a mosaic of patches of grass of different heights (see *Grass harvesting*, *Livestock grazing* and *Burning* above). In many reserves, it is a major challenge to develop the correct management regime, given the need to balance the conservation of all grassland biodiversity (including large mammals and threatened birds) with sustainable use by local communities. Thus, rotational grazing and burning will be difficult to achieve given the huge number of people involved, but it is vital to bring the cutting and burning regime under greater control. Crucially, rotational management can simultaneously increase herbage production and conserve grassland ecosystems, and thus offers a partial remedy to the key problem of the region's growing livestock population and shrinking pastures. In some protected areas current management practices clearly need adjustment, e.g. in Royal Chitwan National Park areas of shorter grassland favoured by Bengal Florican are developing into taller *Narenga*- and *Saccharum*-dominated grassland, and eventually to scrub and forest. Conservation measures are difficult to implement in many protected areas because of poor infrastructural facilities and general lack of training and adequate salaries for protected-area staff. To improve their management, grassland reserves require increased manpower and equipment, along with training and better salaries for staff.

Bengal Florican requires a mosaic of both short grass for foraging and tall grass for cover.

PHOTO: J. C. EAMES

Exploitation of birds

■ *HUNTING*
Trapping and hunting for food remain serious threats in many areas, especially for Swamp Francolin and Bengal Florican. Some 70,000 human families currently rely on the professional trading of wild birds in Uttar Pradesh, India, a country traditionally benign towards wildlife. Socio-economic studies of hunting are required to investigate the impact on bird populations and potential alternative livelihoods. Existing legislation needs to be better enforced in protected areas, with adequate numbers of guards or wardens to patrol the reserves. Control of gun ownership might be a practical method of addressing this problem in rural areas.

■ *WILD BIRD TRADE*
Swamp Francolins are trapped and sold for cock-fighting, while Finn's Weaver is traded as a cage-bird. Between 200 and 300 Finn's Weaver are traded annually in India, even though trapping and trade of the species has been banned there since 1991. Existing laws to control trapping should be more strictly enforced, with conservation awareness programmes initiated near key sites for Finn's Weaver, to highlight the laws and profile the threatened species amongst trappers and local people. There may also be a need for improved training for enforcers responsible for the control of trapping, and the provision of alternative livelihoods to traditional subsistence bird-trappers.

Gaps in knowledge

■ *INADEQUATE DATA ON THREATENED BIRDS*
Further research is required to identify the optimal grazing, burning, cutting and flooding regimes in grasslands (both in general and at particular grassland sites), to suit the needs of people and threatened birds. There are many gaps in knowledge of the distribution and status of threatened birds, and surveys are required to identify further key sites for their conservation. Manipur Bush-quail has not been recorded since 1932, and surveys are needed to relocate it in the Manipur basin (which could run alongside searches for the Slender-billed Babbler which was once common in this state) and elsewhere. The current status of threatened grassland birds is poorly known in Bangladesh, and any surviving areas of natural or semi-natural grassland in Sylhet should be searched for Swamp Francolin, Marsh Babbler and Black-breasted Parrotbill. The distribution of Bengal Florican in the non-breeding season deserves study, and a full winter survey is required of White-throated Bushchat in the northern Gangetic plains and Brahmaputra valley, where very few current wintering sites are known. Surveys are required to locate Finn's Weaver colonies and determine measures needed for their protection, with investigations of the effect of trade on wild populations, and of the possible link between pesticide use and increased nest predation by House Crows (see *Pesticides*).

SOUTH ASIAN ARID HABITATS

G03

THE 12 threatened species of this region are found in a variety of arid and semi-arid habitats in the Indian subcontinent, many of which have been greatly modified by centuries of human usage. They have here been divided into five groups (see coloured outlines on map) according to their habitat requirements and distributions: (G03A) three *Gyps* vulture species which roam over all types of forest and open country; (G03B) four species which inhabit semi-arid plains (rolling short grasslands, open deserts and low shrublands), including two migrants from Central Asia; (G03C) Lesser Florican, which breeds in moderately high grasslands in arid to semi-arid areas; (G03D) three species restricted to dry forests and scrub; (G03E) Green Avadavat, which inhabits rough grasslands, agricultural land and orchards. The conservation needs of each of these five groups is discussed in separate sections below.

- **Key habitats** Grassland, arid and semi-arid woodland and scrub, semi-desert, cultivation and other open habitats.
- **Altitude** 0–c.1,000 m.
- **Countries and territories** **Pakistan**; **India** (Himachal Pradesh; Punjab; Haryana; Delhi; Rajasthan; Gujarat; Uttar Pradesh; Madhya Pradesh; Maharashtra; Goa; Karnataka; Andhra Pradesh; Kerala; Tamil Nadu; Bihar; Orissa; West Bengal; Sikkim; Arunachal Pradesh; Assam; Meghalaya; Nagaland; Manipur); **Nepal**; **Bhutan**; **Bangladesh**.

	Threatened species			
	CR	EN	VU	Total
◉	2	2	4	8
◆	2	—	—	2
✈	—	—	2	2
Total	4	2	6	12

Key: ◉ = breeds only in this grassland region.
◆ = also breeds in other region(s).
✈ = non-breeding visitor from another region.

Vultures play an important role as cleansers, but following the recent crash in their numbers in South Asia they are being replaced by packs of feral dogs as the main scavengers of carcasses. PHOTO: ASAD RAHMANI

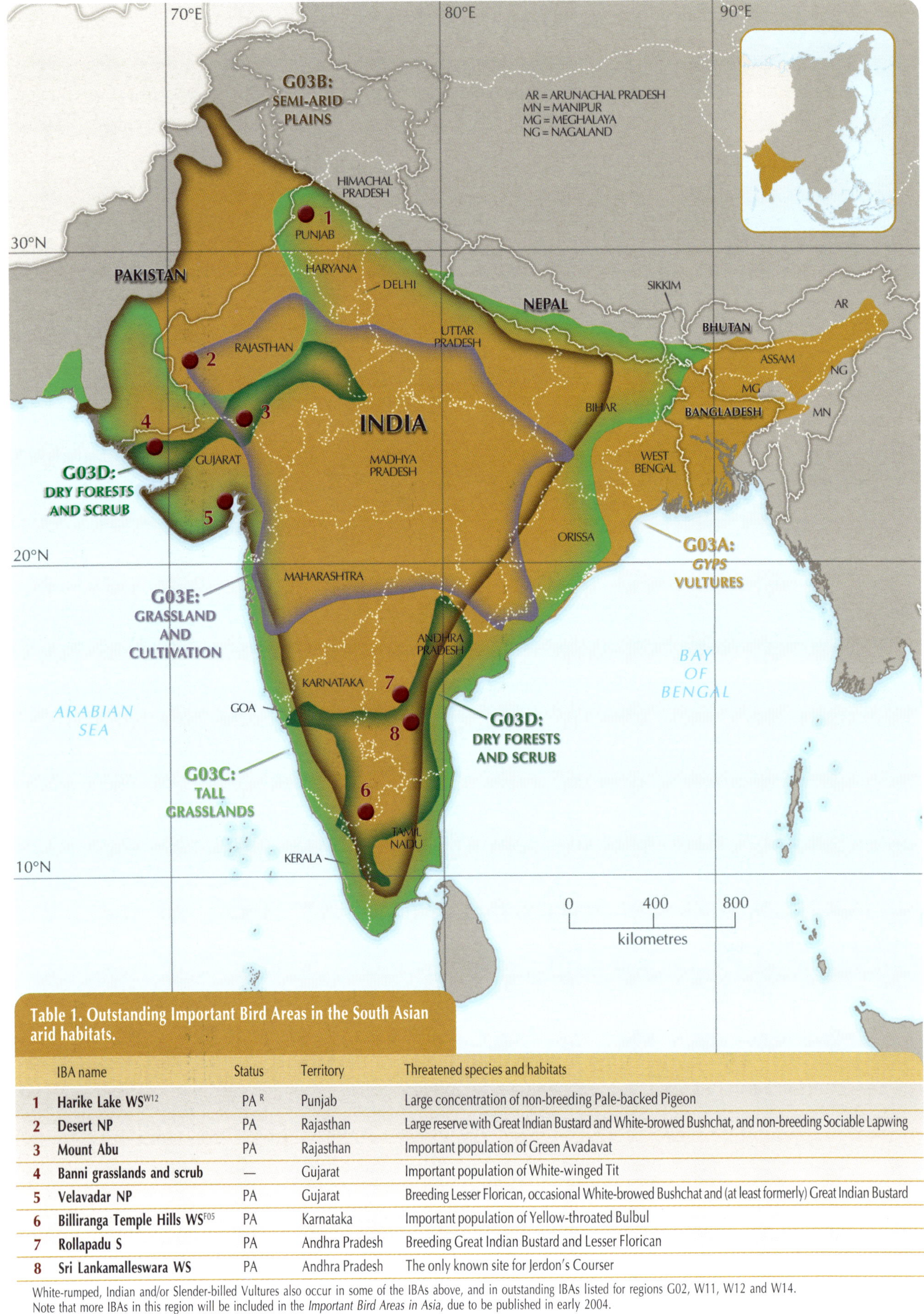

Table 1. Outstanding Important Bird Areas in the South Asian arid habitats.

	IBA name	Status	Territory	Threatened species and habitats
1	**Harike Lake WS**[W12]	PA [R]	Punjab	Large concentration of non-breeding Pale-backed Pigeon
2	**Desert NP**	PA	Rajasthan	Large reserve with Great Indian Bustard and White-browed Bushchat, and non-breeding Sociable Lapwing
3	**Mount Abu**	PA	Rajasthan	Important population of Green Avadavat
4	**Banni grasslands and scrub**	—	Gujarat	Important population of White-winged Tit
5	**Velavadar NP**	PA	Gujarat	Breeding Lesser Florican, occasional White-browed Bushchat and (at least formerly) Great Indian Bustard
6	**Billiranga Temple Hills WS**[F05]	PA	Karnataka	Important population of Yellow-throated Bulbul
7	**Rollapadu S**	PA	Andhra Pradesh	Breeding Great Indian Bustard and Lesser Florican
8	**Sri Lankamalleswara WS**	PA	Andhra Pradesh	The only known site for Jerdon's Courser

White-rumped, Indian and/or Slender-billed Vultures also occur in some of the IBAs above, and in outstanding IBAs listed for regions G02, W11, W12 and W14.
Note that more IBAs in this region will be included in the *Important Bird Areas in Asia*, due to be published in early 2004.

Key *IBA name*: NP = National Park; S = Sanctuary; WS = Wildlife Sanctuary.
Status: PA = IBA is a protected area; (PA) = IBA partially protected; — = unprotected; R = IBA is wholly or partially a Ramsar Site (see pp.31–32); F05 = also supports threatened forest birds of region F05; W12 = also supports threatened waterbirds of region W12.

Table 2. Threatened birds of the South Asian arid habitats.

Species			Distribution and habitat
			G03A: *GYPS* VULTURES
White-rumped Vulture *Gyps bengalensis*	◆	CR	Urban and cultivated areas, light woodland and open habitats in lowland South Asia and South-East Asia
Indian Vulture *Gyps indicus*	●	CR	Urban and cultivated areas, light woodland and open habitats in lowland Pakistan and southern India
Slender-billed Vulture *Gyps tenuirostris*	◆	CR	Cultivated areas, light woodland and open habitats in lowland northern India, southern Nepal, Bangladesh and South-East Asia
			G03B: SEMI-ARID PLAINS
Great Indian Bustard *Ardeotis nigriceps*	●	EN	Range now highly fragmented in semi-desert, grassland, scrub and cultivation in western India and eastern Pakistan
Sociable Lapwing *Vanellus gregarius*	✈	VU	Non-breeding visitor in small numbers to dry plains in Pakistan and north-west India
Pale-backed Pigeon *Columba eversmanni*	✈	VU	Non-breeding visitor to open and sparsely wooded areas in Pakistan and north-west India
White-browed Bushchat *Saxicola macrorhyncha*	●	VU	Local in sandy semi-desert with low scrub in north-west India and (at least formerly) Pakistan
			G03C: TALL GRASSLANDS
Lesser Florican *Sypheotides indica*	●	EN	Breeds in grasslands and open fields in north-west India, outside breeding season disperses widely in India, and sometimes in Pakistan and Nepal
			G03D: DRY FORESTS AND SCRUB
Jerdon's Courser *Rhinoptilus bitorquatus*	●	CR	Known from dry rocky undulating ground with a thin woodland or scrub cover at a handful of localities in Maharashtra and Andhra Pradesh in southern India
Yellow-throated Bulbul *Pycnonotus xantholaemus*	●	VU	Found in sparse thorn scrub interspersed with trees on boulder-strewn inland hills in Karnataka, Andhra Pradesh and Tamil Nadu in southern India
White-naped Tit *Parus nuchalis*	●	VU	Occurs in dry thorny woodland and scrub, with disjunct populations in north-west and southern India
			G03E: GRASSLAND AND CULTIVATION
Green Avadavat *Amandava formosa*	●	VU	Inhabits grassland and scrub, sugarcane fields and orchards in the lowlands of north and central India

● = breeds only in this grassland region; ◆ = also breeds in other region(s); ✈ = non-breeding visitor from another region

OUTSTANDING IBAs FOR THREATENED BIRDS (see Table 1)

Eight IBAs have been selected, which together support populations of almost all of the threatened species of this region. However, most of these birds occur at low densities or have highly fragmented populations, and their conservation is therefore generally best addressed at the landscape level and through the protection of networks of small sites. Many of these will be identified during BirdLife's ongoing regional IBA Project.

G03A: *GYPS* VULTURES

CURRENT STATUS OF HABITATS AND THREATENED SPECIES

White-rumped and Slender-billed Vultures occur in the Indian subcontinent and South-East Asia, but Indian Vulture is found only in India and Pakistan. Early accounts suggest that all three species were once abundant across most of their ranges, having adapted well to modification of their natural habitats by man, although vulture populations declined rapidly in South-East Asia during the twentieth century (see W16 and W18). Vulture numbers remained high in the Indian subcontinent until the last few years of the twentieth century, when an extremely rapid decline was reported, with large numbers of sick birds suggesting that a disease or poisoning may be responsible. The implications are grave, as low reproductive rates and high longevity means that only relatively small increases in adult mortality rates can lead to precipitous population declines. If numbers in the Indian subcontinent continue to fall, the remnant populations of White-rumped and Slender-billed Vultures in South-East Asia will become increasingly important for the survival of these species.

CONSERVATION ISSUES AND STRATEGIC SOLUTIONS (summarised in Table 3a)

Habitat loss and degradation

■ *REDUCED NESTING HABITAT*

These vultures have adapted well to modifications of their habitats by man, and an abundance of suitable habitat remains both in the Indian subcontinent and in South-East Asia. However, some localised declines have been noticed in parts of Bangladesh, mainly around cities, apparently linked to a scarcity of mature trees for nesting and roosting. Where possible mature trees should be left standing, and the provision of artificial nest sites considered, especially where shortage of nest sites is a problem.

■ *PESTICIDES*

Large quantities of pesticides are used in agricultural areas of the Indian subcontinent, and high levels have been found (including of DDT) in tissue samples from cattle and pig carcasses in several parts of India. This may well cause problems to scavengers such as vultures, and there is some evidence that eggs are broken in vulture nests, possibly as a result of eggshell-thinning toxins. However, recent analyses of tissues from dead vultures in India and Pakistan have not found significant concentrations of pesticides; nevertheless, controls on the use of pesticides and other agrochemicals should be tightened throughout the region.

Exploitation of birds

■ *HUNTING AND PERSECUTION*

In the Indian subcontinent vultures are generally regarded as unclean and are not often hunted, and are valued for their role as cleansers. However, efforts have been made to eliminate vulture populations near to some airports to reduce the risk of air-strikes. Vultures are generally unpopular in Asia, a factor that hinders conservation action, and awareness campaigns are required to improve understanding of their current plight, and of the crucial role that they play in the disposal of carrion. In some areas where vultures have declined they are being replaced by packs of feral dogs as the main scavengers of carcasses, which could lead to an increased rabies problem.

■ *POISONING*

Deliberate poisoning of carcasses (e.g. with strychnine), normally to defend against mammalian livestock predators (e.g. leopards), has long affected vultures in Asia. Vultures are communal feeders, meaning that an entire population can be exterminated by feeding at a single poisoned carcass. The use of poisoning baits to control carnivore populations should be prevented throughout.

Gaps in knowledge

■ *INADEQUATE DATA ON THREATENED BIRDS*

Traditionally, very little attention has been paid to the identification, importance and conservation status of vultures. The Indian Vulture and Slender-billed Vulture were considered a single species until recently, and this has hampered any appraisal of recent changes in their distribution and abundance. However, since the vulture crisis began in the late 1990s, numerous surveys and studies of nesting colonies and vulture pathology have been initiated, in India under the coordination of Bombay Natural History Society. This research should be continued and extended, with the most immediate priority to continue laboratory studies to identify the causes of vulture declines in the Indian subcontinent, and to devise appropriate measures for its control. Detailed, standardised surveys are required to monitor the status (including breeding success and incidence of mortality) of the three vulture species in the Indian subcontinent, with survey teams armed with new identification criteria to distinguish Indian Vulture and Slender-billed Vulture; in India, the vast manpower of the Forest Department could facilitate surveys.

Other conservation issues

■ *INCREASED MORTALITY*

Until recently, it was thought that the most likely cause of the recent rapid declines in the numbers of vultures in the Indian subcontinent was infectious disease. Post-mortem studies of vulture carcasses from across India and Pakistan have found evidence of visceral or renal gout, and dying birds in all areas have shown similar clinical signs of sickness, such as neck-drooping. It appears that only vultures of the genus *Gyps* are affected, and it is known that some infectious diseases, especially viral, can exhibit this

Gyps ***vultures in South Asia have suffered a mass mortality in the past decade, but the precise cause is not yet understood.***

PHOTO: OTTO PFISTER

Table 3a. Conservation issues and strategic solutions for the South Asian arid habitats: *Gyps* vultures.

Conservation issues	Strategic solutions
Habitat loss and degradation	
■ REDUCED NESTING HABITAT ■ PESTICIDES	➤ Retain mature trees in areas where a lack of suitable nest and roost sites may be a problem ➤ Tighten controls on the use of pesticides and other agrochemicals
Exploitation of birds	
■ HUNTING AND PERSECUTION ■ POISONING	➤ Conduct awareness campaigns to improve understanding of the current plight of Asia's vultures, and their importance as removers of carrion ➤ Ban the use of poisoned baits to control carnivore populations
Gaps in knowledge	
■ INADEQUATE DATA ON THREATENED BIRDS	➤ Continue research to identify the cause of the vulture decline in the Indian subcontinent, and to devise appropriate measures for its control ➤ Monitor vulture status and distribution using standard survey techniques, including breeding success and mortality at nesting colonies
Other conservation issues	
■ INCREASED MORTALITY	➤ Rapidly implement measures to combat the current cause of decline once these are available ➤ Initiate captive breeding programmes to maintain healthy populations of all three vulture species

type of genus specificity. However, dead birds from three colonies in Pakistan were recently found to contain a drug, Diclofenac, which is used in veterinary medicine, and it is speculated that this could be the main cause of the observed sickness and mortality. Laboratory and field studies need to be continued to confirm whether this drug or another factor, or a combination of factors, is killing vultures. If the drug is confirmed to be the main source of the problem, urgent measures are required to prevent vultures ingesting it with their food, possibly including the development and provision of alternative treatments for sick animals, with awareness campaigns to promote their use. A captive breeding programme has been proposed, to try to maintain healthy captive populations of all three vulture species; until a solution has been found to the problem affecting wild populations, efforts should be continued to breed from any healthy birds in zoos, and construct facilities to hold and breed captive vultures in the Indian subcontinent.

G03B: SEMI-ARID PLAINS

CURRENT STATUS OF HABITATS AND THREATENED SPECIES

Much of eastern Pakistan, and north-western, central and south-eastern India was once a vast expanse of savanna, but this habitat is now much reduced and fragmented. Huge areas have been converted to agriculture over recent decades (often involving large-scale irrigation projects) to meet the demands of a growing human population, while increasing numbers of cattle have caused widespread overgrazing. Both Great Indian Bustard and White-throated Bushchat avoid intensively cultivated areas, so their ranges and numbers have been greatly reduced, and the wintering populations of the other two threatened species have presumably also been negatively affected.

CONSERVATION ISSUES AND STRATEGIC SOLUTIONS (summarised in Table 3b)

Habitat loss and degradation

IRRIGATION AND CONVERSION TO AGRICULTURE

Until the early twentieth century, the semi-arid plains of India and eastern Pakistan were of limited value for agriculture, but a succession of irrigation schemes has led to the conversion of vast areas to intensive cultivation. This continuing loss and fragmentation of grassland and semi-desert habitats, together with associated increases in pesticide use and disturbance, are seriously affecting the populations of threatened birds, particularly Great Indian Bustard. The planned Indira Ghandi Nagar Project (IGNP) will directly irrigate 11% of the Thar Desert, and its canals will bisect the already heavily disturbed Desert National Park; and it could therefore affect large areas of semi-desert habitat. It is vital that irrigation schemes and associated agricultural developments are modified (possibly through environmental impact assessments) near the most important areas for Great Indian Bustard and other threatened birds, to prevent large-scale conversion of their habitats, particularly inside protected areas. Irrigation should be avoided in Desert National Park (and other known centres of bustard populations) by re-routing the IGNP canals, or by prohibiting any irrigation of land (via these canals) within the park boundaries.

The semi-desert habitat of Great Indian Bustard has been rapidly lost and fragmented over recent decades through large-scale irrigation projects and conversion to agriculture.

PHOTO: ASAD RAHMANI

AGRICULTURAL INTENSIFICATION

More intensive forms of agriculture have recently become widespread, linked to the large irrigation schemes. Great Indian Bustards can persist in agricultural areas where the farmers use non-intensive traditional practices, but not in intensively cultivated areas. Even the use of new crops may make a difference, with the substitution of groundnuts for millet and cotton affecting the bustards because this new crop does not provide cover for young. The retention of some areas of farmland where non-intensive traditional agricultural practices are used and bustard-friendly crops are grown should be encouraged, and strictly imposed in buffer zones to reserves. This type of landscape-scale measure is vital as Great Indian Bustards range widely and cannot be conserved within protected areas alone, however important these may be as havens to maximise breeding success.

LIVESTOCK GRAZING AND GRASSLAND MANAGEMENT

Vast numbers of livestock roam Indian semi-deserts and grasslands, chiefly cattle, although sheep- and camel-breeding operations are spreading. Grazing animals may cause disturbance, degrade habitat and destroy the nests of ground breeding birds, although limited grazing is beneficial, preventing vegetation becoming unsuitably tall. At present, many grasslands are poorly managed and hence of limited value for threatened birds, either because excessive grazing or burning has damaged the grasslands and created denuded plains, or because the removal of these factors—as has occurred in some protected areas—has allowed the vegetation to rapidly regenerate into dense scrubland. A balance is needed between these extremes. In general, the perpetuation of traditional grassland management and the introduction of rotational grazing under controlled conditions would be of great benefit, and should be promoted among local communities in and

Grassland reserves such as Rollapadu require careful management of grazing and burning to develop ideal habitat for nesting bustards.

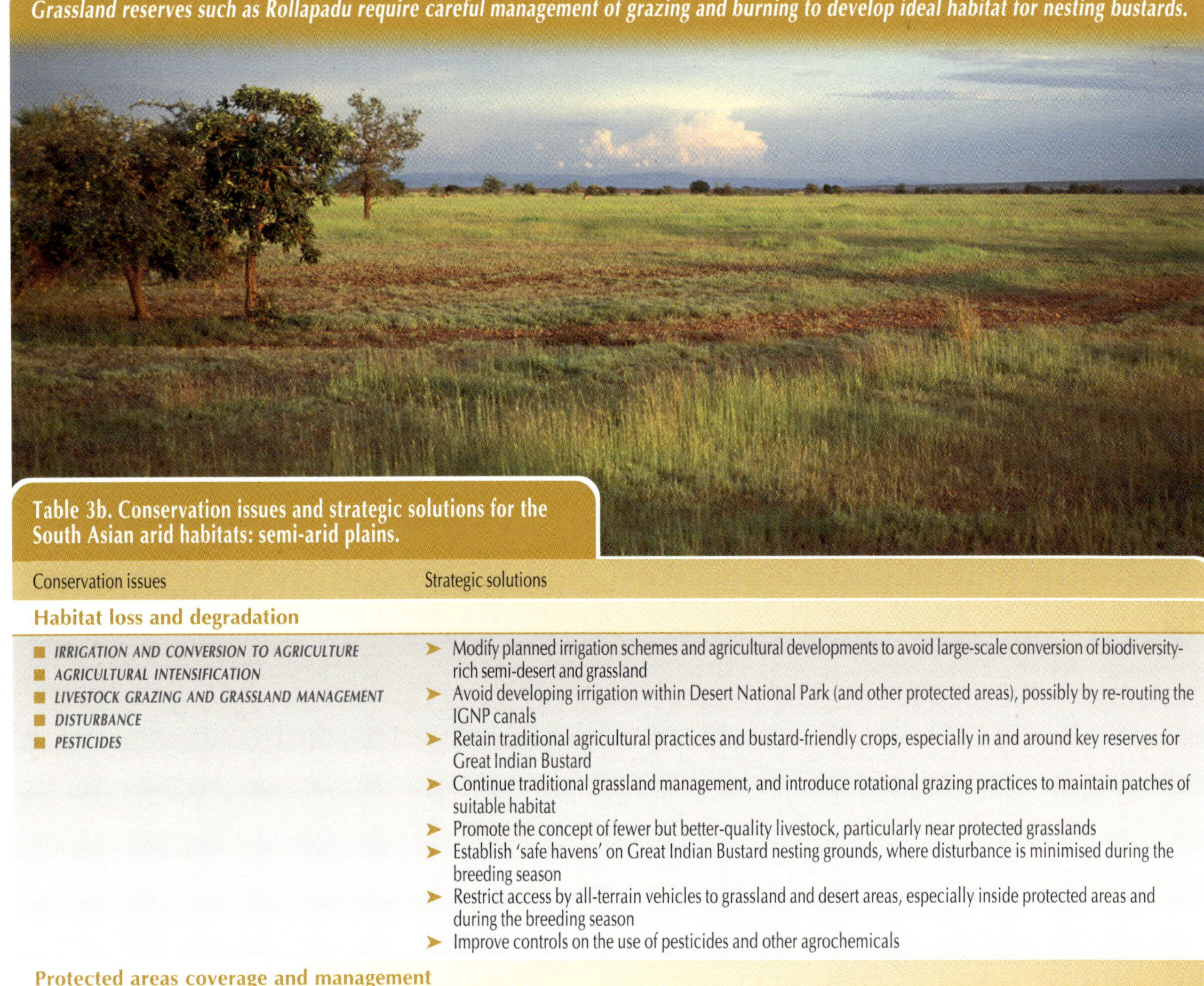

PHOTO: ASAD RAHMANI

Table 3b. Conservation issues and strategic solutions for the South Asian arid habitats: semi-arid plains.

Conservation issues	Strategic solutions
Habitat loss and degradation	
■ *IRRIGATION AND CONVERSION TO AGRICULTURE* ■ *AGRICULTURAL INTENSIFICATION* ■ *LIVESTOCK GRAZING AND GRASSLAND MANAGEMENT* ■ *DISTURBANCE* ■ *PESTICIDES*	➤ Modify planned irrigation schemes and agricultural developments to avoid large-scale conversion of biodiversity-rich semi-desert and grassland ➤ Avoid developing irrigation within Desert National Park (and other protected areas), possibly by re-routing the IGNP canals ➤ Retain traditional agricultural practices and bustard-friendly crops, especially in and around key reserves for Great Indian Bustard ➤ Continue traditional grassland management, and introduce rotational grazing practices to maintain patches of suitable habitat ➤ Promote the concept of fewer but better-quality livestock, particularly near protected grasslands ➤ Establish 'safe havens' on Great Indian Bustard nesting grounds, where disturbance is minimised during the breeding season ➤ Restrict access by all-terrain vehicles to grassland and desert areas, especially inside protected areas and during the breeding season ➤ Improve controls on the use of pesticides and other agrochemicals
Protected areas coverage and management	
■ *GAPS IN PROTECTED AREAS SYSTEM* ■ *WEAKNESSES IN RESERVE DESIGN AND MANAGEMENT*	➤ Increase coverage of threatened birds by establishing new protected areas and extending existing ones ➤ Redemarcate the boundaries of Desert National Park, and redevelop core areas in this and other reserves for breeding Great Indian Bustards ➤ Increase the resources available for protected area management and training of reserve staff, particularly at key sites for threatened species
Exploitation of wildlife	
■ *HUNTING*	➤ Improve enforcement of hunting legislation, by developing an anti-poaching task force, regular patrolling of protected areas, and controls on gun ownership
Gaps in knowledge	
■ *INADEQUATE DATA ON THREATENED BIRDS* ■ *LACK OF AWARENESS*	➤ Study the movements of Great Indian Bustard, possibly using satellite-tracking ➤ Survey White-browed Bushchat (especially in Pakistan), and wintering Sociable Lapwing and Pale-backed Pigeon, to help identify key sites for their conservation ➤ Launch 'Project Bustard', to raise awareness of the plight of India's bustards and to coordinate conservation measures for these birds and their grassland habitats

around important areas for threatened birds. The concept of fewer but better-quality livestock could also be promoted, particularly to reduce grazing pressure near protected grasslands.

■ *DISTURBANCE*

Great Indian Bustards are extremely shy and easily disturbed, especially on their breeding grounds, and the scarcity of secure nesting areas is a significant threat to the species. People, cattle or other ungulates such as blackbuck are often abundant in its nesting habitat, posing a serious risk of disturbance and trampling of nests. Moreover, irrigation projects are set to increase the already considerable human use and disturbance of once-pristine areas such as Desert National Park. The spread of roads and the improvement of all-terrain vehicles has allowed access to even the remotest portions of the Thar desert, leading to increased disturbance and hunting pressure. 'Safe havens' are needed on bustard nesting grounds inside protected areas, where disturbance is minimised during the breeding season, with patrolling by park staff to control human activities. Measures to limit disturbance where bustards nest outside protected areas could be agreed with farmers and landowners, possibly involving payment of compensation if this affects their economic activities. The use of all-terrain vehicles should be restricted in grassland and desert areas, especially inside protected areas and during the breeding season.

■ *PESTICIDES*

The use of pesticides has increased in India in recent decades, linked to irrigation schemes and intensified agriculture. This may directly harm insectivorous birds such as Great Indian Bustard through food-chain toxin accumulation, and indirectly by reducing prey abundance. The use of these chemicals should be carefully controlled, and fodder and crop development schemes that rely on pesticides and fertilisers avoided. In key areas for threatened species, low-intensity traditional agricultural practices, with a low input of pesticides and other agrochemicals, should be promoted, possibly involving compensation schemes.

Protected areas coverage and management

■ *GAPS IN PROTECTED AREAS SYSTEM*

There are many protected areas which contain suitable habitat for Great Indian Bustard and other threatened species, including some very large ones such as Desert National Park in Rajasthan. However, the bustards roam over large areas and make seasonal and nomadic movements, and the only practical plan for their long-term protection is to maximise the area (and the effectiveness of management) of suitable habitat within protected areas, whilst taking measures to improve conditions outside protected areas (see above). Wherever possible, new protected areas are needed to protect important populations of Great Indian Bustard and other threatened species, and some existing reserves need to be enlarged. For example, the Naliya grasslands in Gujarat, which support good populations of Great Indian Bustard and Lesser Florican, are proposed for establishment as a Community Conservation Area (a new category created by a 2002 amendment to the Indian Wildlife Protection Act).

■ *WEAKNESSES IN RESERVE DESIGN AND MANAGEMENT*

Given the pressures on natural habitats, management of protected areas must maximise the value to Great Indian Bustard and the other threatened species. However, in the past there has been inappropriate habitat management in several reserves, for example in Karera Bustard Sanctuary in Madhya Pradesh, where grazing animals were removed from a plot of grassland, which consequently became overgrown and entirely unsuitable for bustards, while protection within a larger area caused the number of blackbuck to increase, which trampled bustard nests and caused considerable damage to villagers' crops. In the huge Desert National Park, the main hope for the fauna of arid India, park authorities have only two vehicles, and inadequate management has allowed hunting and disturbance to take place in core areas.

Improvements in protected area management are therefore needed. Grazing and burning regimes must be carefully managed in protected areas (and ideally also in adjacent lands: see above) to develop ideal bustard habitat, which will require close cooperation with local communities; in some reserves, it may be necessary to control blackbuck populations (through culling or, preferably, translocation). In general, the habitat for Great Indian Bustard can best be protected through the use of large buffer zones where only traditional agriculture and moderate grazing is permitted, surrounding much smaller core areas which are protected from all interference in the breeding season; the boundaries of some reserves, e.g. Desert National Park, may need to be redemarcated and the core areas redeveloped to achieve this zonation. Improved funding is required to manage reserves, increase their staffing levels, and provide better equipment and training. The enforcement capacity of governmental departments responsible for environment and forestry needs to be improved through increased resources and better training.

Exploitation of wildlife

■ *HUNTING*

Hunting of Great Indian Bustards has been a problem for decades, if not centuries, and may also affect wintering Sociable Lapwing and Pale-backed Pigeon. The impact of hunting on the bustard increased greatly during the twentieth century as firearms and (particularly) jeeps became widely available, causing a major decline. Poaching continues in many areas, involving city-dwelling 'sport-hunters', military and police personnel, and local pastoralists; until recently this threat had not affected the species's last stronghold in the interior of the Thar desert, but recent improvements in all-terrain vehicles have made all these areas accessible. The low reproductive rates (only a single egg is laid per clutch) and high longevity of bustards mean that relatively small increases in adult mortality rates lead to significant population declines. The problem of hunting needs urgent attention, and should be tackled through improved enforcement of existing legislation. An anti-poaching task force has been called for to counter persistent local hunters and city-based poachers. Protected areas need to be regularly patrolled to detect hunters, by well-equipped and trained staff, and gun ownership should be more strictly controlled.

Gaps in knowledge

■ *INADEQUATE DATA ON THREATENED BIRDS*

The movements of Great Indian Bustards are poorly understood and require further study, possibly including through satellite-tracking, a project that could potentially be used to raise awareness of the species. Recent studies have improved understanding of White-browed Bushchat in parts of India, but similar work is needed elsewhere, particularly in areas of potential habitat in Pakistan. The distribution and numbers of wintering Sociable Lapwing and (especially) Pale-backed Pigeon, should be investigated to identify key areas for protection.

■ *LACK OF AWARENESS*

There is general ignorance of the plight of Great Indian Bustard in India, and most local people are unaware that it is protected. Given the severe and complex problems facing

White-browed Bushchat has been surveyed in India, but it needs to be looked for in areas of potential habitat in Pakistan.

PHOTO: OTTO PFISTER

this species, Bengal Florican (see G02) and Lesser Florican (see G03C below), an integrated national 'Project Bustard' has been proposed, along the lines of Project Tiger and Project Elephant. This would aim to establish more bustard sanctuaries, to upgrade existing closed areas, to coordinate the management of sanctuaries, to undertake research on the species and their habitats, and to integrate grassland conservation with national grazing policy. Semi-arid habitats are generally considered a low conservation priority in India, and 'Project Bustard' could raise the profile and effectiveness of national grassland conservation, and would benefit a whole suite of other threatened and endemic grassland animals.

G03C: TALL GRASSLANDS

CURRENT STATUS OF HABITATS AND THREATENED SPECIES

Grasslands were once extensive in the main breeding and wintering ranges of Lesser Florican in Gujarat, Maharashtra, Andhra Pradesh and other areas of the Deccan plateau southward to Tamil Nadu, but they have been degraded and fragmented over recent decades. Huge areas have been claimed for agriculture to meet the demands of a growing human population, and the increasing number of cattle has led to widespread overgrazing. Rainfall is crucial to the breeding success of Lesser Florican: in good monsoon years grass regrows strongly after grazing, but in drought years it does not, resulting in habitat inadequate for breeding and dramatic dips in population size. The species appears capable of rapid recovery after drought as long as sufficient habitat is available to support a large enough overall population.

CONSERVATION ISSUES AND STRATEGIC SOLUTIONS (summarised in Table 3c)

Habitat loss and degradation

■ *CONVERSION TO AGRICULTURE*

Tall grassland, the habitat favoured by Lesser Florican, is a highly threatened biome in much of India. In Gujarat, one of the main strongholds of the species, it continues to be encroached illegally by immigrants, who are ploughing up tracts of land and causing the loss of habitat for floricans and local herdsmen. Further conversion to agriculture of traditional fodder-producing grasslands (*vidis*), the main hope for breeding floricans, needs to be minimised, through improved protection and management as outlined below.

Velavadar National Park in Gujarat supports an important breeding population of Lesser Florican.

PHOTO: ASAD RAHMANI

■ *GRASSLAND MANAGEMENT*

The area of habitat available to Lesser Floricans is declining, because *vidis* are diminishing in number and extent, and because many surviving grasslands are poorly managed. The perpetuation of traditional grassland management and the introduction of rotational grazing under controlled conditions would be of great benefit to both floricans and people, as *vidis* are the most economically and ecologically viable land use in this part of India.

■ *LIVESTOCK GRAZING*

Vast numbers of livestock, chiefly cattle, roam Indian grasslands, causing disturbance, habitat degradation and nest loss in ground breeding birds. While too much grazing or burning creates denuded plains, the removal of livestock or burning regimes allows dense scrubland to develop. Controlled or rotational grazing regimes are required, especially in core areas of reserves and other florican breeding areas. The number of grazing animals should be managed, involving the promotion of cattle camps (to reduce indiscriminate grazing pressure) and the concept of fewer but better-quality livestock.

■ *DISTURBANCE*

Lesser Floricans are shy and easily disturbed, especially on their breeding grounds. People, cattle or other ungulates such as blackbuck are often abundant in their nesting habitat, posing a serious risk of disturbance and trampling of nests. Disturbance should be minimised on florican nesting grounds during the breeding season by the exclusion of livestock and control of human access and activities; in protected areas this could be achieved through patrolling by park staff, and outside protected areas through management agreements with farmers and landowners.

■ *PESTICIDES*

The use of pesticides has increased in India in recent decades, which may harm insectivorous birds such as Lesser Florican directly through ingestion with prey, and indirectly by reduced food availability. The use of these chemicals should be more carefully controlled, and fodder and crop development schemes that rely on pesticides and fertilisers avoided.

■ *INVASIVE SPECIES*

The introduced and invasive mesquite tree *Prosopis* poses a serious threat to florican habitat, devaluing grasslands for both birds and fodder production. The Indian Forest Department has recently begun eradication programmes in Rajasthan, and similar initiatives need to be developed in other states.

Protected areas coverage and management

■ *GAPS IN PROTECTED AREAS SYSTEM*

The distribution of Lesser Florican varies annually with rainfall, with birds migrating to the best-watered grassland areas. It is therefore necessary to protect and manage many grasslands within its large potential breeding range in western India, through a combination of secure protected areas and sympathetic grassland management elsewhere. New protected areas should be established (e.g. the proposed Community Conservation Area at Naliya grasslands in Gujarat: see G03B) to increase protection and

Traditional fodder-producing grasslands or vidis are ideal habitat for Lesser Florican, but are diminishing in number and extent.

PHOTO: ASAD RAHMANI

Table 3c. Conservation issues and strategic solutions for the South Asian arid habitats: tall grasslands.

Conservation issues	Strategic solutions
Habitat loss and degradation	
■ CONVERSION TO AGRICULTURE ■ GRASSLAND MANAGEMENT ■ LIVESTOCK GRAZING ■ DISTURBANCE ■ PESTICIDES ■ INVASIVE SPECIES	➤ Minimise further conversion of traditional fodder-producing grasslands (*vidis*) to agriculture ➤ Promote the traditional management of *vidis* ➤ Introduce rotational grazing to maintain patches of florican habitat, and promote the concept of fewer but better-quality livestock at key grasslands ➤ Control access by livestock and people to grasslands where floricans are nesting ➤ Improve controls on the use of pesticides and other agrochemicals ➤ Develop programmes to eradicate mesquite from grasslands
Protected areas coverage and management	
■ GAPS IN PROTECTED AREAS SYSTEM ■ WEAKNESSES IN RESERVE DESIGN AND MANAGEMENT	➤ Establish new protected areas and extend existing reserves in improve coverage of Lesser Florican habitat ➤ Redemarcate the boundaries of reserves to produce large buffer zones, and develop core areas to provide safe havens for breeding floricans ➤ Increase the resources available for protected area management and training of reserve staff, particularly at key sites
Exploitation of wildlife	
■ HUNTING	➤ Improve enforcement of hunting legislation, by developing an anti-poaching task force, regular patrolling of protected areas, and controls on gun ownership
Gaps in knowledge	
■ INADEQUATE DATA ON THREATENED BIRDS ■ LACK OF AWARENESS	➤ Study the migratory movements of Lesser Florican, possibly using satellite-tracking ➤ Launch 'Project Bustard', to raise awareness of the plight of India's bustards and to coordinate conservation measures for these birds and their grassland habitats

improve management of its specialised habitat, and existing reserves enlarged to include additional areas of grassland.

■ *WEAKNESSES IN RESERVE DESIGN AND MANAGEMENT*

Many reserves where Lesser Florican occurs are inadequately protected, being poorly demarcated and advertised, with minimal infrastructure and administrative presence; hence (e.g.), large areas of grasslands in Sailanor Kharmor Sanctuary in Madhya Pradesh were converted to agriculture or leased to graziers during the 1990s. Stricter management of protected grasslands is required, with grazing and burning regimes carefully regulated. The semi-nomadic nature of the Lesser Florican, coupled with the intense pressure on land in India, means that reserves for the species can best be achieved through the development of large buffer zones, surrounding much smaller core areas which are protected from all interference in the breeding season. Improved funding is required to manage reserves, increase their staffing levels, and provide better equipment and training. The enforcement capacity of governmental departments responsible for environment and forestry needs to be improved through increased resources and better training.

Exploitation of wildlife

■ *HUNTING*

The major historical decline in numbers and range of Lesser Florican is linked to severe hunting pressure, and hunting with guns and snares is still widespread. This needs to be addressed through improved enforcement of existing hunting legislation. An anti-poaching task force has been called for to counter persistent hunters in India (see *G03B: Hunting* above). Protected areas should be regularly patrolled by well-equipped and trained staff, and gun ownership strictly controlled.

Gaps in knowledge

■ *INADEQUATE DATA ON THREATENED BIRDS*

The movements and non-breeding range of Lesser Florican are poorly understood and require study, possibly through satellite-tracking; this would help identify key non-breeding sites, where conservation action may be required.

■ *LACK OF AWARENESS*

Given the severe and complex problems facing this species, Bengal Florican (see G02) and Great Indian Bustard, an integrated national 'Project Bustard' has been proposed, along the lines of Project Tiger and Project Elephant (see *G03B: Lack of awareness* for further details).

G03D: DRY FORESTS AND SCRUB

CURRENT STATUS OF HABITATS AND THREATENED SPECIES

Dry forests and scrublands were formerly extensive in Gujarat, Rajasthan and southern India, but have been widely cleared and degraded by a combination of localised clearance for agriculture, villages and other developments, and overgrazing and unsustainable exploitation of forest products. The remaining habitat is now severely fragmented in many areas, and is under continuing pressure from development and exploitation.

CONSERVATION ISSUES AND STRATEGIC SOLUTIONS (summarised in Table 3d)

Forest loss and degradation

■ *CONVERSION TO AGRICULTURE*

Given the semi-aridity and often rocky nature of this habitat, clearance for agriculture is only a localised threat, but has caused significant losses. For example, the Shevaroy hills in Tamil Nadu are almost entirely covered with coffee plantations, confining the Yellow-throated Bulbul to escarpments, and some former sites for White-naped Tit in Gujarat have been cleared for agriculture. Further habitat conversion at key sites for threatened birds needs to be controlled, involving the establishment of a network of dry forest reserves.

■ *EXPLOITATION OF FOREST PRODUCTS*

Throughout this region, especially near settlements, gathering of fuelwood and foliage (for fodder) is heavily degrading or even clearing dry forests and scrub. In Gujarat, wood is collected for illegal charcoal-making and bakeries, removing old trees used by nesting White-naped Tits, and *Acacia* twigs are collected in large numbers to make disposable toothbrushes. In Andhra Pradesh, the habitat of Jerdon's Courser may be threatened by collection of fuelwood, timber and thatch by local villagers, although moderate levels of wood-cutting and livestock grazing may benefit the species by maintaining the open nature of the forest. At key sites for threatened species, social forestry initiatives should be developed, including village forests with communal management, and the regeneration of dry forest on wastelands. Measures are also required to alleviate local demands on forest resources, including the introduction of fuel-efficient stoves to reduce wood-fuel consumption.

■ *LIVESTOCK GRAZING*

Browsing of vegetation by cattle, goats and other livestock is causing localised degradation of dry forests and scrub. At some Yellow-throated Bulbul sites, the hills have been almost totally denuded by intense browsing, and livestock damage the shrubs which provide the bulbul's food. Grazing needs to be more effectively managed, especially in protected areas.

■ *DEVELOPMENT (URBAN, INDUSTRIAL, ETC.)*

Some areas of dry forest are being cleared or fragmented by the expansion of villages and towns, or by quarrying for granite and gypsum. The Somasilla dam caused 57 villages to be relocated within the range of Jerdon's Courser in Andhra Pradesh, and the settlers in the Lankamalai area may pose a serious threat to the species's habitat. Suitable habitat (in excellent condition) for White-winged Tit at the Narayan Sarovar Chinkara Sanctuary has been threatened

Jerdon's Courser habitat in Sri Lankamalleswara Wildlife Sanctuary.

PHOTO: ASAD RAHMANI

by proposed cement factories. Large development projects that could damage important areas of dry forest should be modified (possibly through environmental impact assessments) to minimise loss of habitat, particularly inside protected areas. The biodiversity and economic value of dry forests and scrubland needs to be brought to the attention of government and civil society in India, to reduce the damage to these habitats from development and exploitation.

■ *INVASIVE SPECIES*

The introduced mesquite tree *Prosopis* is thought to be having a serious effect on dry habitats in India. Programmes are needed to eradicate it where it is reducing the quality of dry forest and scrub habitats at important sites for threatened birds.

Protected areas coverage and management

■ *GAPS IN PROTECTED AREAS SYSTEM*

Three wildlife sanctuaries have been established in and around the single known site for Jerdon's Courser. However, Yellow-throated Bulbul and White-winged Tit are both known from very few protected areas, reflecting the low priority that has been given to the conservation of dry forest and scrub. There is an urgent need to establish a network of new sanctuaries (or expanded existing ones) to protect healthy populations of these two species, following surveys to locate their best habitats. These conservation efforts could include sites where habitats exist in good condition owing to religious practices (e.g. temple forests in Pali districts in Rajasthan).

■ *WEAKNESSES IN RESERVE MANAGEMENT*

Stronger protection is likely to be required in many dry forest reserves, because of inadequate infrastructure and administrative presence. Improved funding is required to manage reserves, increase their staffing levels, and provide better equipment and training. The enforcement capacity of governmental departments responsible for environment and forestry needs to be improved through increased resources and better training. The purpose and regulations of protected areas need to be publicised in adjacent communities.

Efforts are underway to locate further sites for Jerdon's Courser, including through analysis of satellite images and field surveys.

PHOTO: SIMON COOK

Gaps in knowledge

■ *INADEQUATE DATA ON THREATENED BIRDS*

Efforts are currently underway to locate further sites for Jerdon's Courser, including through surveys using knowledge of calls and searches for footprints, together with analysis of satellite images to identify potential habitat. These should be continued, together with studies to investigate the courser's habitat requirements, and to clarify how grazing and wood-cutting can be managed to maintain optimum habitat.

Table 3d. Conservation issues and strategic solutions for the South Asian arid habitats: dry forests and scrub.

Conservation issues	Strategic solutions
Habitat loss and degradation	
■ CONVERSION TO AGRICULTURE ■ EXPLOITATION OF FOREST PRODUCTS ■ LIVESTOCK GRAZING ■ DEVELOPMENT (URBAN, INDUSTRIAL, ETC.) ■ INVASIVE SPECIES	➤ Control agricultural expansion at key sites for threatened birds ➤ Develop social forestry initiatives to improve management of dry forests, and promote the regeneration of dry forest on wastelands ➤ Introduce fuel-efficient stoves to reduce wood consumption ➤ Improve the management of grazing, especially inside protected areas ➤ Modify planned development projects to avoid large-scale clearance of dry habitats ➤ Conduct campaigns to raise awareness of the biodiversity and economic value of dry forest and scrub ➤ Initiate programmes to control mesquite
Protected areas coverage and management	
■ GAPS IN PROTECTED AREAS SYSTEM ■ WEAKNESSES IN RESERVE DESIGN AND MANAGEMENT	➤ Develop a network of protected areas for Yellow-throated Bulbul and White-winged Tit, and their dry forest and scrub habitats ➤ Increase the resources available for protected area management and training of reserve staff, particularly at key sites for threatened species
Gaps in knowledge	
■ INADEQUATE DATA ON THREATENED BIRDS	➤ Continue studies of Jerdon's Courser, to locate additional key sites and to improve understanding of its habitat and management requirements ➤ Survey Yellow-throated Bulbul and White-naped Tit, to help clarify the network of sanctuaries required for their protection

G03

Recent surveys of Yellow-throated Bulbul and White-naped Tit have provided a clearer picture of their distribution and abundance, and should be continued to help design the network of sanctuaries required for their protection.

G03E: GRASSLAND AND CULTIVATION

CURRENT STATUS OF HABITATS AND THREATENED SPECIES

The natural habitats of Green Avadavat in central India have been (and continue to be) greatly modified, mainly through conversion for (and intensification of) agriculture. The effects of these changes on the species are unclear, but by far the most important threat that it faces is large-scale trapping for the wild bird trade.

CONSERVATION ISSUES AND STRATEGIC SOLUTIONS (summarised in Table 3e)

Habitat loss and degradation

■ *AGRICULTURAL INTENSIFICATION*

Green Avadavat occurs in a broad range of regenerating and open habitats, and therefore may not be greatly threatened by the continuing conversion of grasslands (that are presumably its natural habitat). However, in some areas cultivation is so intensive (i.e. mustard fields in parts of Rajasthan) that populations of most passerines have declined significantly, probably including Green Avadavat. It eats seeds, and may also be affected by increasing pesticide use, although there is no direct evidence for this. Traditional low-intensity farming practices should be promoted at key sites for the species, especially in and around protected areas.

Several thousand Green Avadavats are traded annually within India, despite a national ban.

PHOTO: GERHARD HOFMANN

Protected areas coverage and management

■ *WEAKNESSES IN RESERVE MANAGEMENT*

Green Avadavat has been recorded in several protected areas, but it appears to be semi-nomadic, and conservation measures within these reserves alone are therefore unlikely to be adequate to ensure its continued survival. However, the protected areas within its range should be managed to provide suitable low scrub and rough grassland, and bird trapping must be prevented, with reserve staff trained and equipped to improve patrolling and law enforcement.

Exploitation of birds

■ *WILD BIRD TRADE*

Green Avadavat is a popular cagebird and is captured in substantial numbers, usually using nets or funnel traps. Several thousand birds were estimated to be traded within India each year during the 1990s, despite a national ban on trapping and trade, and the species still appears in international trade despite being on CITES Appendix II. Strict enforcement of existing laws is therefore needed, including the policing of bird markets, with improved training for enforcers. However, traditional bird-trappers are often severely impoverished, and a rehabilitation programme is perhaps required, including wherever possible the provision of alternative livelihoods. Awareness campaigns are needed to publicise the plight of Green Avadavat and dissuade aviculturists from purchasing it, and governments should be lobbied to strengthen the legislation protecting it from national and international trade.

Gaps in knowledge

■ *INADEQUATE DATA ON THREATENED BIRDS*

The conservation status of Green Avadavat is poorly known, and surveys are required to clarify which are the key sites for its protection, with studies to improve understanding of its habitat requirements, movements and the impact of trade. Monitoring of numbers in trade needs to be continued and improved, to help guide efforts to reduce trapping levels. Socio-economic research is needed into the wild bird trade, to help identify potential alternative livelihoods for trappers.

Table 3e. Conservation issues and strategic solutions for the South Asian arid habitats: grassland and cultivation.

Conservation issues	Strategic solutions
Habitat loss and degradation	
■ *AGRICULTURAL INTENSIFICATION*	➤ Promote traditional low-intensity farming practices at key sites for Green Avadavat
Protected areas coverage and management	
■ *WEAKNESSES IN RESERVE MANAGEMENT*	➤ Maximise the availability of Green Avadavat habitat inside key protected areas ➤ Increase resources for protected area management and staff training, particularly to counter the wild bird trade
Exploitation of wildlife	
■ *WILD BIRD TRADE*	➤ Improve enforcement of laws banning trapping and trade of Green Avadavats, with policing of bird markets and improved training for law enforcers ➤ Develop an awareness campaign to dissuade aviculturists from purchasing Green Avadavats
Gaps in knowledge	
■ *INADEQUATE DATA ON THREATENED BIRDS*	➤ Survey Green Avadavat, to locate key sites for its protection and to improve understanding of the impact of trade on its wild populations ➤ Use socio-economic research of the wild bird trade to help identify potential alternative livelihoods for trappers

ARCTIC TUNDRA

THE Arctic tundra is immensely rich in breeding waterbirds, in particular shorebirds. The Asian section extends across northern Russia from Taymyr to Chukotka and Koryakia, and supports breeding populations of four threatened waterbird species. The breeding ranges of two of these extend westwards outside the Asian region, Lesser White-fronted Goose to northern Europe, and Siberian Crane into western Siberia (although only tiny numbers of the crane breed outside the Asian region). Two species have highly specialised habitat requirements and localised distributions, Siberian Crane, which breeds in lacustrine depressions in north-east Yakutia, and Spoon-billed Sandpiper, which nests only coastal tundra on the Chukotka Peninsula.

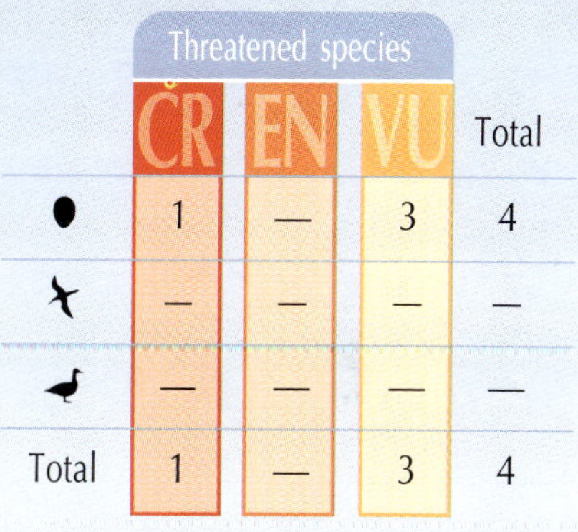

Threatened species

	CR	EN	VU	Total
●	1	—	3	4
(passage migrant)	—	—	—	—
(non-breeding visitor)	—	—	—	—
Total	1	—	3	4

Key: ● = breeding in this wetland region.
= passage migrant.
= non-breeding visitor.

- **Key habitats** Tundra wetlands.
- **Countries and territories** **Russia** (Taymyr, Krasnoyarsk, Yakutia, Chukotka, Koryakia).

Much of this region is remote and relatively undisturbed, and large areas of tundra are still pristine. PHOTO: AXEL BRÄUNLICH

Table 1. Outstanding Important Bird Areas in the Arctic tundra.

	IBA name	Status	Territory	Threatened species
1	**Khroma-Indigirka tundra**	(PA) AP	Yakutia	Large area of tundra with many nesting Lesser White-fronted Goose, Baikal Teal and Siberian Crane (most of the eastern population)
2	**Vankarem lowlands and Kolyuchin bay**	—	Chukotka	Important breeding site for Spoon-billed Sandpiper
3	**Lower Anadyr' plains**	—	Chukotka	Important breeding site for Spoon-billed Sandpiper

Note that more IBAs in this region will be included in the *Important Bird Areas in Asia*, due to be published in early 2004.

Key *Status*: PA = IBA is a protected area; (PA) = IBA partially protected; — = unprotected; AP = IBA is wholly or partially an Asia-Pacific waterbird network site (see p.35).

Spoon-billed Sandpiper breeds only in a few areas of suitable coastal tundra in Chukotka and Koryakia.

PHOTO: CHRIS SCHENK

Almost the entire global population of Siberian Cranes nests in tundra wetlands in north-east Yakutia.

PHOTO: EUGENE POTAPOV

Table 2. Threatened birds of the Arctic tundra.

Species			Distribution and population
Lesser White-fronted Goose *Anser erythropus*	◯	VU	Breeds in wooded tundra in Taymyr, Krasnoyarsk, Yakutia and Chukotka
Baikal Teal *Anas formosa*	?	VU	Breeds (or formerly bred) in tundra from Taymyr to Chukotka, as well as in boreal forest (see F01)
Siberian Crane *Grus leucogeranus*	●	CR	The entire eastern population (>95% of global total) breeds in lake depressions in tundra in Yakutia
Spoon-billed Sandpiper *Eurynorhynchus pygmeus*	●	VU	Breeds on sandy ridges near lakes and marshes in coastal tundra in Chukotka and Koryakia

● = region estimated to support >90% of global breeding population, ◯ = 10–50%, ? = proportion of global breeding population unknown

OUTSTANDING IBAs FOR THREATENED BIRDS (see Table 1)

Three IBAs have been selected, which together support a high proportion of the global population of the highly threatened Siberian Crane, and important populations of the other three threatened species.

CURRENT STATUS OF HABITATS AND THREATENED SPECIES

Much of this region is remote and relatively undisturbed, and large areas of pristine habitat survive. However, human disturbance may be affecting Siberian Cranes and Spoon-billed Sandpipers at some nesting sites, lead shot appears to be poisoning Siberian Cranes and other birds in parts of Yakutia, and there is a potential threat from oil exploration and development. In the 1970s and 1980s, pollution by mining operations greatly reduced the species diversity and abundance of aquatic life-forms in the Khroma river, and similar pollution may still affect this and other rivers in the region.

CONSERVATION ISSUES AND STRATEGIC SOLUTIONS (summarised in Table 3)

Habitat loss and degradation

OIL EXPLORATION AND DEVELOPMENT

Oil has been discovered in and near the areas where Siberian Crane nests in Yakutia, and further exploration and development pose a significant threat. Tundra habitats are fragile and take a long time to (or never) recover if they are damaged: any oil development must be carefully planned and implemented to minimise its impact on the environment.

POLLUTION

Oil development in and near the breeding grounds of Siberian Crane could lead to pollution of the tundra wetlands. The breeding habitat of Spoon-billed Sandpiper is also under threat from oil pollution of the continental shelf (e.g. in the Anadyr' bay and Khatyrka areas). In the recent past, industrial pollution affected the Khroma river, and similar pollution may affect some of the river valleys where Lesser White-fronted Geese nest. Industrial projects need to be carefully managed and monitored to minimise the risk of pollutants being released into the environment.

DISTURBANCE

Although human populations are generally sparse in the tundra, some activities cause significant disturbance to nesting birds. For example, during the spring migration of wild reindeer *Rangifer tarandus* and, especially, when domesticated reindeer are being driven, Siberian Cranes are disturbed and forced to leave their nests, leaving their eggs vulnerable to predation by skuas or large gulls. Nests of Spoon-billed Sandpiper are sometimes destroyed by reindeer herds and herders' dogs or by other human activities. There is a need for awareness campaigns for local villagers and administrators, to help minimise disturbance on the nesting grounds of threatened birds during the breeding season. Reindeer herders should be encouraged to adjust their herding schedule to avoid key areas when there are eggs or young chicks in the nests.

Protected areas coverage and management

■ *GAPS IN PROTECTED AREAS SYSTEM*

There are 13 federal and regional protected areas on the breeding grounds of Siberian Crane in north-east Yakutia; particularly important are Kytalyk Resource Reserve and Chaygurgino and Khroma State Reserves, which together support about half of the global population. However, coverage of this highly threatened species should be increased by extending Kytalyk Resource Reserve in the south (to Kubalah and Alysardah lakes) and in the north-east (to Russkoye Ustie settlement and the Indigirka river), and Chaygurgino State Reserve should be extended southwards to include the Siberian Crane nest sites between the Alazeya and Chukochia rivers. Three new (lower status) reserves should be established in important breeding and passage areas, at Nizhneyansky (the eastern part of the Yana delta and adjacent areas of the Yana-Indigirka lowlands), at Kuoluma (the northern part of the Aldan-Amga) and at Chabda. The reserves established for Siberian Crane also benefit Lesser White-fronted Goose and Baikal Teal, and additional reserves proposed for the goose include its breeding grounds in the upper Ozhogina, Tyung and Ercha rivers. Several local wildlife refuges protect breeding Spoon-billed Sandpipers on the Chukotsk peninsula, but further locally designated protected areas are needed at the most important breeding sites of the species, e.g. Ukouge lagoon, Cape Rekokaurer, Khatyrka estuary and Russkaya Koshka.

Shooting in Russia may have contributed to the decline of Lesser White-fronted Goose, but hunting pressure is much more severe on its wintering grounds in China.

PHOTO: INGAR JOSTEN ØIEN/BIRDLIFE

Exploitation of birds

■ *HUNTING*

Hunting in Yakutia appears to have contributed to the decline of Lesser White-fronted Goose. Although spring hunting of the species has been banned in Yakutia since 1995, hunters are unable to distinguish it from Greater White-fronted Goose *Anser albifrons* (which it is still legal to hunt). To reduce pressure on both species, the timing of the spring hunting season should be adjusted, to open later in the season after the main goose migration.

■ *POISONING WITH LEAD SHOT*

The stomachs of two moribund Siberian Cranes in Yakutia in 1995 contained lead shot, presumably ingested as grit; the birds apparently died of lead poisoning. Subsequent studies have found lead shot in waterfowl and in deposits on the bottom of waterbodies, with the highest levels in the Kolyma lowlands. Alternatives to lead shot should be promoted, and laws enforced to reduce levels of hunting at important sites.

Gaps in knowledge

■ *INADEQUATE DATA ON THREATENED BIRDS*

Numerous ground and aerial surveys have been conducted of the eastern breeding population of Siberian Crane in Yakutia, and its population demography and the status of its breeding habitat should continue to be monitored. Studies are required at the sites most favoured by hunters (usually near the larger settlements) in north-east and central Yakutia, to evaluate the level of threat posed by lead poisoning and the measures required to counter this problem. Further surveys of Spoon-billed Sandpiper are needed on the Chukotka and Koryakia coasts to improve understanding of the species's population, and its numbers regularly monitored at the most important breeding sites.

Table 3. Conservation issues and strategic solutions for birds in the Arctic tundra.

Conservation issues	Strategic solutions
Habitat loss and degradation	
■ OIL EXPLORATION AND DEVELOPMENT ■ POLLUTION ■ DISTURBANCE	➤ Manage oil and other development projects to prevent habitat loss and pollution ➤ Conduct awareness campaigns to minimise disturbance of nesting Siberian Cranes and Spoon-billed Sandpipers
Protected areas coverage and management	
■ GAPS IN PROTECTED AREAS SYSTEM	➤ Increase protected areas coverage of nesting Siberian Cranes in north-east Yakutia ➤ Establish additional local reserves to protect nesting Spoon-billed Sandpipers
Exploitation of birds	
■ HUNTING ■ POISONING WITH LEAD SHOT	➤ Adjust the timing of the spring hunting season to open after the main goose migration ➤ Promote alternatives to lead shot and reduce hunting at important sites
Gaps in knowledge	
■ INADEQUATE DATA ON THREATENED BIRDS	➤ Continue monitoring Siberian Crane population and breeding habitat ➤ Study the impact of lead poisoning on Siberian Crane in Yakutia ➤ Survey Spoon-billed Sandpiper on the Chukotka and Koryakia coasts

SEA OF OKHOTSK and SEA OF JAPAN COASTS

W02

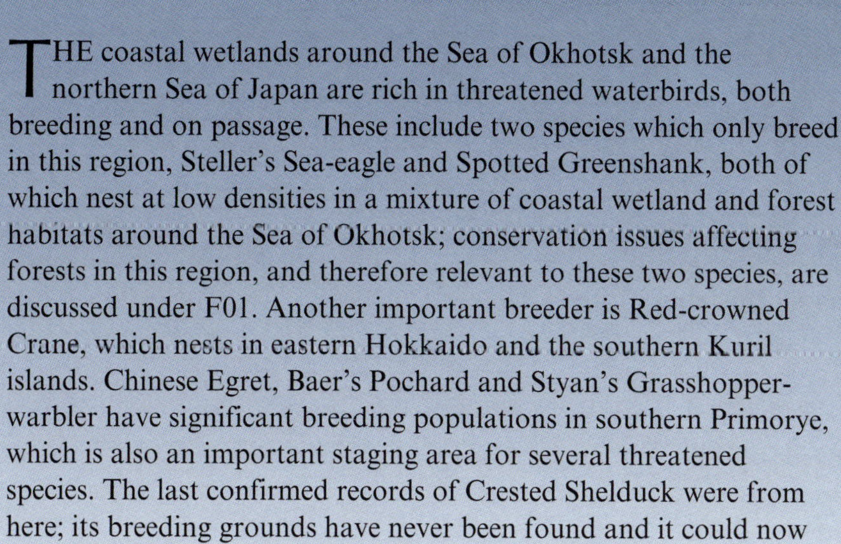

THE coastal wetlands around the Sea of Okhotsk and the northern Sea of Japan are rich in threatened waterbirds, both breeding and on passage. These include two species which only breed in this region, Steller's Sea-eagle and Spotted Greenshank, both of which nest at low densities in a mixture of coastal wetland and forest habitats around the Sea of Okhotsk; conservation issues affecting forests in this region, and therefore relevant to these two species, are discussed under F01. Another important breeder is Red-crowned Crane, which nests in eastern Hokkaido and the southern Kuril islands. Chinese Egret, Baer's Pochard and Styan's Grasshopper-warbler have significant breeding populations in southern Primorye, which is also an important staging area for several threatened species. The last confirmed records of Crested Sheldnon were from here; its breeding grounds have never been found and it could now be extinct.

- **Key habitats** Coastal wetlands, freshwater wetlands on coastal plains.
- **Countries and territories** **Russia** (Chukotka, Koryakia, Kamchatka, Magadan, Khabarovsk, Primorye, Sakhalin); **Japan** (Hokkaido); **North Korea**.

Threatened species

	CR	EN	VU	Total
Breeding	—	3	4	7
Passage	—	1	6	7
Non-breeding	1	1	—	2
Total	1	5	10	16

Key: ● = breeding in this wetland region. ✈ = passage migrant. = non-breeding visitor.

The spectacular Steller's Sea-eagle is unique to this region, nesting in a mixture of coastal wetland and forest habitats around the Sea of Okhotsk. PHOTO: BIRDLIFE

W02

Table 1. Outstanding Important Bird Areas on the Sea of Okhotsk and Sea of Japan coasts.

	IBA name	Status	Territory	Threatened species
1	**Malkachan tundra**	—	Magadan	Breeding Steller's Sea-eagle and Spotted Greenshank, important staging area for waterfowl
2	**Konstantin bay**	—	Khabarovsk	Important breeding site for Spotted Greenshank
3	**Islands in Peter the Great bay**	(PA)	Primorye	Breeding Styan's Grasshopper-warbler, Crested Shelduck was recorded in the 1960s
4	**Tumen estuary**	PA	Primorye	Breeding Baer's Pochard and Chinese Egret (on Furugelm island), important staging area for cranes and shorebirds
5	**North-east Sakhalin lowlands**	—	Sakhalin	Important breeding area for Swan Goose and Spotted Greenshank
6	**Shiretoko peninsula**[F01]	PA	Hokkaido	Important wintering area of Steller's Sea-eagle
7	**Notsuke peninsula**	PA	Hokkaido	Important breeding site for Red-crowned Crane
8	**Furen-ko lake**[F01]	PA	Hokkaido	Important breeding site for Red-crowned Crane, and wintering area of Steller's Sea-eagle
9	**Kushiro marshes**	PA [AP,R]	Hokkaido	Important breeding site for Red-crowned Crane

Note that more IBAs in this region will be included in the *Important Bird Areas in Asia*, due to be published in early 2004.

Key *Status*: PA = IBA is a protected area; (PA) = IBA partially protected; — = unprotected; AP = IBA is wholly or partially an Asia-Pacific waterbird network site (see p.35); R = IBA is wholly or partially a Ramsar Site (see pp.31–32); F01 = also supports a threatened forest bird of region F01.

OUTSTANDING IBAs FOR THREATENED BIRDS (see Table 1)

Nine IBAs have been selected, covering many important breeding and passage populations of threatened waterbirds. Steller's Sea-eagle and Spotted Greenshank breed at low densities, and require conservation at the landscape level, and only a few outstanding IBAs have been selected for these species.

CURRENT STATUS OF HABITATS AND THREATENED SPECIES

Large parts of this region are relatively undeveloped, particularly in the north, and the natural habitats are mainly intact. However, there has been extensive wetland loss on Hokkaido and in parts of south-east Russia, and increased development could lead to further habitat reduction and degradation in many areas. On Hokkaido,

The Red-crowned Cranes on Hokkaido are resident, and virtually dependent on artificial feeding sites in winter.

PHOTO: P. B. TAYLOR/BIRDLIFE

Table 2. Threatened birds of the Sea of Okhotsk and Sea of Japan coasts.

Species			Distribution and population
Chinese Egret *Egretta eulophotes*	○	VU	The only Russian breeding colony is on Furugelm island in southern Primorye
Oriental Stork *Ciconia boyciana*	passage <10%	EN	Occurs on passage in the south of this region
Swan Goose *Anser cygnoides*	○	EN	Nests in northern Sakhalin and the lower Amur in Khabarovsk, and moults on the south-west coast of the Sea of Okhotsk
Lesser White-fronted Goose *Anser erythropus*	passage <10%	VU	Widespread on passage within this region
Crested Shelduck *Tadorna cristata*	EX?	CR	Recorded the 1960s on an island in Peter the Great bay, Primorye
Baikal Teal *Anas formosa*	passage <10%	VU	Widespread on passage within this region
Baer's Pochard *Aythya baeri*	○	VU	Nests in wetlands near the coast of southern Primorye
Scaly-sided Merganser *Mergus squamatus*	? (non-breeding unknown)	EN	Small numbers winter on the coast of Primorye
Steller's Sea-eagle *Haliaeetus pelagicus*	●	VU	Nests in coastal Chukotka, Koryakia, Kamchatka, Magadan, Khabarovsk and Sakhalin, many winter in Japan (mainly Hokkaido)
White-naped Crane *Grus vipio*	passage 10–50%	VU	Occurs on passage in the south of this region, largest numbers in southern Primorye
Hooded Crane *Grus monacha*	passage 10–50%	VU	Occurs on passage in the south of this region
Red-crowned Crane *Grus japonensis*	◍ passage 10–50%	EN	Breeds on Hokkaido and the southern Kuril islands, and occurs on passage in southern Primorye
Swinhoe's Rail *Coturnicops exquisitus*	passage 10–50%	VU	Occurs on passage in the south of this region
Spotted Greenshank *Tringa guttifer*	●	EN	Nests in larch forest by wet coastal meadows and mudflats in Magadan, Khabarovsk, Sakhalin, and presumably Kamchatka
Spoon-billed Sandpiper *Eurynorhynchus pygmeus*	passage >90%	VU	Widespread on passage within this region
Styan's Grasshopper-warbler *Locustella pleskei*	◍	VU	Nests on several islands in Peter the Great Bay, Primorye

Other threatened waterbirds recorded from this region as rare visitors are: Black-faced Spoonbill *Platalea minor*, Siberian Crane *Grus leucogeranus*, Saunders's Gull *Larus saundersi* and Relict Gull *L. relictus*.

● = region estimated to support >90% of global breeding population, ◍ = 10–50%, ○ = <10%; ? = regional proportion of global non-breeding population unknown; passage >90% = region estimated to support >90% of global population on passage, passage 10–50% = 10–50%, passage <10% = <10%; EX? = probably extinct the wetlands favoured by nesting Red-crowned Cranes continue to be lost to development, especially agricultural expansion, river channelisation and road-building; for example, since the 1970s one-third of the 291 km² of marshland in Kushiro has been converted to agricultural, industrial or residential use. On Sakhalin, wetland habitats have been lost because of exploitation of the oil and gas reserves in the north of the island. The coastal meadows and estuaries where Spotted Greenshank nest are usually adjacent to villages, and are therefore subject to disturbance by people, livestock and dogs.

CONSERVATION ISSUES AND STRATEGIC SOLUTIONS (summarised in Table 3)

Habitat loss and degradation

■ *CONVERSION TO AGRICULTURE*

Agricultural expansion and associated river channelisation continue to reduce the breeding habitat of Red-crowned Crane on Hokkaido, and are presumably a potential threat elsewhere in the region.

■ *DEVELOPMENT (URBAN, INDUSTRIAL, ETC.)*

Hokkaido is already industrially developed, and several major coastal ports, new roads and other development projects are planned or underway in eastern Russia. The Tumen River Area Development Programme plans a large seaport and railway, and could lead to habitat loss and increased human disturbance, pollution and other pressures; this may affect breeding populations of Chinese Egret and Styan's Grasshopper-warbler in southern Primorye, as well as migratory waterbirds in the Tumen river estuary. Further north, proposed large-scale coastal developments for the petrol industry in Magadan and other provinces are potential threats to the habitats of Steller's Sea-eagle, and presumably also Spotted Greenshank, and could cause coastal pollution. These projects should be subject to environmental impact assessments, to consider their potential impacts on the environment and threatened species, and to develop plans for mitigation of any negative effects. Management plans need to be prepared for any key areas for threatened birds that could be directly affected by these projects, to ensure that all human activities are compatible with the conservation of the area.

■ *COLLISIONS WITH POWER-LINES*

On Hokkaido, a study in the 1970s found that c.70% of the Red-crowned Cranes found dead in the wild were killed by collision with power-lines, and this is still a major cause of mortality, particularly when cranes are flushed. Techniques need to be developed to minimise collisions, including relocation of power-lines away from crane roosts, improved control of disturbance (especially by tourists), and possibly by marking power-lines to make them more visible to cranes.

■ *DISTURBANCE AND INCREASED PREDATION*

As one of the largest of all raptor species, Steller's Sea-eagle is easily energetically stressed and thus intolerant of more than 5–6 flights a day caused by human disturbance, which may explain its avoidance of human settlements. On Sakhalin, the proportion of abandoned nesting sites has increased in areas where human activity has increased. Disturbance by holiday-makers is a potential threat at the Chinese Egret nesting colony in Primorye. The main predator of Spotted Greenshank nests is the Carrion Crow *Corvus corone*, and the development of settlements near to breeding areas on Sakhalin has led to an increase in the crow population. Industrial and urban development should be controlled around the breeding areas of all of these threatened species, with special measures at some sites to minimise human disturbance during the nesting season.

■ *POISONING*

In Primorye, Steller's Sea-eagle have been poisoned by fish from water polluted by industrial waste, and high levels of airborne pollutants (DDT/DDE and PCBs) in Magadan and Khabarovsk may be affecting the eagles and other

The fishing industry on Hokkaido has until recently provided an abundant supply of waste fish for wintering Steller's Sea-eagles.

PHOTO: JON HORNBUCKLE

wildlife. On Hokkaido, wintering Steller's Sea-eagles are tending to move inland because of changes in the availability of fish and scavenge on sika deer *Cervus nippon* killed by hunters, thus running a high risk of lead poisoning through ingestion of lead shot; at recent rates of poisoning, the eagle population could be halved within c.50 years. In 2001, the Ministry of the Environment of Japan banned the use of lead bullets for deer hunting, requiring hunters to replace them with copper bullets. However, lead bullets can still be used to shoot bears (as hunters claim that copper is not strong enough), making it difficult to police any illegal use of lead bullets to kill deer. This loophole must be removed, possibly by strictly controlling the use of lead shot.

■ *REDUCTION IN FOOD SUPPLY*

On the large salmon rivers near Magadan, the Steller's Sea-eagle population could soon collapse, following the overharvesting of salmon runs during the 1990s, and this is also likely to be a problem elsewhere on the coast and on the large rivers flowing into the Sea of Okhotsk. In Primorye, several fur farms where wintering eagles used to feed on waste have recently closed down, and some of the remaining fur farms have started to recycle the waste, which has deprived the eagles of one of their sources of food. The wintering population on Hokkaido relies heavily on waste fish, but its distribution there has already shifted in response to changes in fishing activities, and future changes are likely to continue to affect it. If food availability on the wintering grounds is low, the breeding success of the eagles may also be lower. Regulations restricting salmon fishing to certain sections of the coast and rivers flowing into the Sea of Okhotsk need to be more strictly enforced. The industrial-scale collection of fish eggs from the spawning grounds to supply salmon hatcheries also needs to be controlled, including through the establishment of new reserves (see below). Given the dependence of some wintering populations of Steller's Sea-eagle on artificial food supplies, supplementary feeding may need to be temporarily provided when these are disrupted, although in the longer term efforts should be made to increase their natural and semi-natural food supplies.

■ OVER-CONCENTRATION CAUSED BY SUPPLEMENTARY FEEDING

The population of Red-crowned Cranes on Hokkaido is resident, despite the fact that only a few wetlands on the island (those kept free of ice by hot springs) are available for them to forage in winter. The cranes are therefore virtually dependent on c.20 artificial feeding sites, and the concentration of wintering birds at these sites might put them at risk from the outbreak of disease or some other catastrophe. In the short term, this could be addressed by providing limited supplies of food at more scattered feeding stations, but in the long term a strategy should be devised to encourage the cranes to migrate southward in winter to the warmer parts of Japan.

Protected areas coverage and management

■ *GAPS IN PROTECTED AREAS SYSTEM*

Many important sites for threatened waterbirds in this region are inside protected areas, but there are some significant gaps in coverage. The breeding grounds of Spotted Greenshank need to be protected in new sanctuaries, including at Konstantin bay in Khabarovsk, where the exploitation of natural resources should be strictly regulated and reindeer herds moved elsewhere during the nesting period. Several sanctuaries have been established in northern Sakhalin for Spotted Greenshank, but additional reserves are needed to protect the breeding and moulting habitats of Swan Goose, particularly those on the north-west coast. New reserves to protect the main salmon spawning grounds that Steller's Sea-eagle relies on for food are vital for its conservation in Russia, for example on the lower Inya river in Khabarovsk. The Tumangan Nature Park has been established on the coastal plains around the Tumen river estuary, but designated hunting areas (for wildfowling and pheasant shooting) close to this park cause disturbance, and the zonation of peripheral

Table 3. Conservation issues and strategic solutions for birds of the Sea of Okhotsk and Sea of Japan coasts.

Conservation issues	Strategic solutions
Habitat loss and degradation	
■ *CONVERSION TO AGRICULTURE* ■ *DEVELOPMENT (URBAN, INDUSTRIAL, ETC.)* ■ *COLLISIONS WITH POWER-LINES* ■ *DISTURBANCE AND INCREASED PREDATION* ■ *POISONING* ■ *REDUCTION IN FOOD SUPPLY* ■ *OVER-CONCENTRATION CAUSED BY SUPPLEMENTARY FEEDING*	➤ Assess the environmental impact of proposed development projects, and prepare management plans for any key areas affected ➤ Develop techniques to minimise collisions by Red-crowned Cranes with power-lines on Hokkaido ➤ Control development and human disturbance around the breeding grounds of threatened species ➤ Stop the use of lead bullets for deer hunting on Hokkaido, to prevent poisoning of Steller's Sea-eagles ➤ Enforce the existing regulations to manage salmon fisheries around the Sea of Okhotsk, and control salmon egg collection from spawning grounds for hatcheries ➤ Provide food to Red-crowned Cranes on Hokkaido at more widely spread feeding stations, and encourage them to migrate to southern Japan in winter
Protected areas coverage and management	
■ *GAPS IN PROTECTED AREAS SYSTEM*	➤ Establish new reserves for nesting Spotted Greenshank and on Sakhalin for Swan Goose, and to protect the salmon stocks required by Steller's Sea-eagle ➤ Improve zonation of Tumangan Nature Park, and establish this site as an international protected area ➤ Extend the boundaries of the Far Eastern Marine Reserve
Exploitation of birds	
■ *WILD BIRD TRADE*	➤ Prohibit trading of Steller's Sea-eagle
Gaps in knowledge	
■ *INADEQUATE DATA ON THREATENED BIRDS*	➤ Survey breeding Baer's Pochard, Spotted Greenshank and Styan's Grasshopper-warbler ➤ Search for Crested Shelduck ➤ Study the effects of pollutants and other potential threats on Steller's Sea-eagle

areas needs to be improved; ideally, an international protected area should be established to include all important habitats for waterbirds around the estuary, including areas in Russia, China and North Korea. Although the breeding colony of Chinese Egrets on Furugelm island is inside the Far Eastern Marine Reserve, the coastal lagoons and bays between the Tumen river mouth and Pos'yet bay, where the birds from the colony feed, are not protected; the reserve boundaries should be extended to include this stretch of coastline, as well as the islets in Peter the Great bay where Styan's Grasshopper-warbler breeds.

Steller's Sea-eagle populations could soon collapse in parts of Russia, following overharvesting of salmon runs during the 1990s.

PHOTO: EUGENE POTAPOV

Exploitation of birds

■ *HUNTING*

Some of the threatened waterbirds are hunted in this region, including Baer's Pochard and Swan Goose, but at levels that do not appear to constitute a major threat.

■ *WILD BIRD TRADE*

Several animal trading companies in eastern Russia have collected chicks and eggs from Steller's Sea-eagle nests, and continue to put pressure on the local Nature Protection Agency in Magadan to allow trade in the species; this could negatively affect the eagle population, particularly as disturbance at nests is believed to affect breeding success adversely, and trade in this protected species should not be allowed.

Gaps in knowledge

■ *INADEQUATE DATA ON THREATENED BIRDS*

There are important gaps in knowledge of the distribution and numbers of several threatened waterbirds in this region. Early summer surveys are required to locate more breeding sites of Spotted Greenshank in potentially suitable habitats to the south-west of Okhotsk and between the Ul'beya river and Cape Onatsevich. A comprehensive survey is needed of Baer's Pochard in Russia. Studies are also required of the islands in Peter the Great bay in Primorye, to locate more populations of Styan's Grasshopper-warbler, and searches are needed to try to locate the little known Crested Shelduck both on these islands and elsewhere in the region. Many studies have been conducted on Steller's Sea-eagle in Russia and Japan, but, given its potential vulnerability to many of the environmental changes taking place or planned within its range, further work is needed, for example to monitor the effects on the eagles of pollutants (DDT/DDE and PCBs) in Russia and lead shot in Japan.

AMUR, USSURI and SUNGARI RIVER BASINS

THIS region includes the floodplains of the middle and lower Amur river and the Ussuri river in south-east Russia, and the Sungari (Songhua) river in north-east China, which includes the Sanjiang (Three Rivers) plain. The extensive lowland wetlands on these floodplains support important breeding populations of several threatened waterbirds, most notably almost the entire global population of Oriental Stork, and high proportions of the global populations of Baer's Pochard and Red-crowned Crane, and probably Swinhoe's Rail. Large numbers of waterbirds occur on passage, notably Baikal Teal and Siberian and Hooded Cranes.

- **Key habitats** Freshwater wetlands on riverine plains.
- **Countries and territories** **Russia** (Khabarovsk, Amur, Jewish Autonomous Region, Primorye); **China** (Heilongjiang, Jilin, Inner Mongolia).

Threatened species

	CR	EN	VU	Total
● (breeding)	—	3	5	8
(passage migrant)	1	1	3	5
(non-breeding visitor)	—	—	—	—
Total	1	4	8	13

Key: ● = breeding in this wetland region. = passage migrant. = non-breeding visitor.

The wetlands at Zhalong National Nature Reserve in China support important populations of several threatened waterbirds, but are affected by drainage and development.
PHOTO: SIMBA CHAN

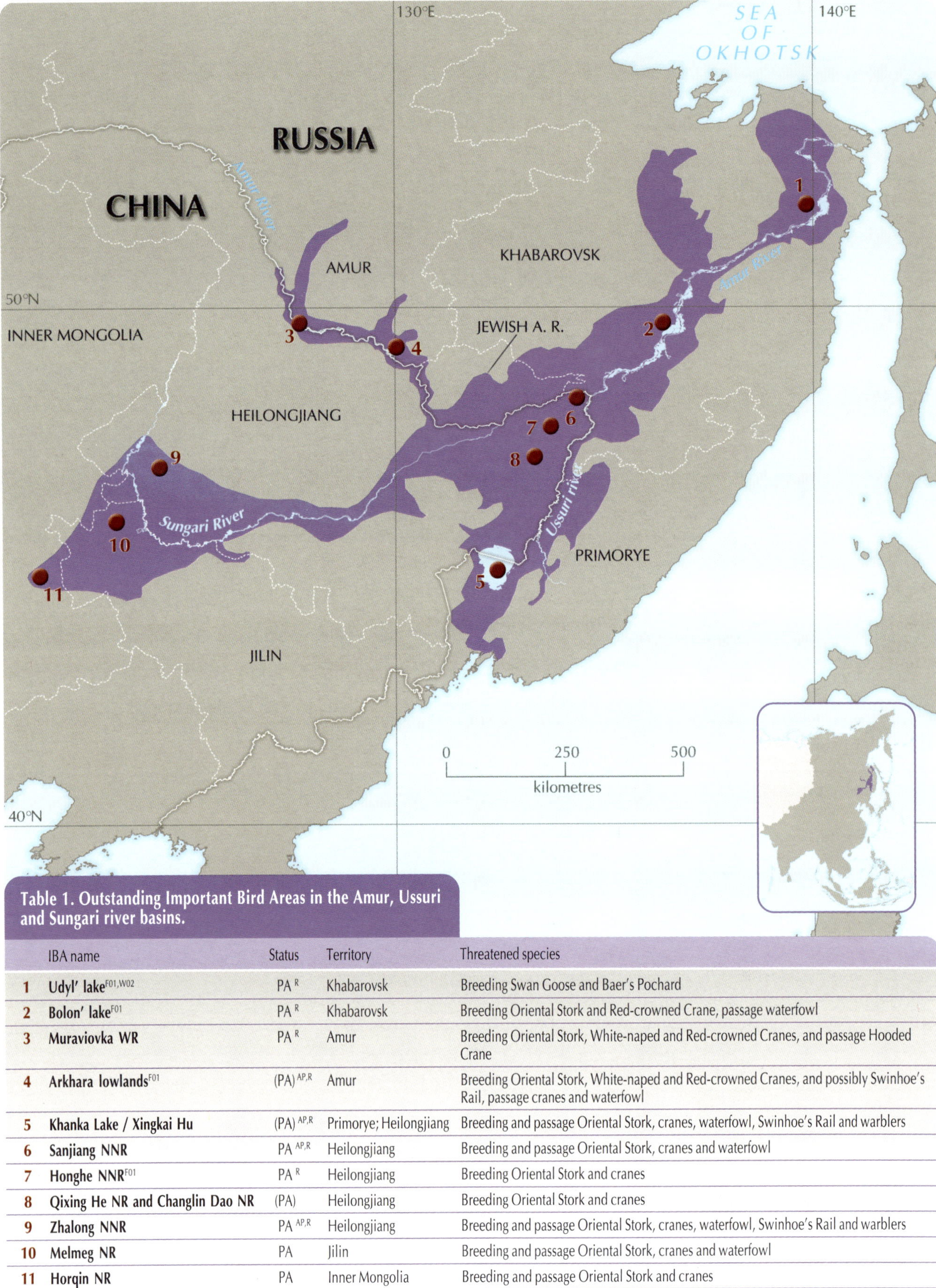

Table 1. Outstanding Important Bird Areas in the Amur, Ussuri and Sungari river basins.

	IBA name	Status	Territory	Threatened species
1	**Udyl' lake**[F01,W02]	PA [R]	Khabarovsk	Breeding Swan Goose and Baer's Pochard
2	**Bolon' lake**[F01]	PA [R]	Khabarovsk	Breeding Oriental Stork and Red-crowned Crane, passage waterfowl
3	**Muraviovka WR**	PA [R]	Amur	Breeding Oriental Stork, White-naped and Red-crowned Cranes, and passage Hooded Crane
4	**Arkhara lowlands**[F01]	(PA) [AP,R]	Amur	Breeding Oriental Stork, White-naped and Red-crowned Cranes, and possibly Swinhoe's Rail, passage cranes and waterfowl
5	**Khanka Lake / Xingkai Hu**	(PA) [AP,R]	Primorye; Heilongjiang	Breeding and passage Oriental Stork, cranes, waterfowl, Swinhoe's Rail and warblers
6	**Sanjiang NNR**	PA [AP,R]	Heilongjiang	Breeding and passage Oriental Stork, cranes and waterfowl
7	**Honghe NNR**[F01]	PA [R]	Heilongjiang	Breeding Oriental Stork and cranes
8	**Qixing He NR and Changlin Dao NR**	(PA)	Heilongjiang	Breeding Oriental Stork and cranes
9	**Zhalong NNR**	PA [AP,R]	Heilongjiang	Breeding and passage Oriental Stork, cranes, waterfowl, Swinhoe's Rail and warblers
10	**Melmeg NR**	PA	Jilin	Breeding and passage Oriental Stork, cranes and waterfowl
11	**Horqin NR**	PA	Inner Mongolia	Breeding and passage Oriental Stork and cranes

Several of the waterbirds of this region breed in the IBAs listed for F01 (Bikin river basin for Oriental Stork, White-naped and Red-crowned Cranes; Iman river basin for Oriental Stork, Baer's Pochard, White-naped and Red-crowned Cranes; and Xianghai NNR for Oriental Stork, Swan Goose, Baer's Pochard, White-naped and Red-crowned Cranes).
Note that more IBAs in this region will be included in the *Important Bird Areas in Asia*, due to be published in early 2004.

Key *IBA name*: NR = Nature Reserve; NNR = National Nature Reserve; WR = Wildlife Refuge.
Status: PA = IBA is a protected area; (PA) = IBA partially protected; — = unprotected; AP = IBA is wholly or partially an Asia-Pacific waterbird network site (see p.35); R = IBA is wholly or partially a Ramsar Site (see pp.31–32); F01 = also supports threatened forest birds of region F01; W02 = also supports threatened waterbirds of region W02.

OUTSTANDING IBAs FOR THREATENED BIRDS (see Table 1)

Eleven IBAs have been selected in this region, covering the most important breeding and passage sites of Oriental Stork, cranes and waterfowl.

CURRENT STATUS OF HABITATS AND THREATENED SPECIES

The extensive plains of this region were originally covered with vast wetlands, but these are now much reduced, fragmented and degraded as a result of conversion for agriculture, fires and other pressures. The rate of wetland conversion has accelerated in recent decades, as the human population has expanded. For example, the area of marshland on the Sanjiang plain in Heilongjiang has been reduced from 54,000 km² in the early 1950s to 14,700 km² at present (from 49% to 13% of the total area of the plain); fortunately, in 2000 the Chinese government announced that agricultural development there will come to an end. There has also been extensive deforestation, and the scarcity of mature trees suitable for nesting is a threat to Oriental Stork. Despite these problems, the region still has many globally and regionally outstanding wetlands, notably the key sites listed in Table 1.

CONSERVATION ISSUES AND STRATEGIC SOLUTIONS (summarised in Table 3)

Habitat loss and degradation

■ *WETLAND DRAINAGE*

Drainage of wetlands for agriculture and pasture continues to reduce the habitat of the region's threatened waterbirds,

Table 2. Threatened birds of the Amur, Ussuri and Sungari river basins.

Species			Distribution and population
Oriental Stork *Ciconia boyciana*	●	EN	Breeds mainly in the Amur and Ussuri basins, with smaller numbers on the Sanjiang plain
Swan Goose *Anser cygnoides*	○	EN	Breeds in the Amur basin and on the Sanjiang plain
Lesser White-fronted Goose *Anser erythropus*	△	VU	Widespread on passage within this region
Baikal Teal *Anas formosa*	◬	VU	Widespread on passage within this region, with large numbers in Primorye on spring migration
Baer's Pochard *Aythya baeri*	◉	VU	Widespread, but localised breeding species within this region
Scaly-sided Merganser *Mergus squamatus*	△?	EN	Scarce passage migrant from its nearby breeding grounds (in F01)
Siberian Crane *Grus leucogeranus*	▲	CR	The eastern population migrates through this region, with flocks staging at several sites in Amur, Heilongjiang and Jilin
White-naped Crane *Grus vipio*	◎	VU	Widespread, but localised breeding species within this region
Hooded Crane *Grus monacha*	◭	VU	Widespread on passage within this region
Red-crowned Crane *Grus japonensis*	◉	EN	Widespread, but localised breeding species within this region
Swinhoe's Rail *Coturnicops exquisitus*	◎	VU	Recorded in the breeding season in Amur, Primorye, Heilongjiang and Jilin
Manchurian Reed-warbler *Acrocephalus tangorum*	◎	VU	Breeds at Khanka lake in Primorye and at Zhalong in Heilongjiang
Marsh Grassbird *Megalurus pryeri*	◎	VU	Breeds at Zhalong in Heilongjiang and recorded in summer at Khanka lake in Primorye

● = region estimated to support >90% of global breeding population, ◉ = 50–90%, ◎ = 10–50%, ○ = <10%; ▲ = region estimated to support >90% of global population on passage, ◭ = 50–90%, ◬ = 10–50%, △ = <10%, △? = proportion of global population on passage unknown

Large areas of wetland in north-east China have been converted to farmland, but there is now a ban on agricultural development on Sanjiang plain.

PHOTO: SIMBA CHAN

Table 3. Conservation issues and strategic solutions for birds of the Amur, Ussuri and Sungari river basins.

Conservation issues	Strategic solutions
Habitat loss and degradation	
■ *WETLAND DRAINAGE* ■ *DEVELOPMENT (URBAN, INDUSTRIAL, ETC.)* ■ *AGRICULTURAL FIRES* ■ *CUTTING OF NESTING TREES* ■ *POLLUTION/PESTICIDES* ■ *DISTURBANCE* ■ *FISHERIES*	➤ Incorporate wetland protection into the regional land-use planning processes in Russia and China ➤ Assess the environmental impact of proposed development projects ➤ Promote alternatives to spring fires, to improve pastureland without damaging waterbird nesting habitat ➤ Protect Oriental Stork nesting trees, plant new trees and provide carefully-sited artificial nest-posts ➤ Enforce laws to control the use of toxic chemicals ➤ Reduce disturbance around the nest sites of threatened birds ➤ Improve management of fisheries at key wetlands
Protected areas coverage and management	
■ *GAPS IN PROTECTED AREAS SYSTEM* ■ *WEAKNESSES IN RESERVE MANAGEMENT*	➤ Expand Lake Khanka Nature Reserve and join the separate blocks into a single area ➤ Designate local sanctuaries to protect nesting storks, waterfowl and cranes ➤ Give nature reserve management offices in China more authority to control land use inside their reserves ➤ Strengthen reserve management in China through improved funding, infrastructure and staff training
Exploitation of birds	
■ *HUNTING*	➤ Ban spring hunting of all waterfowl in eastern Russia and China ➤ Improve enforcement of hunting laws in China, and strictly prohibit the use of toxic chemicals to kill birds
Gaps in knowledge	
■ *INADEQUATE DATA ON THREATENED BIRDS*	➤ Survey breeding Baer's Pochard, Swinhoe's Rail and marshland warblers ➤ Search for Crested Ibis and Crested Shelduck

although there is now a ban on agricultural development on Sanjiang plain (see above). In the Khanka lowlands and north-east China, large areas of marshland have been converted into rice paddies. Drainage of land adjacent to some key sites has affected their hydrology, for example at Zhalong, where the water inflow to the marsh has been reduced in recent years. Some nature reserves on the Sanjiang plain are also affected by water shortage, because the rivers that feed them have been dammed for irrigation projects. Further drainage should be prevented in and around areas supporting significant populations of threatened birds, through the establishment of new (and the effective management of existing) protected areas. Wetland protection needs to be incorporated into the land-use planning processes of the relevant provinces of Russia and China, with input from hydrological and ornithological experts.

■ *DEVELOPMENT (URBAN, INDUSTRIAL, ETC.)*

Infrastructural development can negatively affect wetlands. For example, the construction of National Highway 301 through Zhalong National Nature Reserve has damaged the wetlands. A proposed series of dams in the Amur river basin is likely to have a devastating impact on the wetlands through flooding and increased agricultural development. Environmental impact assessments should be conducted for development projects that could negatively affect wetlands, with the aim of minimising their negative effects and developing appropriate mitigation plans.

■ *AGRICULTURAL FIRES*

The practice of setting fire to agricultural land in spring and autumn, to clear dead vegetation and improve pastures, is widespread. Spring fires are the main problem, as they destroy the nests and young of ground-nesting birds, and also the tall vegetation used to conceal nests. The breeding success of White-naped and Red-crowned Cranes has been seriously affected in many areas, for example on the Khanka plains where 50–90% of potential crane habitat is burned annually, and fires have killed several nesting trees used by Oriental Storks. Alternative techniques need to be devised to improve pastures without destroying crane and stork nests and breeding habitat. These could include fire prevention in spring and, in the areas used for nesting by threatened birds, the cutting of long grass in late summer (after nesting has finished) rather than burning. Awareness campaigns will be required to persuade farmers to adopt these measures.

■ *CUTTING OF NESTING TREES*

Oriental Storks normally use large trees for nesting, but the clearance of suitable trees for timber and firewood has been widespread. For example, in the Khanka lowlands there are now almost no tall trees suitable for nesting and the storks

Oriental Storks normally use large trees for nesting, but many suitable trees have been cut for timber and firewood.

PHOTO: WEN-HSIN HUANG

are forced to use lower trees and man-made structures such as pylons; birds nesting on electricity pylons have sometimes been electrocuted. Artificial nest-posts have been erected and successfully used at several sites in Russia and China, although the provision of artificial posts in a build-up area in Russia was discontinued when birds perished after becoming too confiding. This issue should be addressed by improved protection of nesting trees, provision of carefully sited artificial nest-posts, and planting of elm and willow trees (which are favoured by the storks) to replace the artificial posts in the long term. Oriental Stork is a large and charismatic bird, and new nest sites could be positioned to take advantage of the species's potential to raise conservation awareness.

■ *POLLUTION/PESTICIDES*

Several forms of pollution affect the region's wetlands. Poison baits are used widely, being placed by poachers to kill ducks and geese (see *Hunting* below) and by farmers to control rodents, but they also cause high mortality amongst cranes (and can harm the people who consume poisoned birds). Run-off containing fertiliser or toxic residues leads to eutrophication of wetlands, for example at Zhalong National Nature Reserve, and reduces birds' food supply by killing aquatic organisms. Industrial pollution is also a problem in some areas. For example, in the Amur drainage water pollution by phenols is affecting the quality of fish and the survival of fish fry, and the fishing industry and salmon runs are in danger of collapsing. The laws to control the use of toxic chemicals should be more strictly enforced, with education campaigns to warn users of their adverse effects on wildlife and people.

■ *DISTURBANCE*

Disturbance by people and livestock is a problem for threatened waterbirds in many areas, even within nature reserves, and leads to desertion of nests and increased predation by crows. Measures are required to reduce disturbance of nesting birds, including through regular patrolling of nature reserves.

■ *FISHERIES*

In many wetlands, the intensity of fishing has caused stocks to decline, for example in Zhalong and Xinghai National Nature Reserves, and fishermen now catch smaller fish. This will have reduced the food supply of cranes and other threatened waterbirds, and improved management of fisheries is required for the long-term benefit of both birds and fishermen, with strict enforcement of regulations to prevent illegal fishing.

Protected areas coverage and management

■ *GAPS IN PROTECTED AREAS SYSTEM*

Many of the most important wetlands in this region are protected in nature reserves, but some significant gaps remain. Lake Khanka Nature Reserve (392 km^2, in five separate blocks) should be expanded to c.600 km^2 and the separate blocks joined into a single area, as many threatened waterbirds nest outside the current reserve boundaries. Many nesting storks, waterfowl and cranes are widely distributed at low densities, and their conservation cannot be fully addressed by the creation of one or even several large reserves; the nesting territories of some individual pairs or clusters of nests should be designated as local sanctuaries (known as 'zakazniks' in Russia), where human activities can be regulated while they are nesting.

■ *WEAKNESSES IN RESERVE MANAGEMENT*

Although many important wetlands in this region are officially protected, they are not necessarily secure because of management problems linked to inadequate budgets. In China, many reserves have to generate income for their own operating budget, and business enterprises often operate within them, even in their core areas. For example, Xingkai

Lake Khanka Nature Reserve should be expanded, and the separate blocks joined into a single area, to protect all key areas for threatened waterbirds.

PHOTO: SIMBA CHAN

Hu Nature Reserve in Heilongjiang includes two state farms and one fish farm, where the reserve management office does not have the right to control land use, and other parts of the reserve are leased to individual developers for fisheries and to a mining bureau. A general measure that would greatly benefit conservation would be to give the management offices of nature reserves more authority to control land use inside their reserves. The National Endangered Plant and Wildlife Protection and Nature Reserve Construction Program is a new Chinese government initiative to improve the existing protected area system and establish new reserves, and it provides a mechanism to address the current management problems. It has the potential to provide stable funding for reserves and to improve reserve management, by improving their infrastructure, staff training, staff working conditions and the livelihood of local communities.

Exploitation of birds

■ *HUNTING*

Shooting during the spring and autumn is a major danger to the region's threatened ducks and geese; storks and cranes are also occasionally shot. The threatened waterfowl are often difficult to distinguish from the commoner species, and it has therefore been proposed that the spring hunting of all waterfowl should be banned in the Russian Far East and China. In north-east China, the large-scale collection of eggs for food may be causing a rapid decline in waterfowl populations, inside and outside nature reserves, and improved enforcement of existing laws is required to control this illegal activity. Poison baits are used widely by poachers in China to kill ducks and geese, and these also cause high mortality amongst cranes. The use of toxic chemicals to kill birds should be strictly prohibited, with awareness campaigns to improve understanding of the relevant laws and the possibility of being harmed through the consumption of poisoned birds.

Gaps in knowledge

■ *INADEQUATE DATA ON THREATENED BIRDS*

The distributions of several threatened species are poorly known, including Baer's Pochard and Swinhoe's Rail, and surveys are required to identify key sites for their conservation. Manchurian Reed-warbler and Marsh Grassbird are known from very few sites, where their status is poorly understood, and it is possible that the undiscovered breeding grounds of Streaked Reed-warbler (see W06) are within this region; efforts to locate these skulking species could use tape-recordings and mist-netting. Crested Ibis (see W07) is considered extinct in this region, but searches for remnant populations have been proposed at former sites, and efforts should also be made to locate Crested Shelduck (see W02), including through the distribution of illustrated leaflets.

JAPANESE WETLANDS

THIS region includes the wetlands on the main southern islands of Japan, other than the coastal wetlands of Kyushu and the Nansei Shoto islands which are treated as part of the China Sea coast (W10). Many waterbirds declined substantially in Japan during the past two centuries, notably Crested Ibis, which is extinct there, Oriental Stork, which no longer breeds, and wintering Baikal Teal. However, Honshu still has an important population of Marsh Grassbird, and Izumi on Kyushu supports a remarkable wintering concentration of White-naped and Hooded Cranes.

- **Key habitats** Freshwater and coastal wetlands.
- **Countries and territories** **Japan** (Honshu, Izu islands, Shikoku, Kyushu).

Threatened species

	CR	EN	VU	Total
● (breeding)	—	—	2	2
(passage migrant)	—	2	3	5
(non-breeding visitor)	—	3	6	9
Total	—	5	11	16

Key: ● = breeding in this wetland region.
(passage migrant symbol) = passage migrant.
(non-breeding visitor symbol) = non-breeding visitor.

About 85% of the global population of Hooded Cranes winters at Izumi on Kyushu.
PHOTO: JON HORNBUCKLE

W04

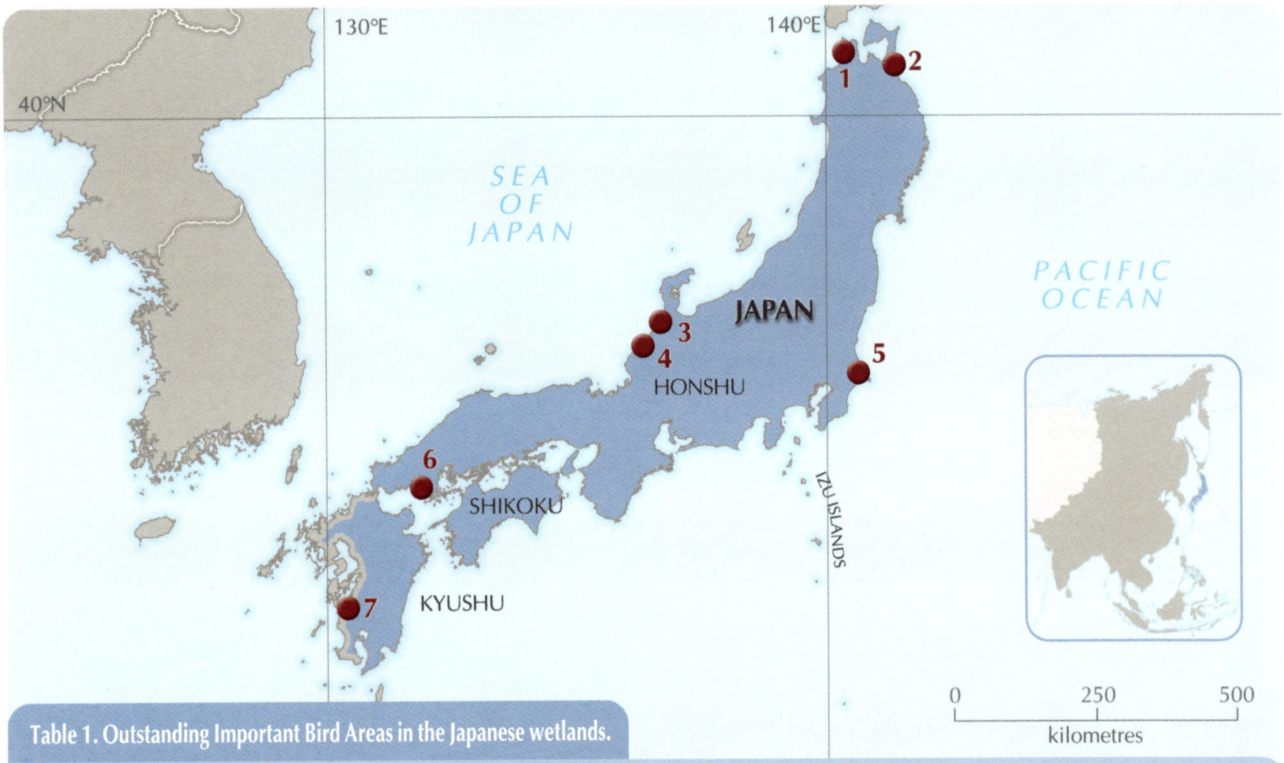

Table 1. Outstanding Important Bird Areas in the Japanese wetlands.

	IBA name	Status	Territory	Threatened species
1	**Iwaki-gawa**	—	Honshu	An important Marsh Grassbird breeding site
2	**Hotoke-numa**	(PA)	Honshu	The most important Marsh Grassbird breeding site in Japan
3	**Kahoku-gata**	—	Honshu	Several hundred Baikal Teal regularly winter
4	**Katano Duck Pond**	PA [AP,R]	Honshu	Several hundred Baikal Teal regularly winter
5	**Tone-gawa**	PA	Honshu	An important Marsh Grassbird breeding site, Swinhoe's Rail recorded in winter
6	**Yashiro**	PA [AP]	Honshu	Supports a small wintering flock of Hooded Crane
7	**Izumi**	PA [AP]	Kyushu	Holds large flocks of wintering White-naped and Hooded Cranes

Note that more IBAs in this region will be included in the *Important Bird Areas in Asia*, due to be published in early 2004.

Key *Status*: PA = IBA is a protected area; (PA) = IBA partially protected; — = unprotected; AP = IBA is wholly or partially an Asia-Pacific waterbird network site (see p.35); R = IBA is wholly or partially a Ramsar Site (see pp.31–32).

Almost half of the global population of White-naped Cranes winters at Izumi.

PHOTO: JACOB WIJPKEMA

OUTSTANDING IBAs FOR THREATENED BIRDS (see Table 1)

Seven IBAs have been selected in this region, most notably Izumi which in winter supports c.85% and c.40% respectively of the global populations of Hooded Crane and White-naped Cranes. The three most important breeding populations of Marsh Grassbird in Japan have also been covered, as well as two wintering sites for Baikal Teal.

CURRENT STATUS OF HABITATS AND THREATENED SPECIES

The lowlands of Japan are densely populated, and many wetlands have been converted for agriculture and urban and industrial development. This habitat loss combined with hunting, disturbance, pesticide use and other pressures led to the extinction of the Japanese breeding populations of Oriental Stork and Crested Ibis, and to major reductions in the numbers of some other waterbirds, including wintering Baikal Teal. Protection and supplementary feeding of White-naped and Hooded Cranes at Izumi on Kyushu has compensated for the lack of suitable wetlands elsewhere, but the abnormal concentration of the entire Japanese wintering populations (and high proportions of the global population) of these two species at a single site makes them vulnerable. Many of the most important wetlands are now well protected, and hunting of waterbirds is no longer a problem.

Marsh Grassbird nests at a handful of sites in Japan, where its habitat needs to be carefully protected and managed.

PHOTO: TAKAO BABA

CONSERVATION ISSUES AND STRATEGIC SOLUTIONS (summarised in Table 3)

Habitat loss and degradation

■ *CONVERSION TO AGRICULTURE*

The Marsh Grassbird requires successional wetlands, and in the past it was able to colonise new sites as the old ones became unsuitable, but nowadays the potential for new wetlands to develop has been greatly reduced in Japan. Several breeding sites were recently under pressure, for example Tone-gawa from a river-widening scheme and Hotoke-numa from conversion into grazing land; however, the project at Tone-gawa is unlikely to go ahead, and the Ministry of the Environment of Japan has decided to create a protected area at Hotoke-numa (although it is not yet clear whether this will cover all of the species's habitat). Hachiro-gata was recently established as a protected area, but the Marsh Grassbird habitat there continues to decline as the vegetation changes through natural succession. The protection of the few areas where this specialised bird is known to breed is critical for its conservation, with management to maintain the necessary vegetation diversity and height, especially to maintain the areas of sedge *Carex*. Increasing the area of suitable successional wetlands within its breeding range, either through management of existing wetlands or even by the creation of new wetlands, will help. Freshwater wetland habitats should also be restored in many parts of Honshu, Shikoku and Kyushu for the benefit of other threatened waterbirds, especially waterfowl and cranes.

Table 2. Threatened birds of the Japanese wetlands.

Species		Distribution and population
Chinese Egret *Egretta eulophotes*	VU	Uncommon migrant at coastal wetlands, which has attempted to nest
Oriental Stork *Ciconia boyciana*	EN	Formerly a common breeding bird, but now a rare migrant and winter visitor
Black-faced Spoonbill *Platalea minor*	EN	Rare migrant at coastal wetlands
Swan Goose *Anser cygnoides*	EN	Rare winter visitor
Lesser White-fronted Goose *Anser erythropus*	VU	Rare winter visitor
Baikal Teal *Anas formosa*	VU	Uncommon winter visitor, but significant concentrations at several localities
Baer's Pochard *Aythya baeri*	VU	Regular but rare winter visitor
Scaly-sided Merganser *Mergus squamatus*	EN	Scarce winter visitor
White-naped Crane *Grus vipio*	VU	Large numbers winter at Izumi on Kyushu, and sometime small numbers in Kochi prefecture on Shikoku
Hooded Crane *Grus monacha*	VU	Large numbers winter at Izumi on Kyushu, and small numbers at Yashiro in western Honshu
Swinhoe's Rail *Coturnicops exquisitus*	VU	Rare migrant and winter visitor
Spotted Greenshank *Tringa guttifer*	EN	Rare migrant at coastal wetlands
Spoon-billed Sandpiper *Eurynorhynchus pygmeus*	VU	Rare but regular autumn migrant, generally along the Pacific coast
Saunders's Gull *Larus saundersi*	VU	Rare migrant and winter visitor at coastal wetlands
Styan's Grasshopper-warbler *Locustella pleskei*	VU	Localised breeding species on the Izu islands and small islands off Honshu
Marsh Grassbird *Megalurus pryeri*	VU	Breeds at six sites in northern and eastern Honshu, and winters along the Pacific side of Japan from central Honshu to Shikoku

= region estimated to support 10–50% of global breeding population; = region estimated to support 50–90% of global non-breeding population, = 10–50%, = <10%, ? = proportion of global non-breeding population unknown; = region estimated to support <10% of global population on passage, ? = proportion of global population on passage unknown

Protection and feeding has greatly benefited wintering cranes at Izumi, but concentration at a single site makes them vulnerable to disease or some other catastrophe.

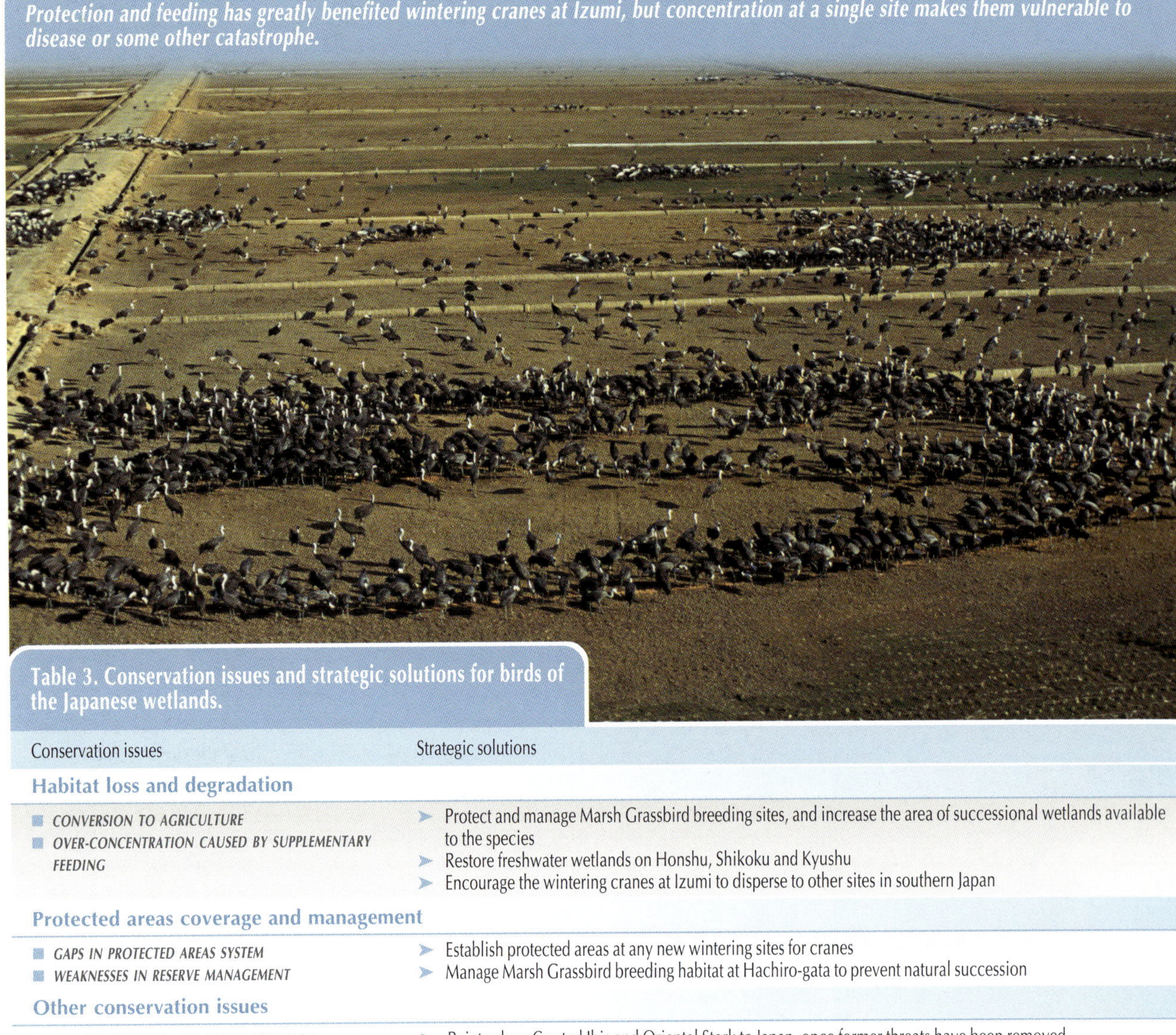

PHOTO: JACOB WIJPKEMA

Table 3. Conservation issues and strategic solutions for birds of the Japanese wetlands.

Conservation issues	Strategic solutions
Habitat loss and degradation	
■ *CONVERSION TO AGRICULTURE* ■ *OVER-CONCENTRATION CAUSED BY SUPPLEMENTARY FEEDING*	➤ Protect and manage Marsh Grassbird breeding sites, and increase the area of successional wetlands available to the species ➤ Restore freshwater wetlands on Honshu, Shikoku and Kyushu ➤ Encourage the wintering cranes at Izumi to disperse to other sites in southern Japan
Protected areas coverage and management	
■ *GAPS IN PROTECTED AREAS SYSTEM* ■ *WEAKNESSES IN RESERVE MANAGEMENT*	➤ Establish protected areas at any new wintering sites for cranes ➤ Manage Marsh Grassbird breeding habitat at Hachiro-gata to prevent natural succession
Other conservation issues	
■ *CAPTIVE BREEDING AND REINTRODUCTION*	➤ Reintroduce Crested Ibis and Oriental Stork to Japan, once former threats have been removed

■ *OVER-CONCENTRATION CAUSED BY SUPPLEMENTARY FEEDING*
Food provision at Izumi on Kyushu has produced an unnaturally large concentration of wintering White-naped and Hooded Cranes, which are at elevated risk from disease or some other catastrophe. Natural habitats and food sources no longer exist there, and the cranes depend completely on intensive management (supported by the Japanese government and local groups), with fresh water being pumped over the agricultural fields where the cranes are fed to cleanse the area. To counter the risk of disease and other threats, the artificial feeding at Izumi should be gradually reduced, and the cranes encouraged to disperse to other potential wintering sites in southern Japan.

Protected areas coverage and management

■ *GAPS IN PROTECTED AREA SYSTEM*
Should the attempts to attract White-naped and Hooded Cranes to other potential wintering sites in Japan be successful, these sites should be protected, possibly including compensation to farmers whose land is used by the cranes.

■ *WEAKNESSES IN RESERVE MANAGEMENT*
The Marsh Grassbird breeding site at Hachiro-gata has recently been established as a protected area, but the reserve needs to be managed to maintain the vegetation diversity and height required by the species.

Other conservation issues

■ *CAPTIVE BREEDING AND REINTRODUCTION*
Several captive Crested Ibises from the population in China have been sent to Japan. When the captive population in Japan has increased sufficiently, a reintroduction programme at carefully selected sites could be considered there, but only once the former threats have been removed. There are also plans to re-establish a breeding population of Oriental Storks in Japan using captive-bred birds.

STEPPE WETLANDS

THERE are many large, rich wetlands associated with the steppe grasslands that extend from eastern Europe through central and western Asia to north-east Asia. Large numbers of waterbirds breed in the east Asian part of the steppes, including several threatened species, most notably the entire world population of Relict Gull and high proportions of the global populations of Swan Goose and White-naped Crane. This region is also important for several threatened grassland birds, which are covered in G01.

- **Key habitats** Freshwater and saline wetlands.
- **Countries and territories** **Russia** (Krasnoyarsk, Khakassia, Tuva, Irkutsk, Buryatia, Chita); **Mongolia**; **China** (Inner Mongolia, Xinjiang, Gansu); outside the Asia region, the steppes extend through central and west Asia to eastern Europe.

Threatened species

	CR	EN	VU	Total
● [breeding][1]	—	3	6	9
[passage migrant]	—	—	2	2
[non-breeding visitor]	1	—	—	1
Total	1	3	8	12

Key: ● = breeding in this wetland region.
[1] The Conservation Dependent Dalmatian Pelican also breeds in this region.
[passage migrant symbol] = passage migrant.
[non-breeding visitor symbol] = non-breeding visitor.

The steppes of eastern Mongolia are one of the main breeding grounds of White-naped Crane.
PHOTO: UTE BRADTER

W05

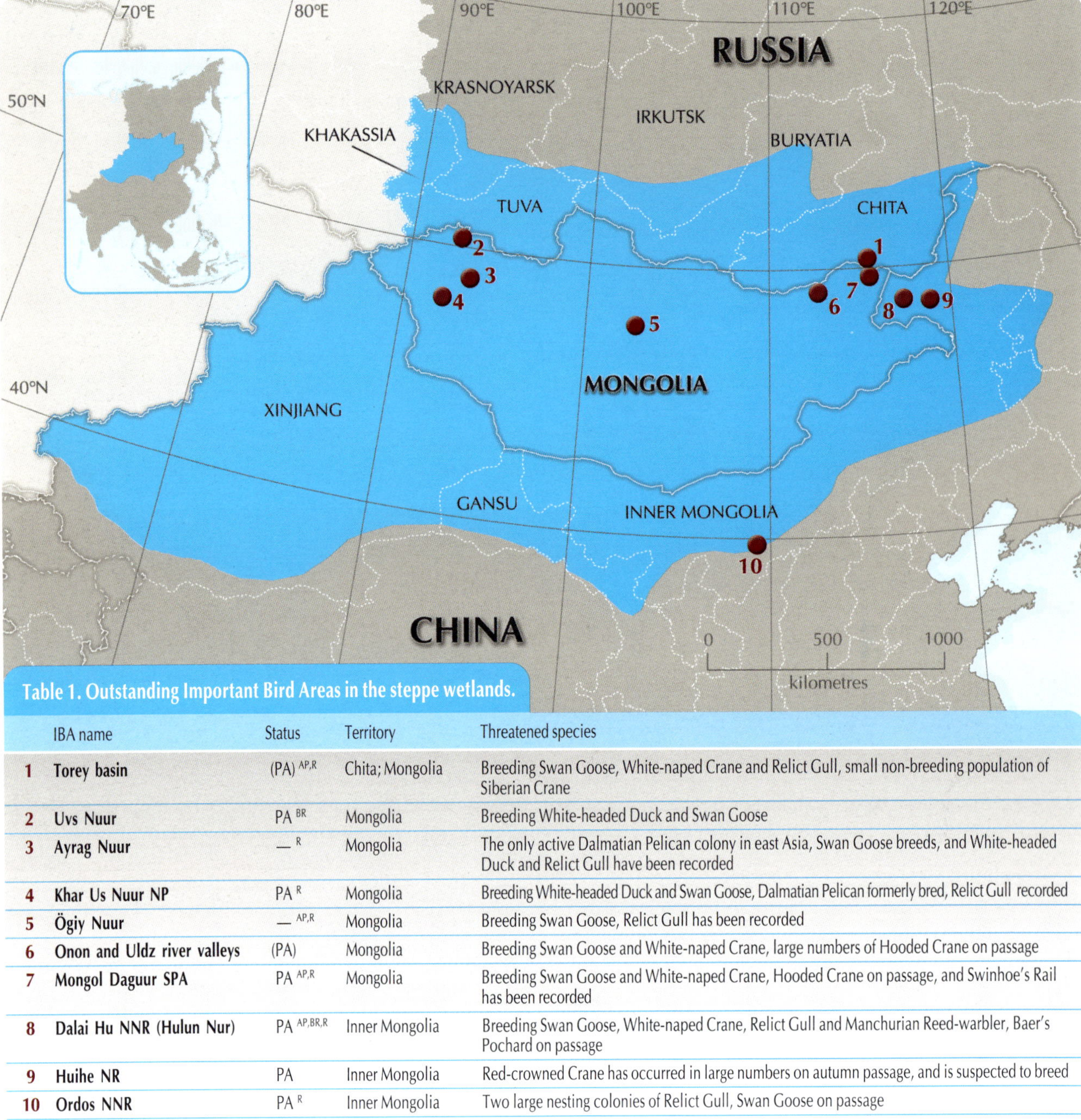

Table 1. Outstanding Important Bird Areas in the steppe wetlands.

	IBA name	Status	Territory	Threatened species
1	**Torey basin**	(PA) AP,R	Chita; Mongolia	Breeding Swan Goose, White-naped Crane and Relict Gull, small non-breeding population of Siberian Crane
2	**Uvs Nuur**	PA BR	Mongolia	Breeding White-headed Duck and Swan Goose
3	**Ayrag Nuur**	— R	Mongolia	The only active Dalmatian Pelican colony in east Asia, Swan Goose breeds, and White-headed Duck and Relict Gull have been recorded
4	**Khar Us Nuur NP**	PA R	Mongolia	Breeding White-headed Duck and Swan Goose, Dalmatian Pelican formerly bred, Relict Gull recorded
5	**Ögiy Nuur**	— AP,R	Mongolia	Breeding Swan Goose, Relict Gull has been recorded
6	**Onon and Uldz river valleys**	(PA)	Mongolia	Breeding Swan Goose and White-naped Crane, large numbers of Hooded Crane on passage
7	**Mongol Daguur SPA**	PA AP,R	Mongolia	Breeding Swan Goose and White-naped Crane, Hooded Crane on passage, and Swinhoe's Rail has been recorded
8	**Dalai Hu NNR (Hulun Nur)**	PA AP,BR,R	Inner Mongolia	Breeding Swan Goose, White-naped Crane, Relict Gull and Manchurian Reed-warbler, Baer's Pochard on passage
9	**Huihe NR**	PA	Inner Mongolia	Red-crowned Crane has occurred in large numbers on autumn passage, and is suspected to breed
10	**Ordos NNR**	PA R	Inner Mongolia	Two large nesting colonies of Relict Gull, Swan Goose on passage

Note that more IBAs in this region will be included in the *Important Bird Areas in Asia*, due to be published in early 2004.

Key *IBA name*: NP = National Park; NR = Nature Reserve; NNR = National Nature Reserve; SPA = Strictly Protected Area.
Status: PA = IBA is a protected area; (PA) = IBA partially protected; — = unprotected; AP = IBA is wholly or partially an Asia-Pacific waterbird network site (see p.35); BR = IBA is wholly or partially a Biosphere Reserve (see pp.34–35); R = IBA is wholly or partially a Ramsar Site (see pp.31–32).

OUTSTANDING IBAs FOR THREATENED BIRDS (see Table 1)

Ten large wetland IBAs with important breeding and passage populations of threatened waterbirds have been selected. Several of them also support some of the threatened grassland birds of G01.

CURRENT STATUS OF HABITATS AND THREATENED SPECIES

The steppes support a relatively low human population compared to many other parts of Asia, and the grasslands and their associated wetlands are largely intact. However, over-grazing, too frequent steppe fires, the indiscriminate use of pesticides and other pressures have reduced the quality of many wetlands. Together with hunting, this has led to declines in the populations and ranges of several threatened waterbird species.

CONSERVATION ISSUES AND STRATEGIC SOLUTIONS (summarised in Table 3)

Habitat loss and degradation

CONVERSION TO AGRICULTURE

The conversion of the remaining grasslands for agriculture is a threat in many areas, notably in Mongolia where the government has plans for large-scale agricultural development in the steppe zone. This has the potential to greatly reduce the habitat of species such as White-naped Crane (see G01 for details and recommendations).

Relict Gulls nest only on steppe lakes in eastern Asia.

PHOTO: HE FENQI

Table 2. Threatened birds of the steppe wetlands.

Species			Distribution and population
Dalmatian Pelican *Pelecanus crispus*	○	CD	Currently known to breed at a single lake in western Mongolia, having abandoned colonies elsewhere in Mongolia and in Xinjiang
White-headed Duck *Oxyura leucocephala*	○	EN	Small numbers breed at lakes in western Mongolia and adjacent parts of Russia; reported to have bred in Xinjiang
Swan Goose *Anser cygnoides*	◕	EN	A widespread breeding bird
Baikal Teal *Anas formosa*	✧	VU	A widespread passage migrant
Baer's Pochard *Aythya baeri*	○	VU	Breeds in Chita and Inner Mongolia, and possibly in eastern Mongolia
Pallas's Fish-eagle *Haliaeetus leucoryphus*	◔	VU	A widespread breeding bird
Siberian Crane *Grus leucogeranus*	✦	CR	Small numbers of non-breeding birds oversummer in Chita and eastern Mongolia
White-naped Crane *Grus vipio*	◕	VU	Breeds in Chita, north-east Mongolia and Inner Mongolia
Hooded Crane *Grus monacha*	✦	VU	A widespread passage migrant
Red-crowned Crane *Grus japonensis*	○	EN	A scarce passage and breeding species in Inner Mongolia, has bred in Chita and probably in eastern Mongolia
Swinhoe's Rail *Coturnicops exquisitus*	◔	VU	This poorly known species has bred in Chita, and occurred in eastern Mongolia
Relict Gull *Larus relictus*	●	VU	Breeds very locally on steppe lakes in Chita, Mongolia and Inner Mongolia, and just outside the Asian region in Kazakhstan
Manchurian Reed-warbler *Acrocephalus tangorum*	◔	VU	Recorded during the breeding season at Dalai Hu lake in Inner Mongolia

Other threatened waterbirds recorded from this region as rare visitors are: Oriental Stork *Ciconia boyciana* and Marsh Grassbird *Megalurus pryeri*.

● = region estimated to support >90% of global breeding population, ◕ = 50–90%, ◔ = 10–50%, ○ = <10%; ✦ = region estimated to support 10–50% of global population on passage, ✧ = <10%

■ *DEVELOPMENT (URBAN, INDUSTRIAL, ETC.)*

Steppe habitats are being affected by development in parts of eastern Russia and northern China, and the government of Mongolia is considering a number of industrial and infrastructural projects in the steppe zone (see G01 for details and recommendations).

■ *STEPPE FIRES*

Steppe fires, which are usually set by man in spring and early summer, are an important means of maintaining grassland quality, but they can negatively affect threatened waterbirds. For example, they sometimes destroy White-naped Crane eggs and young, and presumably affect other ground-nesting species (see G01 for details and recommendations).

■ *FLUCTUATING WATER LEVELS*

On the Mongolian breeding grounds of Dalmatian Pelican, fluctuations in water level frequently reduce nesting success. Elsewhere in the species's range, the provision of floating rafts at colonies has proved effective in reducing the problem of flooding. Plans for artificial nest platforms at Ayrag Nuur should be implemented. This technique could also be used to try to attract breeding birds back to the abandoned colonies at Khar Us Nuur and elsewhere in Mongolia and China.

■ *LIVESTOCK GRAZING*

Livestock levels in many areas exceed the carrying capacity of the grasslands, and overgrazing affects the long grass and marshland vegetation which is the nesting habitat of some threatened waterbirds (see G01 for details and recommendations).

■ *PESTICIDES*

Pesticides have been used in parts of the steppes to try to control periodic outbreaks in the numbers of Brandt's voles *Microtus brandti*, and other 'pest' species, but White-naped Cranes have been reported to have been poisoned by pesticides, and it is very likely that some of the other threatened waterbirds and birds of prey have also been affected (see G01 for details and recommendations).

■ *DISTURBANCE*

Disturbance by people and domestic animals is a significant threat to nesting waterbirds at many steppe wetlands, leading to desertion of nests and increased predation. For example, disturbance by people, dogs and livestock seems to be a significant problem in Buriatia (see G01 for details and recommendations).

Swan Goose is a widespread breeding bird in the steppes, but has declined because of large-scale hunting on its wintering grounds in China.

PHOTO: FRANK TODD

Protected areas coverage and management

■ *GAPS IN PROTECTED AREAS SYSTEM*

Many of the most important steppe wetlands are protected in nature reserves, but some significant gaps remain. For example, in Russia the Daursky Nature Reserve should be enlarged to include Khotogor bay at Barun-Torey lake.

■ *WEAKNESSES IN RESERVE MANAGEMENT*

Given the plans to develop the steppes of eastern Mongolia (see *Development (urban, industrial, etc.)* above), it is important to ensure the effective protection of critical biodiversity in that part of the country by strengthening the management of the nine existing and 12 proposed protected areas, and by supporting biodiversity conservation and sustainable alternative livelihoods in the buffer zones of these reserves.

Exploitation of birds

■ *HUNTING*

Hunting of waterbirds is a problem in some parts of this region. For example, cranes are sometimes killed in Chita, foreign hunting parties have recently visited some steppe lakes in Mongolia, and Dalmatian Pelicans are reported to be hunted for their bills (which are used traditionally as sweat wipes for horses) in Mongolia (see G01 for details and recommendations).

Gaps in knowledge

■ *INADEQUATE DATA ON THREATENED BIRDS*

The ranges of most threatened waterbirds are poorly known in this region, especially in Mongolia. Surveys are required to investigate the distributions and numbers of Swan Goose, Pallas's Fish-eagle, Swinhoe's Rail and Relict Gull, to identify additional key sites. The non-breeding range of Relict Gull is poorly understood, and satellite-tracking of individuals may help locate the main passage and wintering areas (and possibly also undiscovered breeding colonies).

Table 3. Conservation issues and strategic solutions for birds of the steppe wetlands.

Conservation issues	Strategic solutions
Habitat loss and degradation	
■ CONVERSION TO AGRICULTURE	➤ Reconsider plans to convert steppe to agricultural land in Mongolia and China
■ DEVELOPMENT (URBAN, INDUSTRIAL, ETC.)	➤ Assess the environmental impact of proposed development projects in Mongolia
■ STEPPE FIRES	➤ Improve fire management, to avoid harm to nesting birds and their habitats
■ FLUCTUATING WATER LEVELS	➤ Provide artificial nest platforms for Dalmatian Pelican at Ayrag Nuur, and possibly also at former nesting colonies
■ LIVESTOCK GRAZING	➤ Develop ecological management of grazing, encouraging nomadic herders to reduce their herds
■ PESTICIDES	➤ Promote alternatives to pesticides for the control of vole outbreaks
■ DISTURBANCE	➤ Control disturbance in key breeding areas of threatened birds, and provide direct protection to nest sites if required
Protected areas coverage and management	
■ GAPS IN PROTECTED AREAS SYSTEM	➤ Enlarge Daursky Nature Reserve
■ WEAKNESSES IN RESERVE MANAGEMENT	➤ Strengthen the management of protected areas in Mongolia
Exploitation of birds	
■ HUNTING	➤ Improve enforcement of hunting laws, through training of law enforcers and conservation awareness work with hunters and the public ➤ Control gun ownership
Gaps in knowledge	
■ INADEQUATE DATA ON THREATENED BIRDS	➤ Survey threatened waterbirds to identify additional key sites ➤ Use satellite tracking to help locate the migration routes and wintering grounds of Relict Gull

YELLOW SEA COAST

W06

THE coastal wetlands around the Yellow and Bohai Seas are of immense importance for threatened waterbirds. They support the entire known global breeding populations of Black-faced Spoonbill and Saunders's Gull, and almost all breeding Chinese Egrets; the spoonbill and egret nest on small islands, mostly off western Korea, and the gull in coastal saltmarshes. High proportions of the populations of Swan Goose and Red-crowned Cranes winter on the coast of Jiangsu in China, and most of the world's Baikal Teal winter in South Korea. Large numbers of threatened waterbirds move through on passage, notably Spotted Greenshank and Spoon-billed Sandpiper, for which the region's intertidal mudflats provide vital feeding habitat.

- **Key habitats** Coastal wetlands, freshwater wetlands near the coast.
- **Countries and territories** **North Korea**; **South Korea**; **China** (Liaoning, Hebei, Tianjin, Shandong, Jiangsu, Shanghai).

Threatened species

	CR	EN	VU	Total
●	—	1	3	4
✈	1	2	7	10
🦆[1]	—	3	4	7
Total	1	6	14	21

Key: ● = breeding in this wetland region.
✈ = passage migrant.
🦆 = non-breeding visitor.
[1] The Conservation Dependent Dalmatian Pelican is also a non-breeding visitor to this region.

Islets off the west coast of Korea support almost the entire global breeding populations of Black-faced Spoonbill and Chinese Egret. PHOTO: HAIXIANG ZHOU

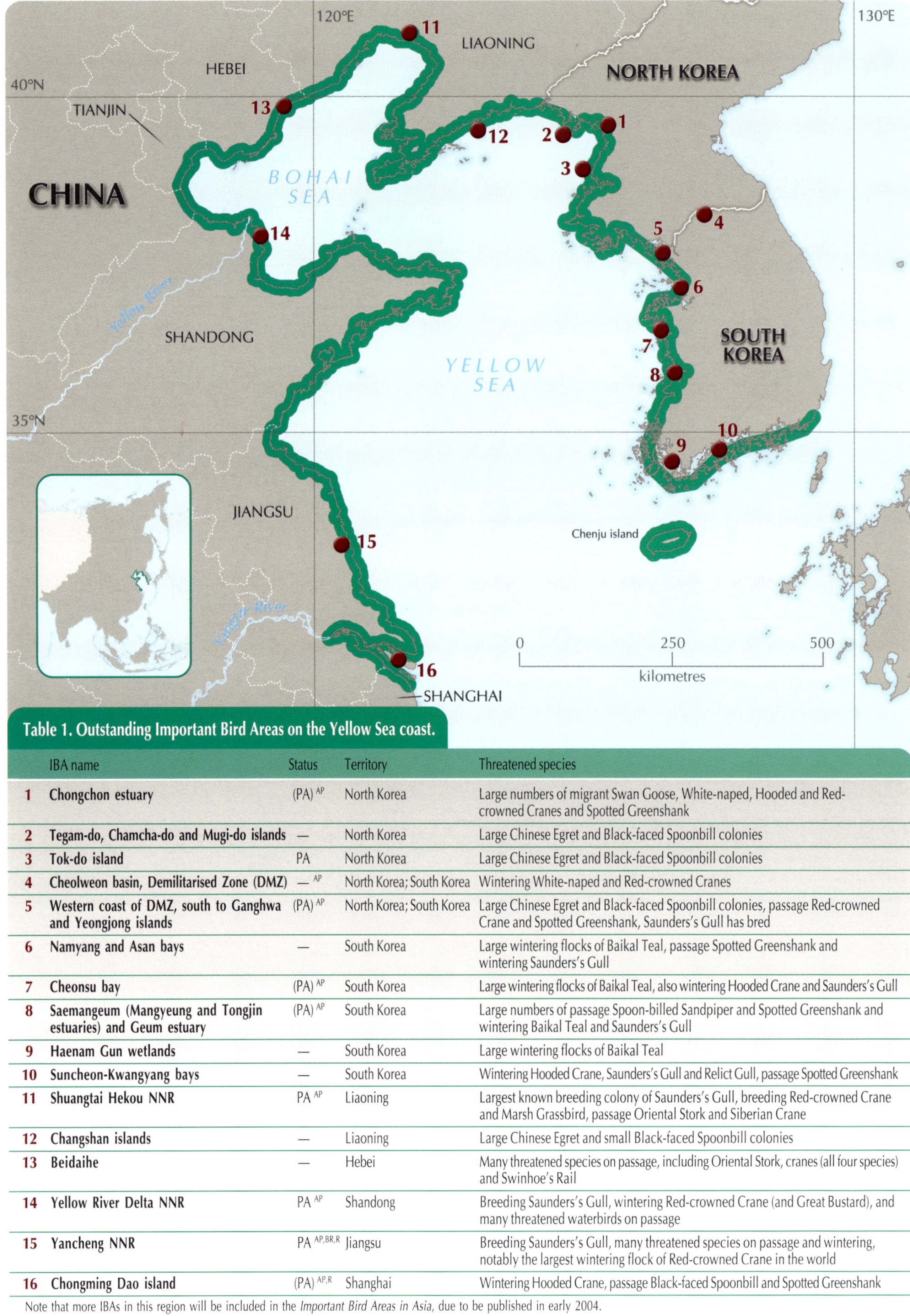

Table 1. Outstanding Important Bird Areas on the Yellow Sea coast.

	IBA name	Status	Territory	Threatened species
1	**Chongchon estuary**	(PA) AP	North Korea	Large numbers of migrant Swan Goose, White-naped, Hooded and Red-crowned Cranes and Spotted Greenshank
2	**Tegam-do, Chamcha-do and Mugi-do islands**	—	North Korea	Large Chinese Egret and Black-faced Spoonbill colonies
3	**Tok-do island**	PA	North Korea	Large Chinese Egret and Black-faced Spoonbill colonies
4	**Cheolweon basin, Demilitarised Zone (DMZ)**	— AP	North Korea; South Korea	Wintering White-naped and Red-crowned Cranes
5	**Western coast of DMZ, south to Ganghwa and Yeongjong islands**	(PA) AP	North Korea; South Korea	Large Chinese Egret and Black-faced Spoonbill colonies, passage Red-crowned Crane and Spotted Greenshank, Saunders's Gull has bred
6	**Namyang and Asan bays**	—	South Korea	Large wintering flocks of Baikal Teal, passage Spotted Greenshank and wintering Saunders's Gull
7	**Cheonsu bay**	(PA) AP	South Korea	Large wintering flocks of Baikal Teal, also wintering Hooded Crane and Saunders's Gull
8	**Saemangeum (Mangyeung and Tongjin estuaries) and Geum estuary**	(PA) AP	South Korea	Large numbers of passage Spoon-billed Sandpiper and Spotted Greenshank and wintering Baikal Teal and Saunders's Gull
9	**Haenam Gun wetlands**	—	South Korea	Large wintering flocks of Baikal Teal
10	**Suncheon-Kwangyang bays**	—	South Korea	Wintering Hooded Crane, Saunders's Gull and Relict Gull, passage Spotted Greenshank
11	**Shuangtai Hekou NNR**	PA AP	Liaoning	Largest known breeding colony of Saunders's Gull, breeding Red-crowned Crane and Marsh Grassbird, passage Oriental Stork and Siberian Crane
12	**Changshan islands**	—	Liaoning	Large Chinese Egret and small Black-faced Spoonbill colonies
13	**Beidaihe**	—	Hebei	Many threatened species on passage, including Oriental Stork, cranes (all four species) and Swinhoe's Rail
14	**Yellow River Delta NNR**	PA AP	Shandong	Breeding Saunders's Gull, wintering Red-crowned Crane (and Great Bustard), and many threatened waterbirds on passage
15	**Yancheng NNR**	PA AP,BR,R	Jiangsu	Breeding Saunders's Gull, many threatened species on passage and wintering, notably the largest wintering flock of Red-crowned Crane in the world
16	**Chongming Dao island**	(PA) AP,R	Shanghai	Wintering Hooded Crane, passage Black-faced Spoonbill and Spotted Greenshank

Note that more IBAs in this region will be included in the *Important Bird Areas in Asia*, due to be published in early 2004.

Key *IBA name*: NR = Nature Reserve; NNR = National Nature Reserve.
Status: PA = IBA is a protected area; (PA) = IBA partially protected; — = unprotected; AP = IBA is wholly or partially an Asia-Pacific waterbird network site (see p.35); BR = IBA is wholly or partially a Biosphere Reserve (see pp.34–35); R = IBA is wholly or partially a Ramsar Site (see pp.31–32).

The inter-tidal habitats around the Yellow Sea are vital for Spotted Greenshank and other migrant shorebirds, but are severely threatened by reclamation projects.

PHOTO: JOHN HOLMES

OUTSTANDING IBAs FOR THREATENED BIRDS (see Table 1)

The Yellow Sea coast is exceptionally important for threatened waterbirds, and sixteen IBAs have been selected to cover the most important breeding, passage and wintering sites for them.

CURRENT STATUS OF HABITATS AND THREATENED SPECIES

About 600 million people (c.10% of the world's population) live in the river catchments draining into the Yellow Sea, and the huge human pressure on the region has had a major impact on the environment. Large areas of coastline continue to be reclaimed for agriculture, industry, urban expansion and other development, with this part of China estimated to have lost c.37% of its intertidal areas since 1950, South Korea c.43% since 1917, as well as large areas in North Korea. The quality of the remaining wetlands has been reduced by pollution, unsustainable fishing and human disturbance. Despite these problems, the region still has many globally and regionally important wetlands, notably the outstanding IBAs listed in Table 1.

Table 2. Threatened birds of the Yellow Sea coast.

Species		Distribution and population
Dalmatian Pelican *Pelecanus crispus*	CD	Rare passage migrant, small wintering population in Jiangsu
Chinese Egret *Egretta eulophotes*	VU	Almost the entire global population breeds on small islands off the west coast of the Korean Peninsula and off Liaoning and (probably) Shandong
Oriental Stork *Ciconia boyciana*	EN	Most of global population migrates through the region, small numbers winter in Jiangsu
Black-faced Spoonbill *Platalea minor*	EN	The entire global population breeds on small islands off the west coast of the Korean Peninsula and off Liaoning
Swan Goose *Anser cygnoides*	EN	Passage and winter visitor, with the largest wintering concentrations in Jiangsu
Lesser White-fronted Goose *Anser erythropus*	VU	Passage and winter visitor, with significant winter counts in Shandong and Jiangsu
Baikal Teal *Anas formosa*	VU	Most of the global population winters in South Korea, mainly near the coast
Baer's Pochard *Aythya baeri*	VU	Widespread passage and winter visitor, largest counts in Tianjin, Shandong and Jiangsu
Scaly-sided Merganser *Mergus squamatus*	EN	Scarce passage and winter visitor
Siberian Crane *Grus leucogeranus*	CR	The eastern population migrates through the region, small numbers stage in Liaoning and Hebei
White-naped Crane *Grus vipio*	VU	A flock winters at the DMZ in Korea, and the Yangtze basin and Japanese wintering populations migrate through this region
Hooded Crane *Grus monacha*	VU	Flocks winter in South Korea and Shanghai, and the Yangtze basin and Japanese wintering populations migrate through this region
Red-crowned Crane *Grus japonensis*	EN	Large numbers winter in Jiangsu, with smaller flocks in Shandong and at the DMZ in Korea
Swinhoe's Rail *Coturnicops exquisitus*	VU	Occurs on passage, recorded annually at Beidaihe in Hebei in recent years
Spotted Greenshank *Tringa guttifer*	EN	Recorded widely on migration in inter-tidal habitats, important concentrations at several sites in South Korea, Shandong and Jiangsu
Spoon-billed Sandpiper *Eurynorhynchus pygmeus*	VU	Recorded widely on migration in small numbers in intertidal habitats, with the most important concentration in the Saemangeum area in South Korea
Saunders's Gull *Larus saundersi*	VU	The entire global population nests in coastal saltmarshes at a few sites in Liaoning, Hebei, Shangdong, Jiangsu and South Korea
Relict Gull *Larus relictus*	VU	Flocks recorded on passage and in winter at a few sites, the main wintering grounds of this poorly known species may prove to be in this region
Styan's Grasshopper-warbler *Locustella pleskei*	VU	Breeds on small islands off the south and west coasts of South Korea
Streaked Reed-warbler *Acrocephalus sorghophilus*	VU	This poorly known species has been recorded on passage in Hebei and Shanghai
Manchurian Reed-warbler *Acrocephalus tangorum*	VU	Recorded on passage in Liaoning and Hebei
Marsh Grassbird *Megalurus pryeri*	VU	Probably breeds in Liaoning and possibly in Shanghai, recorded on passage in South Korea and Hebei

In addition to the waterbirds, Great Bustard *Otis tarda* (VU; see G01) occurs on migration and in winter on the coastal plains of this region, and Greater Spotted Eagle *Aquila clanga* (VU; see F01) and Imperial Eagle *A. heliaca* (VU; see G01) both occur on migration.

● = region estimated to support >90% of global breeding population, ◎ = 10–50%, ○ = <10%; = region estimated to support >90% of global non-breeding population, = 50–90%, = 10–50%, = <10%, ? = proportion of global non-breeding population unknown; = region estimated to support >90% of global population on passage, = 50–90%, = 10–50%

CONSERVATION ISSUES AND STRATEGIC SOLUTIONS (summarised in Table 3)

Habitat loss and degradation

■ *COASTAL RECLAMATION*

Large areas of intertidal habitat have been lost around the Yellow Sea, and there are many large-scale ongoing projects and plans for further reclamation, for industrial development, agricultural land, salt works, urban expansion, aquaculture and freshwater reservoirs. In the past, sediment carried to the sea by large rivers, notably the Yellow and Yangtze Rivers, led to the growth in area of intertidal mudflats, which to some extent compensated for the habitat lost through reclamation and erosion. However, in the past decade the water and sediment flows in the Yellow River have declined drastically because of increased water extraction from the river for agricultural, industrial and domestic purposes. It is predicted that once the Three Gorges Dam is in operation there will also be a significant decline in the sediment discharged into the sea by the Yangtze.

China plans to reclaim a further 45% of its existing Yellow Sea mudflats, South Korea a further 34% (with many additional smaller projects proposed by local governments), while in North Korea it is believed that substantial investments already made in seawalls and irrigation channels will cause additional wetland habitat loss. Many IBAs listed in Table 1 are affected, including Saemangeum in South Korea, which is subject to the largest ongoing reclamation project in the world (at 401 km^2) that aims to landfill both estuaries following the construction of a 33 km dyke, due for completion in 2005. Although recent opposition to many of the ongoing and planned reclamation projects in South Korea appears to have caused a shift away from this type of project, the South Korean government recently decided to continue to implement the Saemangeum project.

There is a need to review and, if necessary, revise plans (at the national, provincial and local levels) for coastal reclamation around the Yellow Sea in North Korea, South Korea and China, in order to reconcile the needs of nature conservation and economic development. Environmental impact assessments should be conducted to review these projects, fully taking into account the value of intertidal wetlands for biodiversity and the ecological services that they provide (e.g. coastal protection, spawning grounds for fish). Given the importance of the Saemangeum area for biodiversity conservation and fisheries, the South Korean government should continue to reconsider this reclamation project, in line with internationally held obligations (e.g. Ramsar Resolution 7.21, Enhancing the conservation and wise use of inter-tidal areas).

■ *CHANGING AGRICULTURAL PRACTICES*

In South Korea, paddy fields are being replaced with other crops (e.g. vegetables and watermelons) grown under vinyl or in greenhouses, because of changing dietry habits and competition with cheap rice imports from the USA and Australia. This is reducing the foraging habitat available to cranes and other waterbirds, and needs to be offset by improved protection and management of the key remaining wetlands.

■ *POTENTIAL DEVELOPMENT OF THE DEMILITARISED ZONE (DMZ) IN KOREA*

The land in and around the DMZ (see Table 1), which currently divides North Korea and South Korea, is relatively undeveloped and subject to very little human disturbance. The DMZ is afforded protection by the current security situation in Korea, but could be opened up for development (with resulting increased disturbance) should the situation change in the future. Various options for the conservation of key areas for biodiversity need to be considered in advance, taking into account the extremely complex and delicate political situation surrounding the DMZ. These might include, for example, proposals to establish wetland conservation areas and national parks, where access and development would be controlled.

■ *POLLUTION/PESTICIDES*

The Yellow Sea is subject to major pollution from industrial effluent and domestic sewage, with long-term monitoring in China and South Korea showing that the seawater quality is steadily declining. This is a potential threat to waterbirds

Table 3. Conservation issues and strategic solutions for birds on the Yellow Sea coast.

Conservation issues	Strategic solutions
Habitat loss and degradation	
■ COASTAL RECLAMATION ■ CHANGING AGRICULTURAL PRACTICES ■ POTENTIAL DEVELOPMENT OF THE DEMILITARISED ZONE (DMZ) IN KOREA ■ POLLUTION/PESTICIDES ■ DISTURBANCE	➤ Review long-term national, provincial and local plans for reclamation on the Yellow Sea coast ➤ Assess the environmental impact of all proposed reclamation projects, and reconsider the Saemangeum project in South Korea ➤ Prepare plans to ensure the protection of key sites for threatened birds within the DMZ ➤ Continue to develop pollution control programmes, including systems to prevent spills at oilfields ➤ Control access to waterbird nesting colonies and roost sites
Protected areas coverage and management	
■ GAPS IN PROTECTED AREAS SYSTEM ■ WEAKNESSES IN RESERVE MANAGEMENT	➤ Establish new wetland nature reserves, notably on the west coast of South Korea ➤ Give nature reserve management offices in China more authority to control land use inside their reserves ➤ Strengthen reserve management in China through improved funding, infrastructure and staff training
Exploitation of birds	
■ HUNTING ■ EGG COLLECTING	➤ Strictly enforce laws banning the use of poisons for hunting in China, and patrol markets to prevent the sale of waterbirds ➤ Improve wardening at important waterbird nesting colonies
Gaps in knowledge	
■ INADEQUATE DATA ON THREATENED BIRDS	➤ Survey offshore islands in Korea and China to locate colonies of Chinese Egret and Black-faced Spoonbill ➤ Study the distribution and management needs of Great Bustards at the Yellow River delta ➤ Monitor populations of large waterbirds through counts on migration

Systems and contingency plans need to be put in place to prevent and control oil spills at major oilfields on the Chinese coast.

PHOTO: SIMBA CHAN

through direct poisoning or a reduction in the biomass of their prey. Shuangtai Hekou Nature Reserve is under threat of pollution from operations in the Liaohe oilfield, and oilfields at the Yellow River delta, near Tianjin and elsewhere are also potential sources of pollution. The use of agrochemicals and pesticides by farmers is also likely to be affecting some threatened waterbirds. Efforts are already underway to reduce pollution levels, for example by the Chinese government which is providing major funding to the Yellow Sea coastal provinces to install pollution control facilities. These measures need to be continued and extended to all parts of the region, with systems in place to prevent oil spills at Shuangtai Hekou Nature Reserve, the Yellow River delta and in the Tianjin area.

DISTURBANCE

In this densely populated region, human disturbance is a serious problem at many wetlands, with activities such as fishing, shellfish harvesting, and eel fry and lugworm collection disturbing the feeding and roosting grounds of threatened waterbirds (and nesting colonies of Saunders's Gull). Disturbance of nesting Chinese Egrets and Black-faced Spoonbills by photographers is thought to have led to increased predation by gulls, and to have been a major factor in egret population collapse at a colony in South Korea. Human usage of sensitive areas of coastal wetlands needs to be managed for biodiversity conservation, by working with local people for the sustainable use of wetland resources. Access to the most sensitive areas needs to be controlled, particularly at nesting colonies and roost sites, with clear guidelines prepared for visitors.

Protected areas coverage and management

GAPS IN PROTECTED AREAS SYSTEM

Several of the most important coastal wetlands around the Yellow Sea are officially protected, but there are some significant gaps, notably in South Korea where (despite recent efforts by the relevant government departments) there are still no comprehensively protected coastal wetlands. New nature reserves are needed on the west and south coasts of South Korea, and possibly at key wetlands in coastal North Korea and China. For example, a coastal wetland reserve should be considered at Beidaihe, because there is great potential for conservation education and the promotion of birdwatching at this popular tourist resort. Some existing reserves in China need expansion, for example Chongming and Xinglong Dongsha.

WEAKNESSES IN RESERVE MANAGEMENT

A major problem in China (and presumably also in North Korea) is the difficulty of managing existing nature reserves in the face of massive pressures from development, pollution and disturbance. A general beneficial measure would be to give the management offices of nature reserves more authority to control land use and development inside their reserves, notably at Yancheng, Yellow River Delta and Chongming Dongtan Nature Reserves, with good supervision from central government and the public to ensure improved management. The National Endangered Plant and Wildlife Protection and Nature Reserve Construction Program is a new Chinese government initiative to improve the existing protected area system and

The fisheries of the Yellow Sea provide a vital source of protein to the many people who live near the coast.

PHOTO: HAIXIANG ZHOU

establish new reserves, and it provides a mechanism to address the current management problems. It has the potential to provide stable funding for reserves and to improve reserve management, by improving their infrastructure, staff training, staff working conditions and the livelihood of local communities.

Exploitation of birds

HUNTING

Hunting is a threat to waterbirds in parts of this region. The use of poison baits is widespread in China and (to a lesser extent) South (and presumably also North) Korea, either deliberately placed by poachers to kill ducks and geese, or used by farmers to control rodents, but they also cause high mortality amongst cranes. The laws to prevent the use of these poisons should be more strictly enforced, and an education campaign launched to warn users of their adverse effects on wildlife and people. Markets in the region should be patrolled, to prevent the illegal sale of threatened waterbirds.

EGG COLLECTING

Local people and fishermen in China collect the eggs of the colonial Chinese Egret, Black-faced Spoonbill and Saunders's Gull for food. At Yancheng Nature Reserve, Saunders's Gull eggs have also been collected for zoos for captive breeding. Improved wardening of colonies is required to prevent these illegal activities, with associated education campaigns.

Gaps in knowledge

INADEQUATE DATA ON THREATENED BIRDS

Recent studies have added greatly to knowledge of threatened birds in this region, but many gaps remain. Some breeding colonies of Chinese Egrets and Black-faced Spoonbills remain to be located, and surveys are required in Korea, particularly in the western DMZ (which is currently very difficult to access because of the security situation) and in China, on uninhabited offshore islands in Liaoning, Shandong and possibly further south; these surveys should also search for nesting colonies of Chinese Crested-tern *Sterna bernsteini* (see S01). In addition to the nesting islands, surveys should identify the main feeding areas used by the nesting egrets and spoonbills, as these intertidal habitats on the mainland are likely to be more seriously threatened than the breeding sites. Structured interviews with fishermen in coastal ports could help locate offshore islands and remote coastal localities with breeding colonies of egrets and other birds. The Yellow River delta supports an important wintering population of Great Bustards, and their distribution needs to be studied to allow improved management of nature reserves for their benefit. The entire global or continental Asian populations of several threatened waterbirds migrate through the Yellow Sea, and changes in their numbers could therefore be monitored through regular systematic counts, for example at Beidaihe, where virtually the entire global populations of Oriental Stork and Siberian Crane have been counted on migration.

Chinese Egrets nest on offshore islets, but need inter-tidal flats on the nearby mainland for feeding.

PHOTO: HAIXIANG ZHOU

CENTRAL CHINESE WETLANDS

THIS region extends across central China, from Shaanxi in the west to Shandong and northern Anhui in the east. It is notable for supporting the only known wild population of Crested Ibis; the species was feared extinct, but was rediscovered in Shaanxi in 1981. There are also some important sites for wintering waterfowl and migrant cranes, mainly in the Yellow River valley.

- **Key habitats** Freshwater wetlands on riverine plains, and traditionally managed agricultural land.
- **Countries and territories** **China** (Shaanxi, Shanxi, Shandong, Henan, Anhui, Jiangsu).

Threatened species

	CR	EN	VU	Total
●	—	1	—	1
(passage migrant)	1	1	2	4
(non-breeding visitor)	—	2	3	5
Total	1	4	5	10

Key: ● = breeding in this wetland region.
(passage migrant symbol) = passage migrant.
(non-breeding visitor symbol) = non-breeding visitor.

Crested Ibis was feared extinct, but in 1981 a small wild population was discovered in Yang Xian county, Shaanxi. PHOTO: XI ZHINONG/BP

Table 1. Outstanding Important Bird Areas in the central Chinese wetlands.

	IBA name	Status	Territory	Threatened species
1	**Yang Xian county**	(PA)	Shaanxi	The only wild population of Crested Ibis in the world
2	**Yubei Huanghe Gudao NNR**	PA	Henan	Wintering Swan Goose and Lesser White-fronted Goose, and cranes (all four species) on passage
3	**Sanmenxia NR**	PA	Henan	Wintering Swan Goose, Lesser White-fronted Goose and Baer's Pochard, passage Red-crowned Crane

Note that more IBAs in this region will be included in the *Important Bird Areas in Asia*, due to be published in early 2004.

Key *IBA name*: NR = Nature Reserve; NNR = National Nature Reserve.
Status: PA = IBA is a protected area; (PA) = IBA partially protected; — = unprotected.

Crested Ibis was formerly widespread in North-East Asia, but Yang Xian county is now one of the few areas with tall trees suitable for nesting adjacent to unpolluted paddyfields.

PHOTO: XI ZHINONG/BP

OUTSTANDING IBAs FOR THREATENED BIRDS (see Table 1)

Three IBAs have been selected, covering the only wild population of Crested Ibis and two important sites for wintering waterfowl and migrant cranes in the Yellow River valley.

CURRENT STATUS OF HABITATS AND THREATENED SPECIES

The wetlands of this region have been utilised by man for several thousand years, and greatly altered by human activities. Large areas have been converted for agriculture or urbanised and virtually all large trees in the lowlands have been cut, and land reclamation, water pollution and hunting continue to affect the region's wetlands. Water is generally in short supply, and the levels in both the Yellow and Huai He rivers are now very low because of excessive water extraction, with the lower reaches of the Yellow River almost totally dried up in some years.

Despite these pressures, the last stronghold of Crested Ibis is in a remote part of this region. This species was widespread and locally common in North-East Asia until the late nineteenth century, but its population then

Farmers in Yang Xian county are given compensation for not using fertilisers and pesticides.

PHOTO: XI ZHINONG/BP

Table 2. Threatened birds of the central Chinese wetlands.

Species			Distribution and population
Oriental Stork *Ciconia boyciana*	non-breeding	EN	Rare passage and winter visitor
Crested Ibis *Nipponia nippon*	●	EN	The only known wild population is in Yang Xian county in Shaanxi, where c.200 wild birds were estimated at the end of the 2002 nesting season
Swan Goose *Anser cygnoides*	non-breeding	EN	Winters in the Yellow River valley and probably on the plains between the Yellow and Huai He rivers
Lesser White-fronted Goose *Anser erythropus*	non-breeding	VU	Winters in the Yellow River valley
Baikal Teal *Anas formosa*	non-breeding	VU	Winter visitor (at least formerly) to Nanxi Hu in Shandong
Baer's Pochard *Aythya baeri*	non-breeding	VU	Winters in the Yellow River valley and recorded at Nanxi Hu in Shandong
Siberian Crane *Grus leucogeranus*	passage	CR	Occurs on passage in the Yellow River valley
White-naped Crane *Grus vipio*	passage	VU	Occurs on passage in the Yellow River valley, and some birds may winter
Hooded Crane *Grus monacha*	passage	VU	Occurs on passage in the Yellow River valley
Red-crowned Crane *Grus japonensis*	passage	EN	Occurs on passage in the Yellow River valley, and some birds may winter

Other threatened waterbirds recorded from this region as rare visitors are: Scaly-sided Merganser *Mergus squamatus*, Pallas's Fish-eagle *Haliaeetus leucoryphus* and Relict Gull *Larus relictus*, and the Conservation Dependent Dalmatian Pelican *Pelecanus crispus* has also occurred. In addition to the waterbirds, Great Bustard *Otis tarda* (VU; see G01) occurs on migration and in winter on the riverine plains of this region.

● = region estimated to support >90% of global breeding population; (non-breeding symbol) = region estimated to support <10% of global non-breeding population; (passage symbol) = region estimated to support <10% of global population on passage

crashed—probably as a result of logging of nesting trees, wetland conversion for agriculture, hunting, agrochemicals and changes in agricultural practices—and it was feared to be extinct in the wild when the last wild birds in Japan were taken into captivity in 1981; however, in the same year a small population was discovered in Yang Xian county. By the 1980s, this was one of few areas where the combination of habitat features that the species requires still remained, with tall trees suitable for nesting adjacent to unpolluted paddyfields. Its numbers have increased steadily since its rediscovery, as a result of the prohibition of logging, firearms for hunting and agrochemicals in paddyfields, with compensation being given to farmers for not using fertilisers and pesticides, and protection of nests.

CONSERVATION ISSUES AND STRATEGIC SOLUTIONS (summarised in Table 3)

Habitat loss and degradation

■ *WETLAND CONVERSION AND AGRICULTURAL CHANGE*

Wetland loss and degradation are continuing in this region. For example, in Heyang county in Shaanxi wetlands are being converted into fishponds and being damaged by tourism development. Wetland protection and management therefore needs to be incorporated into the land-use planning processes of the relevant Chinese provinces and counties. In Yang Xian county, the wetlands have been protected and managed for the benefit of Crested Ibis, and great efforts are made to protect the nesting pairs and help maximise their breeding success, but problems remain. The area of paddies has been declining on the breeding grounds since the early 1980s, which has forced parent birds to forage more widely, and malnutrition of chicks has been noted in some areas. There are a limited number of large trees suitable for nesting near the wetlands in Yang Xian county, which has forced young breeding birds to nest in relatively small trees more accessible to predators. The special measures to protect the Crested Ibises and their habitats need to be continued, and initiated in new areas when the species expands its range.

■ *POLLUTION/PESTICIDES*

Fertilisers and pesticides are widely used, although their application is carefully controlled in Crested Ibis areas. Pollution from industrial effluents is also a problem, including on the wintering grounds of Crested Ibis along the Han Shui river, where several birds have been killed by the ingestion of poisoned materials. Improved laws and their enforcement are required to reduce pollution, together with campaigns to inform farmers about the wise use of fertilisers and pesticides, to minimise their impact on the environment. As the numbers and range of Crested Ibis increase, the use of agrochemicals in newly occupied areas will have to be minimised.

Protected areas coverage and management

■ *GAPS IN PROTECTED AREAS SYSTEM*

A Crested Ibis Conservation and Observation Station has been established in Yang Xian county (managed by Shaanxi Forestry Bureau), which protects ibis nest sites and feeding habitats, and local communities there have been well informed about wildlife conservation and the importance of the species. All of Yang Xian county has been proposed as a Non-Hunting Area, as accidental killing of this species sometimes occurs, and nature reserves are advocated in the wetlands along the Han Shui river, which have become an important feeding ground for Crested Ibis in recent years. Elsewhere in the region, wetland habitat is included in several protected areas, including Sanmenxia Nature Reserve in Henan. However, new protected areas should be considered (following surveys to assess their importance) at the other wetlands in the Yellow River valley and on the Yellow River–Huai He plain.

Great efforts are made in Yang Xian county to protect the nesting ibises and help maximise their breeding success.

PHOTO: XI ZHINONG/BP

Exploitation of birds

■ *HUNTING*

Illegal hunting, including the use of poisoned baits, is widespread, and improved enforcement of hunting legislation is required, backed up by conservation awareness work. Although the protected status of Crested Ibis is now well known in Yang Xian county, and the government has confiscated hunters' guns, small numbers of ibises continue to be accidentally poisoned by hunters at the feeding grounds along the Han Shui river. Full enforcement of hunting laws is needed, especially if the population continues to expand and the ibises move into new areas.

Gaps in knowledge

■ *INADEQUATE DATA ON THREATENED BIRDS*

Information on waterbirds is generally sparse, and surveys are required to clarify the status of the threatened species and identify important sites for their conservation, especially on the Yellow River and Yellow River–Huai He plain. For example, it is unclear whether Red-crowned Cranes winter in the Yellow River valley or only occur there on migration. Further field surveys should be conducted in parts of the Crested Ibis's former range in mainland China, particularly in remoter regions where low-intensity agriculture is still practised and may be providing suitable habitat for this species, for example in southern Gansu. Several studies have been conducted on the ecology of Crested Ibis in Yang Xian county, but there is still scope for a major ecological study of the population. The long-term management of the species will depend on the highest-quality scientific information, and every effort should be made to determine as soon as possible the optimal ecological conditions for this last population.

Other conservation issues

■ *CAPTIVE BREEDING AND REINTRODUCTION*

It has been suggested that captive Crested Ibis (now over 100 birds in the Protection and Rearing Centre in Yang Xian county, and more in Beijing Zoo and Japan) should be reintroduced to parts of the species' former range. However, great care should be taken to follow IUCN guidelines, for example to avoid the danger of the transmission of disease from captive to wild birds.

Table 3. Conservation issues and strategic solutions for birds of the central Chinese wetlands.

Conservation issues	Strategic solutions
Habitat loss and degradation	
■ *WETLAND CONVERSION AND AGRICULTURAL CHANGE* ■ *POLLUTION/PESTICIDES*	➤ Incorporate wetland protection and management into regional land-use planning processes ➤ Continue special measures to protect Crested Ibis habitats and nests, and initiate in new areas if they expand their range ➤ Improve enforcement of laws to reduce pollution, and inform farmers about the wise use of fertilisers and pesticides
Protected areas coverage and management	
■ *GAPS IN PROTECTED AREAS SYSTEM*	➤ Designate all of Yang Xian county as a Non-Hunting Area, and create new reserves along the Han Shui river ➤ Establish new wetland reserves along the Yellow River and on the Yellow River–Huai He plain
Exploitation of birds	
■ *HUNTING*	➤ Improve enforcement of hunting laws
Gaps in knowledge	
■ *INADEQUATE DATA ON THREATENED BIRDS*	➤ Conduct wetland surveys to locate key sites on the Yellow River and Yellow River–Huai He plain ➤ Search for undiscovered populations of Crested Ibis in its former range in mainland China ➤ Study the ecology of Crested Ibis, to determine the optimal conditions for its survival
Other conservation issues	
■ *CAPTIVE BREEDING AND REINTRODUCTION*	➤ Consider the reintroduction of Crested Ibis to parts of its former range in China

LOWER YANGTZE BASIN

THE extensive lakes and marshlands in the lower Yangtze basin support huge numbers of wintering waterbirds. These include many threatened species, notably almost the entire global populations of Oriental Stork and Siberian Crane, and significant proportions of the populations of Swan Goose, Lesser White-fronted Goose, White-naped Crane, Hooded Crane and Marsh Grassbird. The lower Yangtze basin is also thought to be the main wintering area of the poorly known Swinhoe's Rail.

- **Key habitats** Freshwater wetlands on riverine plains.
- **Countries and territories** **China** (Hubei, Anhui, Jiangsu, Jiangxi, Hunan).

Threatened species

	CR	EN	VU	Total
● (breeding)	—	—	—	—
(passage migrant)	—	—	—	—
(non-breeding visitor)[1]	1	3	7	11
Total	1	3	7	11

Key: ● = breeding in this wetland region.
= passage migrant.
= non-breeding visitor.

[1] The Conservation Dependent Dalmatian Pelican is also a non-breeding visitor to this region.

Almost the entire global population of Siberian Crane winters at Poyang Hu lake in Jiangxi.
PHOTO: RON SALDINO

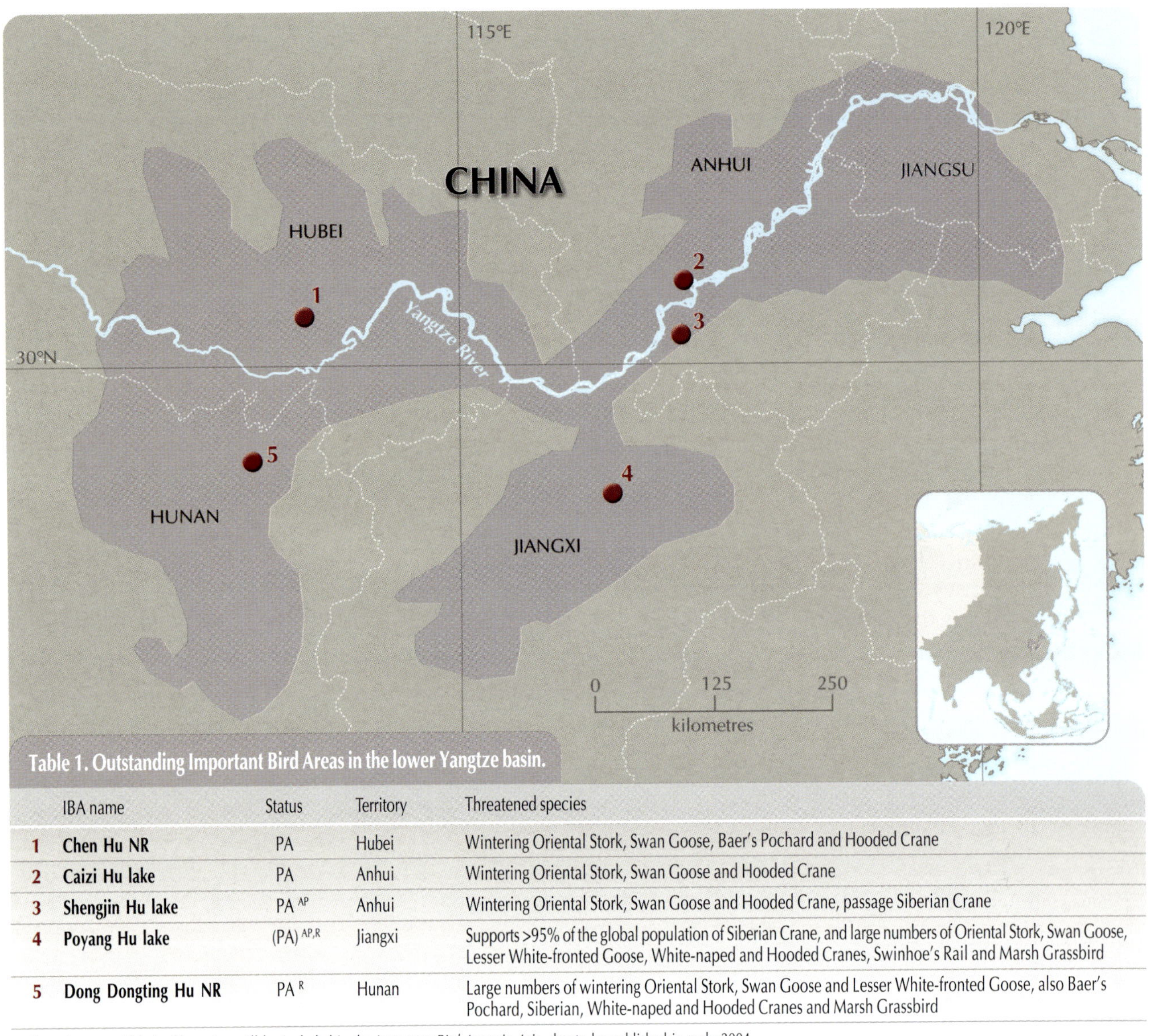

Table 1. Outstanding Important Bird Areas in the lower Yangtze basin.

	IBA name	Status	Territory	Threatened species
1	**Chen Hu NR**	PA	Hubei	Wintering Oriental Stork, Swan Goose, Baer's Pochard and Hooded Crane
2	**Caizi Hu lake**	PA	Anhui	Wintering Oriental Stork, Swan Goose and Hooded Crane
3	**Shengjin Hu lake**	PA [AP]	Anhui	Wintering Oriental Stork, Swan Goose and Hooded Crane, passage Siberian Crane
4	**Poyang Hu lake**	(PA) [AP,R]	Jiangxi	Supports >95% of the global population of Siberian Crane, and large numbers of Oriental Stork, Swan Goose, Lesser White-fronted Goose, White-naped and Hooded Cranes, Swinhoe's Rail and Marsh Grassbird
5	**Dong Dongting Hu NR**	PA [R]	Hunan	Large numbers of wintering Oriental Stork, Swan Goose and Lesser White-fronted Goose, also Baer's Pochard, Siberian, White-naped and Hooded Cranes and Marsh Grassbird

Note that more IBAs in this region will be included in the *Important Bird Areas in Asia*, due to be published in early 2004.

Key *IBA name*: NR = Nature Reserve.
Status: PA = IBA is a protected area; (PA) = IBA partially protected; — = unprotected; AP = IBA is wholly or partially an Asia-Pacific waterbird network site (see p.35); R = IBA is wholly or partially a Ramsar Site (see pp.31–32).

OUTSTANDING IBAs FOR THREATENED BIRDS (see Table 1)

Five IBAs have been selected, all lake systems with large wintering populations of threatened waterbirds. Poyang Hu and Dongting Hu lakes are exceptionally important for several species, most notably the wintering population of Siberian Crane at Poyang Hu.

CURRENT STATUS OF HABITATS AND THREATENED SPECIES

The lower Yangtze basin supports more than 300 million people, and the wetlands there have been much reduced and degraded by economic activities, principally reclamation for agriculture. The total area of lakes has been reported to have declined by 61.5% in c.30 years, from 17,198 km^2 in the 1950s to only 6,605 km^2 in the 1980s. More than 1,100 lakes have been totally reclaimed, notably in Hubei province where there were 1,066 lakes with a total surface area of 8,300 km^2 in the 1950s, but only 83 lakes with a total surface of 2,484 km^2 in the 1980s. The surface area of Poyang Hu lake has been reduced from 5,000 km^2 to 3,600 km^2, and of Dongting Hu lake from 4,350 km^2 to 2,740 km^2, while the area of farmland in Jianghan plain and Dongting Hu lake area was c.8,660 km^2 in 1949, with an agricultural population of about seven million, but the farmland area has now increased to c.15,300 km^2 and the population of farmers to 15 million. Although the total area of wetlands is still large, their quality has been greatly reduced by development, pollution, overfishing and human disturbance, and the wintering waterbirds are mainly concentrated in the relatively few remaining suitable areas of shallow wetland.

CONSERVATION ISSUES AND STRATEGIC SOLUTIONS (summarised in Table 3)

Habitat loss and degradation

■ *WETLAND CONVERSION AND AGRICULTURAL CHANGE*

Conversion and degradation of wetlands is continuing in many parts of the Yangtze basin. Paddyfields inside

The lower Yangtze basin is thought to be the main wintering area of the poorly known Swinhoe's Rail.

PHOTO: PETER LOS/BIRDLIFE

Table 2. Threatened birds of the lower Yangtze basin.

Species			Distribution and population
Dalmatian Pelican *Pelecanus crispus*	<10%	CD	Small, declining wintering population
Oriental Stork *Ciconia boyciana*	>90%	EN	Almost the entire global population winters in this region
Swan Goose *Anser cygnoides*	50–90%	EN	A high proportion of the global population winters in this region
Lesser White-fronted Goose *Anser erythropus*	50–90%	VU	A high proportion of the Asian (and global) population winters in this region
Baikal Teal *Anas formosa*	<10%	VU	Formerly numerous in this region, but numbers now greatly reduced
Baer's Pochard *Aythya baeri*	<10%	VU	Concentrations of hundreds winter at several sites
Scaly-sided Merganser *Mergus squamatus*	?	EN	Scarce and local winter visitor
Siberian Crane *Grus leucogeranus*	>90%	CR	The entire eastern population winters in the region, mainly at Poyang Hu lake
White-naped Crane *Grus vipio*	50–90%	VU	Large numbers winter at Poyang Hu and Dongting Hu lakes
Hooded Crane *Grus monacha*	10–50%	VU	Flocks winter at several wetlands
Swinhoe's Rail *Coturnicops exquisitus*	>90%?	VU	The main wintering grounds of this poorly known species appear to be in this region
Marsh Grassbird *Megalurus pryeri*	50–90%	VU	The main wintering grounds of the continental population of this species appear to be in this region

Other threatened waterbirds recorded from this region as rare visitors are: Chinese Egret *Egretta eulophotes*, Black-faced Spoonbill *Platalea minor*, Red-crowned Crane *Grus japonensis* and Saunders's Gull *Larus saundersi*. In addition to the waterbirds, Great Bustard *Otis tarda* (VU; see G01) occurs in winter on the riverine plains of this region.

[>90%] = region estimated to support >90% of global non-breeding population, [50–90%] = 50–90%, [10–50%] = 10–50%, [<10%] = <10%, [?] = proportion of global non-breeding population unknown

Longgan Hu Nature Reserve in Hubei were converted into cotton fields and lotus ponds during the 1990s, leading to a decline in the number of wintering Hooded Cranes, and conversion of paddies to cotton fields has also reduced wetland habitat inside Dong Dongting Hu Nature Reserve in Hunan. The increased cultivation of perennial crops such as sugarcane means that fewer fallow fields are available for foraging waterbirds, and China's recent entry into the World Trade Organization could lead to many other changes in cropping patterns. At Chen Hu lake in Hubei local people are still illegally reclaiming parts of the lake for fishponds, despite the establishment of a nature reserve there.

Following the devastating floods of 1998, the Chinese government announced measures to protect wetlands and to stop reclamation; the rehabilitation and restoration of the middle and lower Yangtze basin is a priority project of the *National Wetland Action Plan for China*. At some locations, including Poyang Hu and Dong Dongting Hu lakes, many thousands of people are being relocated out of flood-prone areas, and low-lying farmlands and settlements are to be converted back to lake and wetland. The impacts of these flood control measures and of agricultural change need to be monitored, to identify threats to key wetlands and opportunities to improve waterbird habitats. Any further encroachment into natural wetlands should be prevented, both inside and outside protected areas.

CHANGES IN RIVER FLOW

The construction and operation of the Three Gorges Dam will change the seasonal flow of water in the Yangtze River and could negatively affect the wetlands downstream. Artificially maintaining low water levels during the summer flood season and raising them (by an estimated average of one metre) in the winter may well change the character of the wetlands, and the shallow wetlands that most wintering waterbirds require for feeding will be greatly reduced in extent. Siberian Cranes may be particularly badly affected, through the reduced availability of the roots and tubers of aquatic plants that they eat, leading to starvation in late winter and early spring. High water levels caused by releases from the dam in March may force them to move closer to

The Three Gorges Dam is being constructed to control flooding and generate power, but once in operation it could negatively affect the wetlands downstream.

PHOTO: DI YUN/CHINA FEATURES

Table 3. Conservation issues and strategic solutions for birds of the lower Yangtze basin.

Conservation issues	Strategic solutions
Habitat loss and degradation	
WETLAND CONVERSION AND AGRICULTURAL CHANGE CHANGES IN RIVER FLOW POLLUTION/PESTICIDES REDUCED FOOD SUPPLY DISTURBANCE	➤ Prevent encroachment into natural wetlands, and monitor the impacts of ongoing flood control measures and agricultural changes ➤ Develop a programme to monitor the effects of the Three Gorges Dam on wetlands and waterbirds ➤ Enforce laws to reduce pollution, and inform farmers about the wise use of fertilisers and pesticides ➤ Improve management of fisheries, particularly in protected areas
Protected areas coverage and management	
GAPS IN PROTECTED AREAS SYSTEM WEAKNESSES IN RESERVE MANAGEMENT	➤ Establish new wetland reserves at key sites in Hubei, Anhui and Jiangsu ➤ Expand the boundaries and core area of Poyang Hu Nature Reserve ➤ Give nature reserve management offices more authority to control land use inside their reserves ➤ Strengthen reserve management through improved funding, infrastructure and staff training
Exploitation of birds	
HUNTING	➤ Formulate new national and local regulations to improve management of hunting ➤ Increase the protection status of Swan Goose, Lesser White-fronted Goose and Baikal Teal in China ➤ Strictly enforce hunting legislation, particularly to prevent the sale of illegally hunted waterbirds in markets
Gaps in knowledge	
INADEQUATE DATA ON THREATENED BIRDS	➤ Survey poorly known species, notably Swinhoe's Rail and Marsh Grassbird ➤ Conduct regular, coordinated counts at major wetlands, to monitor the numbers of threatened birds and the effects of hunting and other pressures

human settlements and make them more susceptible to poison baits. More predictable water levels may also allow increased encroachment for agriculture and other economic activities. A coordinated programme is therefore needed to monitor (a) the water levels and condition of the key wetlands in the lower Yangtze basin and (b) the numbers of waterbirds and the effects of habitat changes on their distribution and behaviour. The dam authorities should be kept fully informed of the conditions required by Siberian Cranes and other threatened species at Poyang Hu lake and elsewhere in the Yangtze basin, and alerted when any decline in the quality of these wetlands is attributable to water flow.

■ *POLLUTION/PESTICIDES*

Industrial wastes, oil leakage, pesticides and fertilisers have entered the water system and severely polluted wetlands in the Yangtze basin. The ongoing conversion from rice to cotton results in pesticides used in the cotton fields polluting the adjacent wetlands. Improved laws and their enforcement are required to reduce pollution, together with campaigns to inform farmers about the wise use of fertilisers and pesticides, to minimise their environmental impact.

■ *REDUCED FOOD SUPPLY*

Fishermen and eel-catchers now use very fine mesh nets, leaving almost no food for the waterbirds in many wetlands.

Fishing needs to be better controlled inside protected areas, and the management of fisheries improved throughout the region, with conservation awareness campaigns targeted at local communities and governments pointing out that such intensive exploitation (particularly of small fish) cannot be sustainable.

■ *DISTURBANCE*

Disturbance is a problem at most wetlands in the lower Yangtze basin, caused by poaching, fishing, lotus-rhizome digging and other human activities, as well as burning and grazing. This needs to be reduced through improved protected area management (see below), to ensure that undisturbed areas are always available for feeding and roosting birds.

Protected areas coverage and management

■ *GAPS IN PROTECTED AREAS SYSTEM*

Many of the most important wetlands in the lower Yangtze valley are officially protected, but some new nature reserves need to be established, e.g. at Shaobo Hu lake in Jiangsu. Several of these wetlands are degraded, and habitat restoration may be required following their designation as reserves to enhance their value for waterbirds. The nature reserve at Poyang Hu lake only includes a small part of the lake, and it should be expanded to include Nan Hu lake in Yongxiu and Xingzi counties, Dalianzi Hu lake in Boyang county, Saicheng Hu lake in Jiujiang county and Tangyin in Duchang county, and the core area of the reserve should be enlarged. Similar changes may be required at some other wetland protected areas, when surveys show that the current boundaries and reserve zonation are inadequate.

■ *WEAKNESSES IN RESERVE MANAGEMENT*

Although many important wetlands in this region are officially protected, they are not necessarily secure because of inadequate budgets, lack of experienced personnel and poor communication between agencies (agriculture, fisheries, irrigation, etc.). A general beneficial measure would be to give the management offices of nature reserves more authority to control land use inside their reserves. The National Endangered Plant and Wildlife Protection and Nature Reserve Construction Program is a new Chinese government initiative to improve the existing protected area system and establish new reserves, and it provides a mechanism to address the current management problems. It has the potential to provide stable funding for reserves and to improve reserve management, by improving their infrastructure, staff training, staff working conditions and the livelihood of local communities. Specific measures proposed at Poyang Hu Nature Reserve (also relevant to many other wetland reserves) are that ownership of the core area should be transferred to the nature reserve management office and a development plan drafted. There is a need to improve enforcement of wildlife conservation laws and regulations in the reserve, and for education of the general public to help prevent poaching. The water level of the lake should be artificially controlled: two different levels should be used (if necessary by legal enforcement) to maintain the wetland ecosystem at Poyang, a flooding cycle to submerge the grasslands under 17–18 m of water for 50–100 days, and then the maintenance of optimal water levels for feeding of Siberian Cranes and other waterbirds in winter. Control of fishery practices should be improved in the reserve, including compensation of fishermen for not draining the lakes, and regulation of the size of fish that can be taken.

Exploitation of birds

■ *HUNTING*

A study of hunting pressure in the middle and lower basins of the Yangtze River in 1987–1992 estimated that c.50% of the total wintering waterfowl were killed each year by local hunters, using netting, shooting and poisoning; most threatened waterbird species were found in hunters' bags during the study. The numbers of waterfowl in the lower Yangtze basin have declined greatly in the last 10 years, possibly indicating that hunting pressure may have intensified in recent years, and this is linked to the recent declines in the Swan Goose and the eastern population of Lesser White-fronted Goose. Poachers have recently been arrested for killing several hundreds of these two geese, many of which were openly on sale in local markets. Although ducks and geese are the main targets of the hunters, the poisons (rice soaked in strong pesticides) and mist-nets used to capture them also kills storks, cranes and other birds.

National and local hunting regulations should be formulated in China (using the results of scientific investigations), to control the length of the hunting season, to limit the number of hunters and the bag size of each hunter, and to ban inappropriate hunting methods such as the use of poisons. Most threatened waterbirds are already protected in China, but, given the hunting pressure on waterfowl, Swan Goose, Lesser White-fronted Goose and Baikal Teal should all be upgraded to a higher protection status. Increased efforts should be made to enforce the existing (and any new) hunting legislation, to control hunting at key sites in the lower Yangtze basin, and prevent the sale of illegally hunted waterbirds in markets. Appropriate training will therefore be needed for local police and officials. It is important to back up existing and new hunting legislation with publicity to support its vigorous enforcement, including the dangers that the use of

The eastern population of Lesser White-fronted Goose is affected by large-scale commercial hunting.

PHOTO: WEN-HSIN HUANG

Once the Three Gorges Dam is in operation, the numbers of cranes and other threatened waterbirds in the lower Yangtze basin will need to be closely monitored.

PHOTO: BIRDLIFE

poisons poses to wildlife and people. This publicity should target people in the remotest and poorest areas, where it will be necessary to create alternative job opportunities for local hunters.

Gaps in knowledge

■ *INADEQUATE DATA ON THREATENED BIRDS*

Waterbird counts are periodically conducted at the wetlands in the lower Yangtze basin, notably of cranes and storks. However, the distribution and numbers of several threatened waterbirds are incompletely known, and surveys are required, notably for Swinhoe's Rail and Marsh Grassbird. Regular, coordinated counts are required at all of the major wetlands, to monitor the effects of the various threats to waterbirds and their habitats. Along with regular checks on the numbers of birds being sold for food in markets, this would help clarify the levels of hunting and the effects on waterbird populations. Once the Three Gorges Dam is in operation, these counts will be important for monitoring the numbers (and any changes in distribution) of Siberian Crane and other threatened waterbirds (see *Changes in river flow* above).

TIBETAN PLATEAU

W09

THE many high-altitude lakes and marshlands on the Tibetan plateau support one unique waterbird species, Black-necked Crane, which is widely distributed during the breeding season but moves to the relatively low eastern and southern parts of the plateau for the winter. Baer's Pochard and Pallas's Fish-eagle also occur on the southern and eastern fringes of the plateau.

- **Key habitats** High altitude lakes and marshland.
- **Countries and territories** **China** (Xinjiang, Tibct, Qinghai, Gansu, Sichuan, Yunnan, Guizhou); **India** (Jammu and Kashmir [Ladakh], Sikkim, Arunachal Pradesh); **Bhutan**.

Threatened species

	CR	EN	VU	Total
● (breeding)	—	—	2	2
(passage migrant)	—	—	—	—
(non-breeding visitor)	—	—	1	1
Total	—	—	3	3

Key: ● = breeding in this wetland region.
= passage migrant.
= non-breeding visitor.

Black-necked Crane is unique to the high altitude wetlands of the Tibetan plateau.
PHOTO: OTTO PFISTER

OUTSTANDING IBAs FOR THREATENED BIRDS (see Table 1)

Nine IBAs have been selected, including three important breeding areas and six important non-breeding concentrations of Black-necked Crane.

CURRENT STATUS OF HABITATS AND THREATENED SPECIES

The human population density on the Tibetan plateau is very low, and many areas are relatively undisturbed. However, the wetlands are locally under pressure from wetland drainage, overgrazing, peat mining, reservoir construction, pesticide use and changes in agricultural practices, particularly in the relatively low-altitude areas in the south and east of the plateau.

CONSERVATION ISSUES AND STRATEGIC SOLUTIONS (summarised in Table 3)

Habitat loss and degradation

WETLAND DRAINAGE

Some of the high-altitude wetlands used by Black-necked Cranes are being converted into wet grasslands, and then gradually into steppe and arid land. Habitat pressures are heaviest in the species's wintering range, because it is at relatively low altitudes and more densely populated, with many areas in Yunnan and Guizhou being affected by drainage and damming. This wetland drainage is mainly to create new pastureland, but wetlands around Lhasa are being converted to croplands and urbanised. There is therefore a need for improved protection and management of the crane's main breeding and wintering sites, to minimise further conversion of their wetland habitat.

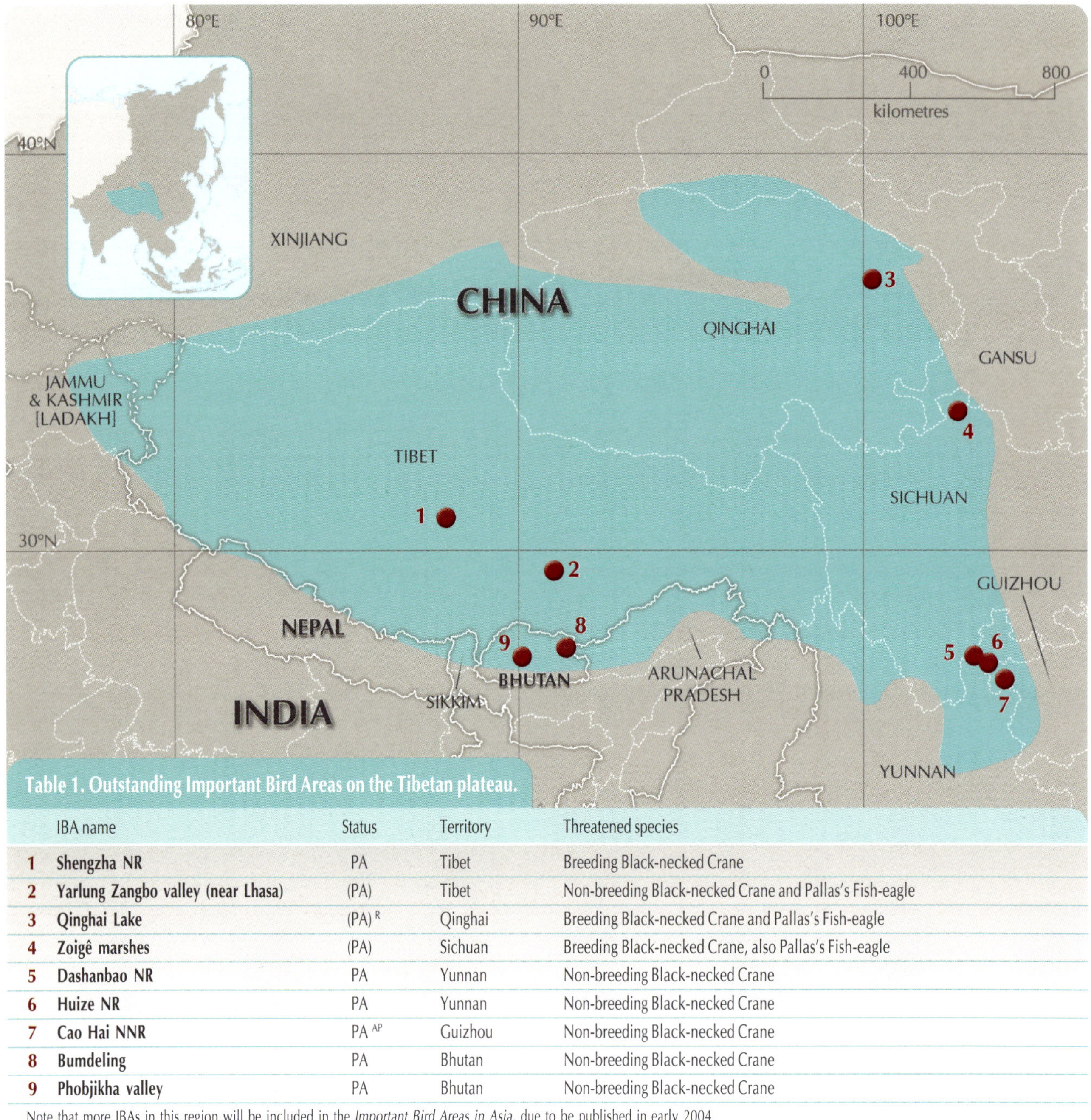

Table 1. Outstanding Important Bird Areas on the Tibetan plateau.

	IBA name	Status	Territory	Threatened species
1	**Shengzha NR**	PA	Tibet	Breeding Black-necked Crane
2	**Yarlung Zangbo valley (near Lhasa)**	(PA)	Tibet	Non-breeding Black-necked Crane and Pallas's Fish-eagle
3	**Qinghai Lake**	(PA) R	Qinghai	Breeding Black-necked Crane and Pallas's Fish-eagle
4	**Zoigê marshes**	(PA)	Sichuan	Breeding Black-necked Crane, also Pallas's Fish-eagle
5	**Dashanbao NR**	PA	Yunnan	Non-breeding Black-necked Crane
6	**Huize NR**	PA	Yunnan	Non-breeding Black-necked Crane
7	**Cao Hai NNR**	PA AP	Guizhou	Non-breeding Black-necked Crane
8	**Bumdeling**	PA	Bhutan	Non-breeding Black-necked Crane
9	**Phobjikha valley**	PA	Bhutan	Non-breeding Black-necked Crane

Note that more IBAs in this region will be included in the *Important Bird Areas in Asia*, due to be published in early 2004.

Key *IBA name*: NR = Nature Reserve.
Status: PA = IBA is a protected area; (PA) = IBA partially protected; — = unprotected; AP = IBA is wholly or partially an Asia-Pacific waterbird network site (see p.35); R = IBA is wholly or partially a Ramsar Site (see pp.31–32).

Table 2. Threatened birds of the Tibetan plateau.

Species			Distribution and population
Baer's Pochard *Aythya baeri*		VU	Rare winter visitor to the eastern edge of the Tibetan plateau
Pallas's Fish-eagle *Haliaeetus leucoryphus*	○	VU	Occurs on the southern and eastern edges of the Tibetan plateau
Black-necked Crane *Grus nigricollis*	●	VU	Widely distributed on the Tibetan plateau during the breeding season, in winter tends to concentrate in the relatively low eastern and southern parts

Other threatened waterbirds recorded from this region as rare visitors are: Black-faced Spoonbill *Platalea minor*, Scaly-sided Merganser *Mergus squamatus*, Hooded Crane *Grus monacha* and Red-crowned Crane *G. japonensis*.

● = region estimated to support >90% of global breeding population, ○ = <10%; = region estimated to support <10% of global non-breeding population

PEAT EXTRACTION

Peat is mined near Zoigê marsh in Sichuan and elsewhere in the region, which could lead to localised loss of Black-necked Cranes breeding and wintering habitat. This activity needs to be carefully controlled at important sites for the species.

CHANGING AGRICULTURAL PRACTICES

In south-central Tibet and north-east Yunnan, many farmers now prefer a higher-yield winter wheat to the traditional barley, spring wheat and broadbean, and in winter this wheat has no grain for the cranes to eat. With the increased emphasis on autumn ploughing for the cultivation of winter wheat and to control insects, less waste grain and other surface residues are available for the cranes, perhaps causing the birds to switch to the winter wheat seedlings. Traditional farming practices should be promoted in the main wintering sites of the cranes, possibly with compensation to farmers for any lost income. Well-managed ecotourism could provide an alternative income for local people at some wintering sites, and an incentive to leave the fields unploughed in winter.

LIVESTOCK GRAZING

High-altitude vegetation is slow-growing and sensitive to disturbance, and increasing livestock densities have degraded grasslands in some areas, for example in Ladakh, and this constant diminution of undisturbed foraging and breeding areas could lead to the extinction of the small Indian population. Improved management of grazing is required at the crane's main breeding and wintering sites.

DEVELOPMENT (URBAN, INDUSTRIAL, ETC.)

The human population is increasing rapidly in some areas, for example in Ladakh and near Lhasa, and associated development could negatively affect the Black-necked Cranes. Road building is causing disturbance and opening up remote areas, while telephone and electricity wires pose a hazard to flying cranes. In north-east Yunnan, most wintering areas are in wetlands near reservoirs built for irrigation, but in other areas reservoir construction is reducing the shallow water areas needed for wading birds like cranes. For example, a proposed dam on the Lhasa River could have a tremendous impact on the 1,000+ cranes that winter there. Environmental impact assessments should be conducted for development projects near the crane's main breeding and wintering sites.

POLLUTION/PESTICIDES

In Tibet, pesticides used by farmers have caused crane mortalities in at least one area, and intensive use of pesticides may have affected this species on its breeding grounds at Zoigê marshes. Pesticides used to control rodents in China have also poisoned Pallas's Fish-eagles. Industrial pollution from new zinc furnaces has increased in the Cao Hai lake watershed in Guizhou, and refuse disposal and sewage from nearby towns are a problem there and in Ladakh. The use of pesticides and herbicides should be legally regulated, with measures to control other forms of pollution around the key sites for cranes.

DISTURBANCE

As they are often fairly tame, Black-necked Cranes are perhaps less susceptible to disturbance than other crane species. However, as human populations increase in their wintering and breeding sites, problems are arising with greater frequency. Increased disturbance from tourists may be a factor in the decline in summering cranes at Qinghai Hu lake, and the development of new tourist destinations in Ladakh may disrupt important crane breeding and feeding areas. In Bhutan, most crane roost sites are accessible to increasing numbers of tourists wanting to approach the birds too closely. Human and livestock activity keeps cranes off their nests in Ladakh, and, in combination with an increase in egg-predating Common Ravens *Corvus corax* around human settlements, has reduced the breeding success of the cranes. There is a need to control human access and activities at the main breeding and wintering sites of the cranes, particularly near nests and roost sites.

Protected areas coverage and management

GAPS IN PROTECTED AREAS SYSTEM

In China, most Black-necked Cranes nest outside protected areas, but a few nature reserves hold small breeding populations, and several hold important non-breeding concentrations. Some of the cranes in Ladakh breed within the Changthang Cold Desert Wildlife Sanctuary, and the

The human population is increasing rapidly in some parts of this region, for example near Lhasa, and development could negatively affect Black-necked Crane.

PHOTO: JOHN HOLMES

main wintering sites in Bhutan are inside protected areas. Pallas's Fish-eagle is known from several protected areas, but it is unclear whether any reserves support significant populations. More nature reserves are needed to protect breeding Black-necked Cranes in China and India, including in the Xiamen region of the Zoigê marshes and in several areas of Ladakh. New reserves should also be considered at unprotected wintering sites in Yunnan. However, the establishment of community-based monitoring and protection schemes may be more appropriate at sites where the cranes winter on agricultural land.

■ *WEAKNESSES IN RESERVE MANAGEMENT*
Many existing protected areas are remote and have little or no infrastructure. To improve their effectiveness, management plans need to be developed and implemented, with sufficient funding to develop the infrastructure and staffing required to address all threats.

Exploitation of birds

■ *HUNTING AND EGG COLLECTION*
Buddhist beliefs in Tibet, Qinghai, western Yunnan, western Sichuan, Ladakh and Bhutan preclude the hunting of wildlife, including Black-necked Cranes which are regarded as supernatural spirits throughout much of their range, being frequently depicted in religious imagery and considered symbols of good luck and happiness. Nevertheless, hunting has recently become a significant threat in several areas owing to increases in firearms and improved access to remote areas. Illegal crane hunting and egg-collection occurs at Zoigê marshes in Sichuan and elsewhere, crane hunting has been observed at several wintering sites in Tibet, and birds are killed (sometimes by poisoned bait) for food by farmers at some wintering sites in Yunnan. Wintering cranes sometimes cause damage to crops, which may explain the recent increase in hunting by farmers, and crane wings are sometimes used for scarecrows. In some parts of China, Pallas's Fish-eagle is hunted for its tail feathers. Black-necked Crane and Pallas's Fish-eagle are legally protected in most parts of this region. All breeding and wintering populations need to be protected through improved law enforcement, with special emphasis in parts of Yunnan where this problem is most relevant; rewards should be offered for reporting poaching incidents.

Gaps in knowledge

■ *INADEQUATE DATA ON THREATENED BIRDS*
Recent surveys and winter counts have greatly improved knowledge of Black-necked Crane, but many breeding areas still await discovery. The status of Pallas's Fish-eagle on the Tibetan plateau is poorly understood, and studies are required there, possibly including satellite tracking to investigate its seasonal movements.

Other conservation issues

■ *PREDATION BY FERAL DOGS*
Predation of eggs and chicks by feral dogs has severely affected the small breeding population in India, especially as they can swim across water to reach nests; of 61 eggs monitored in Ladakh in the 1990s, 35 (57%) did not survive to fledging, and dogs were responsible for 19 (54%) of these failures. A coordinated culling programme appears necessary, but care is needed to respect local traditions on the sanctity of animal life. Alternatively herders' dogs could have a front leg attached to their collars so that they cannot pursue fleeing wildlife, or a dog sterilisation programme might be conducted.

Predation of eggs and chicks by feral dogs is affecting Black-necked Crane in India.

PHOTO: OTTO PFISTER

Table 3. Conservation issues and strategic solutions for birds on the Tibetan plateau.

Conservation issues	Strategic solutions
Habitat loss and degradation	
■ WETLAND DRAINAGE	➤ Minimise conversion of Black-necked Crane habitat for agricultural and urban expansion
■ PEAT EXTRACTION	➤ Limit peat extraction at key sites for Black-necked Crane
■ CHANGING AGRICULTURAL PRACTICES	➤ Promote traditional farming practices which provide food for wintering cranes
■ LIVESTOCK GRAZING	➤ Improve the management of grazing at key sites
■ DEVELOPMENT (URBAN, INDUSTRIAL, ETC.)	➤ Assess the environmental impact of development projects near key sites
■ POLLUTION/PESTICIDES	➤ Legally regulate the use of pesticides and herbicides, and control pollution near key sites
■ DISTURBANCE	➤ Regulate human activities at key sites, particularly near nests and roost sites
Protected areas coverage and management	
■ GAPS IN PROTECTED AREAS SYSTEM	➤ Establish new protected areas for breeding and wintering Black-necked Cranes in China and India
■ WEAKNESSES IN RESERVE MANAGEMENT	➤ Develop community-based monitoring and protection schemes at sites where Black-necked Cranes winter on agricultural land
	➤ Prepare management plans for existing and new reserves, and develop the infrastructure and staffing required for their implementation
Exploitation of birds	
■ HUNTING AND EGG COLLECTION	➤ Improve enforcement of hunting laws, particularly in Yunnan
Gaps in knowledge	
■ INADEQUATE DATA ON THREATENED BIRDS	➤ Investigate the status of Pallas's Fish-eagle on the Tibetan plateau, possibly using satellite tracking
Other conservation issues	
■ PREDATION BY FERAL DOGS	➤ Conduct a coordinated dog culling and/or sterilisation programme in Ladakh

CHINA SEA COAST

W10

THE coasts of southern Japan, southern China, Taiwan and northern Vietnam are fringed with wetlands, mainly estuaries with their associated freshwater and saltwater marshes, mangroves and intertidal mudflats. These wetlands are vital for waterbirds migrating from North-East Asia to their wintering grounds to the south, including substantial numbers of Chinese Egret, Spotted Greenshank and Spoon-billed Sandpiper. They are also important for wintering waterbirds, most notably almost the entire global population of Black-faced Spoonbills and large numbers of Saunders's Gull, while the main wintering grounds of Styan's Grasshopper-warbler also appear to be in this region.

- **Key habitats** Coastal wetlands.
- **Countries and territories** **Japan** (Kyushu, Nansei Shoto); **China** (*mainland*: Zhejiang, Fujian, Guangdong, Guangxi, Hainan; *Hong Kong*; *Macau*; *Taiwan*); **Vietnam**.

	Threatened species			
	CR	EN	VU	Total
● (breeding)	—	—	1	1
passage migrant	—	—	3	3
non-breeding visitor[1]	—	5	6	11
Total	—	5	10	15

Key: ● = breeding in this wetland region.
= passage migrant.
= non-breeding visitor.
[1] The Conservation Dependent Dalmatian Pelican is also a non-breeding visitor to this region.

The China Sea coast region overlaps with part of Conservation International's Indo-Burma Hotspot (see pp.20–21).

Mai Po and Inner Deep Bay is an important passage and wintering area for many threatened waterbirds. PHOTO: RAY TIPPER

OUTSTANDING IBAs FOR THREATENED BIRDS (see Table 1)

Nine IBAs have been selected to cover the most important concentrations of threatened waterbirds in the region, particularly for the wintering populations of Black-faced Spoonbill and Saunders's Gull.

CURRENT STATUS OF HABITATS AND THREATENED SPECIES

The coastal lowlands of southern Japan, southern China and northern Vietnam support a very large human population. Many wetlands have been converted to agricultural land or shrimp- and fish-ponds, or reclaimed for industrial, infrastructural and urban development. For example, several large tidal flats have been reclaimed on Kyushu and c.1,700 km^2 of intertidal mudflats have been reclaimed in Zhejiang province since 1950. The region still has some extensive intertidal wetlands of major importance for threatened birds, but they are under great pressure from further development, human disturbance and pollution.

CONSERVATION ISSUES AND STRATEGIC SOLUTIONS (summarised in Table 3)

Habitat loss and degradation

■ *COASTAL RECLAMATION*

There is huge demand for land in this region because of human population pressure and very rapid economic

Table 1. Outstanding Important Bird Areas on the China Sea coast.

	IBA name	Status	Territory	Threatened species
1	**Sone tidal flats**	—	Kyushu	Wintering Saunders's Gull
2	**Hakata bay**	—	Kyushu	Wintering Black-faced Spoonbill
3	**Ariake bay**	— AP	Kyushu	Wintering Black-faced Spoonbill and Saunders's Gull
4	**Wenzhou bay**	—	Zhejiang	Wintering Dalmatian Pelican and Saunders's Gull, passage Black-faced Spoonbill
5	**Min Jiang estuary**	—	Fujian	Wintering Black-faced Spoonbill and Swan Goose
6	**Mai Po and Inner Deep Bay**	PA AP,R	Hong Kong	Passage and wintering Dalmatian Pelican, Oriental Stork, Black-faced Spoonbill, Spotted Greenshank, Spoon-billed Sandpiper and Styan's Grasshopper-warbler
7	**Taipa-Colones wetland**	—	Macau	Wintering Black-faced Spoonbill
8	**Tsengwen estuary**	PA	Taiwan	Supports almost half of the global wintering population of Black-faced Spoonbill
9	**Xuan Thuy NP**	PA R	Vietnam	Passage and wintering Chinese Egret, Black-faced Spoonbill, Spotted Greenshank, Spoon-billed Sandpiper and Saunders's Gull

Note that more IBAs in this region will be included in the *Important Bird Areas in Asia*, due to be published in early 2004.

Key *IBA name*: NP = National Park.
Status: PA = IBA is a protected area; (PA) = IBA partially protected; — = unprotected; AP = IBA is wholly or partially an Asia-Pacific waterbird network site (see p.35); R = IBA is wholly or partially a Ramsar Site (see pp.31–32).

development. Large areas of wetland have been reclaimed for agriculture and aquaculture, and in recent decades for industrial and urban land. Reclamation of tidal flats has been widespread on Kyushu in Japan, while ongoing and proposed reclamation projects threaten to reduce the extent and quality of wetland habitat still further. A sea wall built at Isahaya in 1997 dried out Japan's largest tidal flat, such that it no longer attracts shorebirds. An airport has been constructed near Daijukarami tidal flat in Saga prefecture, and another is planned near Sone tidal flat at Kitakyushu. A 4 km^2 artificial island at Wajiro in Fukuoka prefecture is now under construction. Similar reclamation has taken place on Taiwan, particularly on the west coast, and the Tsengwen estuary, the most important wintering site for Black-faced Spoonbill, has recently been under considerable pressure for industrial development. In southern China, coastal provinces are experiencing one of the highest rates of economic development of any region in the world, placing wetland habitats under great pressure. In Zhejiang province, large areas of mudflats have already been lost, and there are plans to reclaim a further 659 km^2 in the next 15 years. The wetlands of Deep Bay near Hong Kong have been degraded by housing estates and container storage facilities, as well as by the Shenzhen Special Economic Zone in Guangdong province, which has greatly reduced the area of Futian Nature Reserve. The wintering habitats of Black-faced Spoonbill have been lost in Guangxi through the construction of port facilities.

All future coastal reclamation and development projects in the region should be reviewed and revised, in order to

W10

Large numbers of Saunders's Gulls winter on the China Sea coast.

PHOTO: RAY TIPPER

Table 2. Threatened birds of the China Sea coast.

Species			Distribution and population
Dalmatian Pelican *Pelecanus crispus*		CD	Small wintering populations in Hong Kong and Zhejiang
Spot-billed Pelican *Pelecanus philippensis*		VU	Very small and declining non-breeding population in northern Vietnam
Chinese Egret *Egretta eulophotes*		VU	Widespread on passage in significant numbers
Oriental Stork *Ciconia boyciana*		EN	Irregular winter visitor, mainly to Hong Kong
Black-faced Spoonbill *Platalea minor*		EN	Almost the entire global population winters, with important concentrations in Taiwan, Hong Kong, Vietnam, Japan and Macau
Swan Goose *Anser cygnoides*		EN	Scarce winter visitor
Baikal Teal *Anas formosa*		VU	Scarce winter visitor
Baer's Pochard *Aythya baeri*		VU	Scarce winter visitor
Scaly-sided Merganser *Mergus squamatus*	?	EN	Scarce winter visitor
Swinhoe's Rail *Coturnicops exquisitus*		VU	Scarce winter visitor
Spotted Greenshank *Tringa guttifer*		EN	Widespread on passage, a few overwinter in northern Vietnam
Spoon-billed Sandpiper *Eurynorhynchus pygmeus*		VU	Widespread on passage, a few overwinter in northern Vietnam
Saunders's Gull *Larus saundersi*		VU	Widespread in winter, with important concentrations at several sites
Styan's Grasshopper-warbler *Locustella pleskei*	○ ?	VU	Nests on islets off Kyushu in Japan, all known wintering sites are in southern China, mostly in Hong Kong
Streaked Reed-warbler *Acrocephalus sorghophilus*	?	VU	Recorded on migration
Manchurian Reed-warbler *Acrocephalus tangorum*	?	VU	Recorded on migration

Other threatened waterbirds recorded from this region as rare visitors are: Lesser White-fronted Goose *Anser erythropus*, White-naped Crane *Grus vipio*, Hooded Crane *G. monacha*, Red-crowned Crane *G. japonensis* and Relict Gull *L. relictus*. In addition to the waterbirds, Greater Spotted Eagle *Aquila clanga* (VU; see F01) and Imperial Eagle *A. heliaca* (VU; see G01) occur on migration and in winter.

○ = region estimated to support <10% of global breeding population; = region estimated to support >90% of global non-breeding population, = 50–90%, = <10%, ? = proportion of global non-breeding population unknown: = region estimated to support >90% of global population on passage, = 50–90%, = 10–50%, = <10%

reconcile the needs of nature conservation and economic development. Environmental impact assessments should be used, and the management of coastal wetlands needs to be better coordinated between the relevant sectors. New protected areas should be established at some important wetlands for threatened birds. Conservation awareness initiatives are required to inform local authorities and communities about the ecological services that coastal wetlands provide (e.g. water purification, fish spawning grounds), as well as their importance for biodiversity.

■ *CONVERSION TO AQUACULTURE*

In recent decades, large areas of natural wetland in this region have been converted to shrimp- and fish-ponds. In Vietnam, this is causing the loss of intertidal mudflats in the Red River delta, although the extra accretion of sediment south of the Red River mouth may compensate for this. Reclamation for aquaculture should be controlled, but because shrimp- and fish-ponds that use traditional, extensive practices can provide valuable habitat for waterbirds, this type of sustainable aquaculture should be promoted.

■ *DISTURBANCE*

The feeding and roosting sites of waterbirds are disturbed by human activities at many coastal wetlands. Tourists and shellfish collectors cause major disturbance at Dongzhaigang Nature Reserve on Hainan and in Zhejiang, and fishermen from mainland China cause disturbance to Inner Deep Bay. People also collect molluscs and crabs in the intertidal zone of the Red River delta, Vietnam, disturbing foraging waterbirds. Indeed, a certain degree of conflict may occur between people and Black-faced Spoonbills at various sites, as *Tellina* (a bivalve thought to be important in the diet of spoonbills) is increasingly collected as food for people, and domestic ducks and crabs. Human access to coastal wetlands needs to be managed, and access to the most sensitive sites limited, particularly around feeding and roost sites.

■ *POLLUTION*

Pollution—in the form of industrial effluent, domestic sewage and agrochemical run-off—is generally severe. For example, there is a serious pollution problem linked to the rapid development of Shenzhen and Hong Kong around the Inner Deep Bay area, as 90% of untreated sewage from Shenzhen is directly released and water treatment plants for this and industrial effluent will not be available for the next 10 years. The water quality of Inner Deep Bay has declined sharply as a result, and in summer 1996 the dissolved oxygen content dropped to almost 0%; there were also indications that the numbers of mudskippers, crabs and worms were declining. The levels of pollution are higher still around Shanghai, and southwards along the coasts of Zhejiang and Fujian. In Vietnam, large quantities of pesticides and fertilisers are used throughout the Red River delta, with agricultural run-off, mixing with untreated human and animal waste from Hanoi and Hai Phong, draining into the Red, Thai Binh, Day and other major tributaries discharging into the Gulf of Tonkin. Long-term programmes are required to reduce pollution levels throughout the region, using stronger and more enforceable environmental legislation, and specific projects to address the major sources of pollution.

Protected areas coverage and management

■ *GAPS IN PROTECTED AREAS SYSTEM*

There are several large and vitally important wetland reserves in this region, including Mai Po Marshes Nature Reserve in Hong Kong and Xuan Thuy National Park in Vietnam, but many key wetlands are not officially protected, and new reserves are needed. Several sites on Kyushu deserve protection, including Ariake bay and Sone tidal flat in Fukuoka and Daijyu-garami in Saga. In southern China, new nature reserves are needed, at (e.g.) Wenzhou and Yueqing bays in Zhejiang. In Vietnam, new protected areas should be promoted in the Red River delta and the coastal zone of Quang Ninh province. Although Vietnam was early to ratify the Ramsar Convention, it has

Large areas of coastal wetland have been converted to shrimp- and fish-ponds, but traditional, extensive management of these ponds can provide valuable habitat for waterbirds.

PHOTO: JOHN HOLMES

About half of the global population of Black-faced Spoonbills winters at the Tsengwen estuary on Taiwan.

PHOTO: MARTIN HALE

Table 3. Conservation issues and strategic solutions for birds on the China Sea coast.

Conservation issues	Strategic solutions
Habitat loss and degradation	
■ *COASTAL RECLAMATION* ■ *CONVERSION TO AQUACULTURE* ■ *DISTURBANCE* ■ *POLLUTION*	➤ Assess the environmental impact of proposed reclamation and development projects ➤ Minimise conversion of key wetlands for aquaculture ➤ Promote traditional, extensive aquacultural practices, to maximise the value of shrimp- and fish-ponds for waterbirds ➤ Control access to important waterbird feeding and roosting sites ➤ Develop long-term programmes to reduce coastal pollution levels
Protected areas coverage and management	
■ *GAPS IN PROTECTED AREAS SYSTEM* ■ *WEAKNESSES IN RESERVE MANAGEMENT*	➤ Establish new protected areas at key coastal wetlands in Japan, mainland China, Taiwan, and Vietnam ➤ Designate more wetlands in Vietnam under the Ramsar Convention ➤ Strengthen reserve management through improved funding, infrastructure and staff training ➤ Stop afforestation of intertidal flats with mangroves in China and the Red River delta
Exploitation of birds	
■ *HUNTING*	➤ Strengthen hunting laws and their enforcement, particularly inside protected areas ➤ Control gun and mist-net ownership
Gaps in knowledge	
■ *INADEQUATE DATA ON THREATENED BIRDS*	➤ Survey poorly known areas, to locate key sites for threatened birds ➤ Continue coordinated counts of Black-faced Spoonbill, and study the impact of human resource use on its distribution and numbers ➤ Investigate the wintering range of Styan's Grasshopper-warbler

so far only nominated a single site, and the relevant government agency should list more of the countries' outstanding wetlands under this convention. Other important roosting or foraging sites for waterbirds should be considered for designation as local nature reserves, where reclamation is not allowed and disturbance and hunting are controlled.

■ *WEAKNESSES IN RESERVE MANAGEMENT*

There are major problems in managing some nature reserves because of pressures from development, disturbance and pollution. At some sites, protected area status is virtually ignored by local people, requiring improvements in institutional capacity, management, manpower and equipment. However, it should be recognised that effective protection of the highly productive coastal wetlands in these reserves would be of great benefit to the livelihood of the local people, particularly through increased fish stocks. At Mai Po Marshes Nature Reserve in Hong Kong, protection of the buffer zones should be strengthened, and development controlled wherever possible. The Tsengwen estuary on Taiwan, the main wintering site for Black-faced Spoonbill, was designated an 'Important Wildlife Area' in 2002; a management plan for this new reserve needs to be prepared and implemented, and proposed developments in the adjacent areas should be carefully assessed and monitored. In the Red River delta in Vietnam, the reserve managers are pursuing inappropriate afforestation schemes, promoted by central government and some development NGOs, causing the loss of intertidal mudflats by planting them with mangroves. This practice is probably the main reason for recent declines in the numbers of wintering Black-faced Spoonbills and Saunders's Gulls at sites in the delta, and should be stopped.

Exploitation of birds

■ *HUNTING*

Hunting is a serious threat on the coasts of mainland China and Vietnam, even inside protected areas. All waterbirds are hunted in China, and in some areas mist-netting of shorebirds for food is very common. Fish-farmers sometimes shoot herons and egrets (for which Black-faced Spoonbills could be mistaken) as pests. Similarly, in Vietnam, trapping and shooting of shorebirds and wildfowl (using mist-nets, air-guns and shot-guns) for both local consumption and export to major cities and to China pose a severe threat to migratory waterbirds in the Red River delta. Laws designed to control hunting should be strengthened, and increased efforts made to enforce the existing (and any new) legislation, particularly inside protected areas. Conservation awareness campaigns are required to make hunters more aware of hunting laws and of the pressure that they are putting on threatened species. However, given the difficulties of implementing hunting and trapping regulations, resources might most effectively focus on control of gun and mist-net ownership, especially near key wetlands.

Black-faced Spoonbills have recently been satellite-tracked from Hong Kong, which has identified some important stopover and breeding sites.

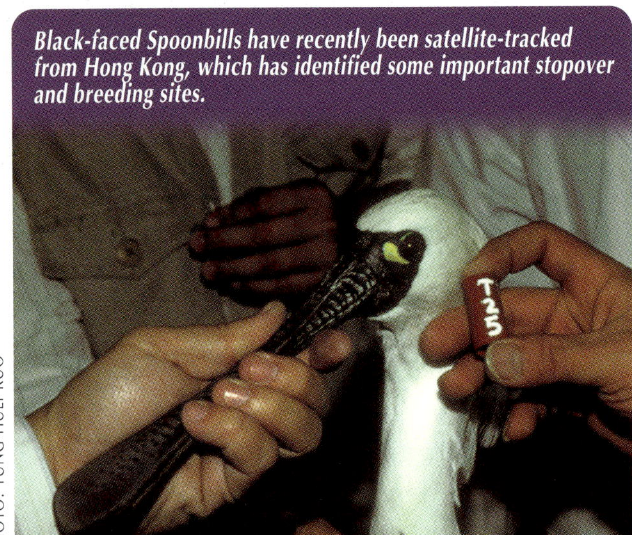

PHOTO: TUNG-HUEI KUO

Gaps in knowledge

■ *INADEQUATE DATA ON THREATENED BIRDS*

The coastal wetlands in parts of this region are poorly known, for example in Fujian and Zhejiang, China, and surveys are needed to identify important sites for waterbird conservation which may warrant designation as new nature reserves. Recent surveys in the poorly known coastal zone of Quang Ninh province in Vietnam found this area to be of greater importance than was previously believed, and located several important sites. The numbers of wintering Black-faced Spoonbills have been counted annually in this region for several years, and this monitoring should continue; studies to investigate the effects of human resource use on the distributions and numbers of this species and other threatened birds would be valuable. An outbreak of avian botulism killed 73 Black-faced Spoonbill at the Tsengwen estuary on Taiwan in winter 2002/2003; the causes of this disease need to be investigated, and plans developed to try to prevent any further outbreaks. The only wintering records of Styan's Grasshopper-warbler are from southern China, mainly in Hong Kong, and systematic surveys are required to clarify of its non-breeding range, and locate additional wintering sites.

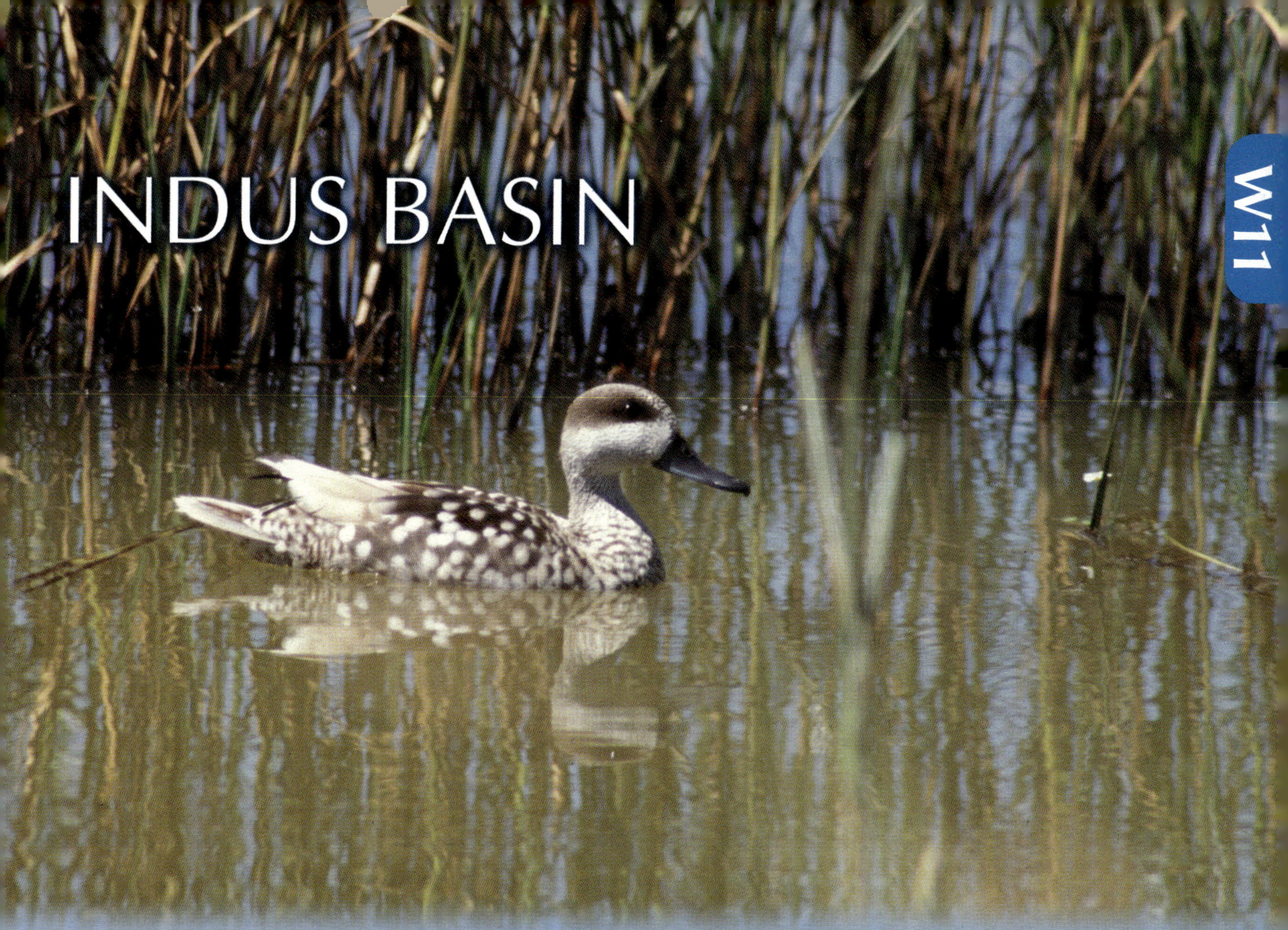

INDUS BASIN

W11

THE wetlands associated with the Indus river and its tributaries are the stronghold for two species of migratory duck in the Asia region, White-headed Duck and Marbled Teal. Populations of the former are concentrated in north-east Pakistan, in Punjab province, while the latter visits shallow wetlands in the lowlands of Sind in southern Pakistan. An important non-breeding population of Dalmatian Pelican also occurs here, mostly on large lakes near the Indus delta. Jerdon's Babbler breeds very locally in reedbeds and other tall riverine grasslands along the Indus river and its tributaries. Breeding populations of Pallas's Fish-eagle and Indian Skimmer were once large, but have declined to low levels over recent decades.

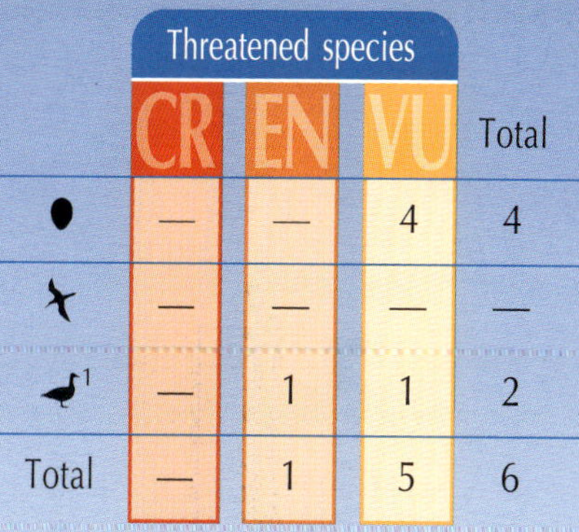

Threatened species	CR	EN	VU	Total
● (breeding)	—	—	4	4
(passage migrant)	—	—	—	—
(non-breeding visitor)[1]	—	1	1	2
Total	—	1	5	6

Key: ● = breeding in this wetland region.
= passage migrant.
= non-breeding visitor.
[1] The Conservation Dependent Dalmatian Pelican is also a non-breeding visitor to this region.

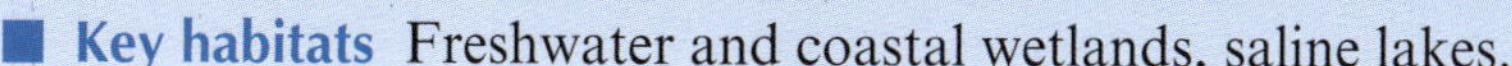

■ **Key habitats** Freshwater and coastal wetlands, saline lakes.

■ **Countries and territories** **Pakistan.**

An important population of Marbled Teals winters in southern Pakistan.
PHOTO: RAY TIPPER

OUTSTANDING IBAs FOR THREATENED BIRDS (see Table 1)

Three IBAs have been selected, which cover some of the most important wintering concentrations of White-headed Duck and Marbled Teal, and populations of Dalmatian Pelican and Pallas's Fish-eagle.

CURRENT STATUS OF HABITATS AND THREATENED SPECIES

Pakistan once supported enormous waterbird populations, especially in winter, but these declined dramatically during the twentieth century. Many natural wetlands disappeared as a result of irrigation and drainage projects to provide more land for agriculture and habitation, although at the same time new lakes and marshes were created upstream of dams and barrages, or as a result of faulty drainage systems or overspill from irrigation canals. Although this overall loss of habitat was undoubtedly a factor in waterbird declines, the main cause was probably the high levels of hunting and disturbance throughout much of the Indus watershed. These pressures continue to depress waterbird numbers, but if they could be controlled there is potential for population recoveries.

CONSERVATION ISSUES AND STRATEGIC SOLUTIONS (summarised in Table 3)

Habitat loss and degradation

■ *CONVERSION TO AGRICULTURE*

Wetlands in Pakistan continue to be affected by irrigation and drainage schemes, usually associated with agricultural development. These activities may alter the water depth, and hence the suitability of wetlands for different waterbird species, and they leave wetlands vulnerable to desiccation and conversion. The potential impact of proposed

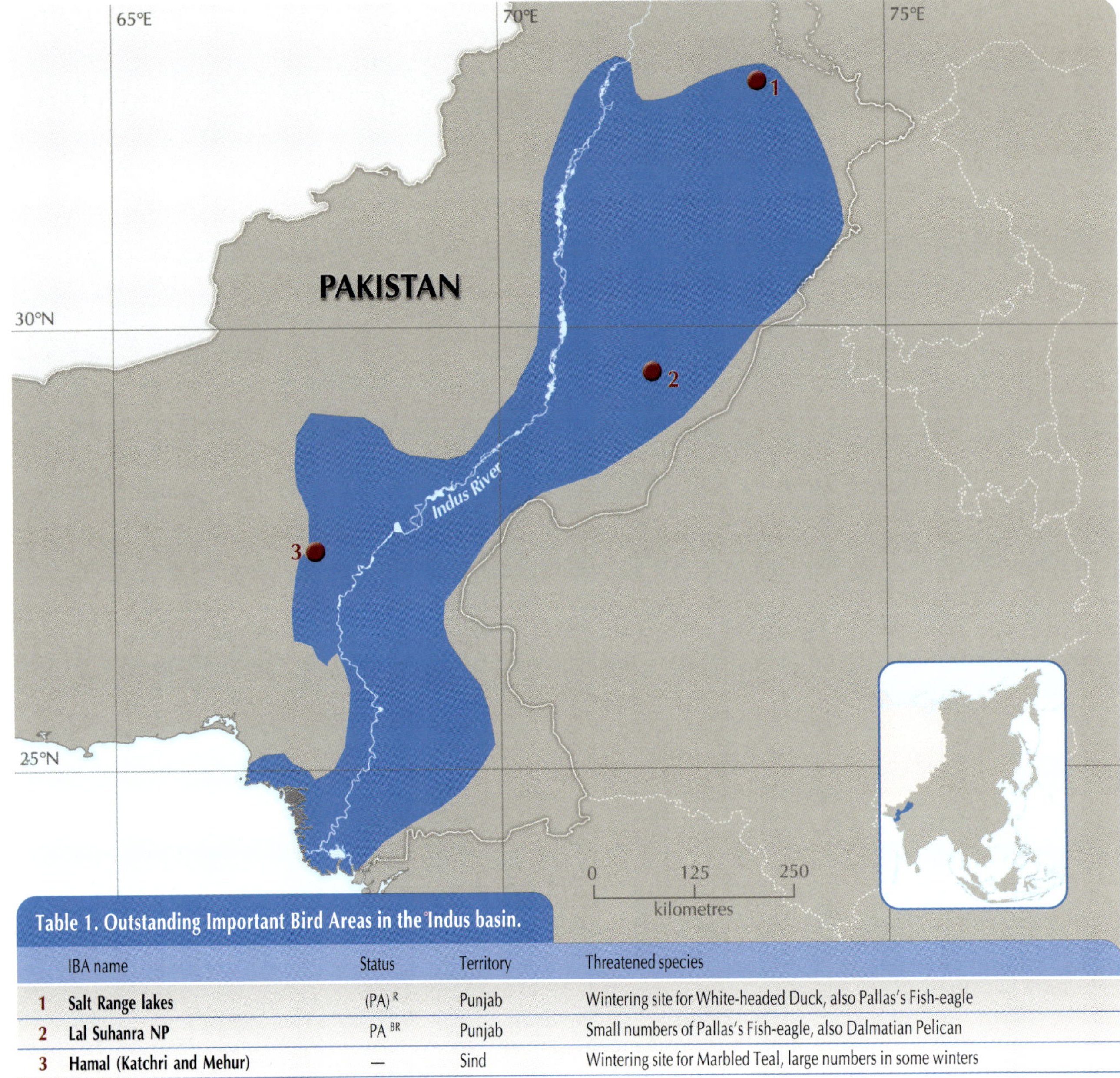

Table 1. Outstanding Important Bird Areas in the Indus basin.

	IBA name	Status	Territory	Threatened species
1	**Salt Range lakes**	(PA) R	Punjab	Wintering site for White-headed Duck, also Pallas's Fish-eagle
2	**Lal Suhanra NP**	PA BR	Punjab	Small numbers of Pallas's Fish-eagle, also Dalmatian Pelican
3	**Hamal (Katchri and Mehur)**	—	Sind	Wintering site for Marbled Teal, large numbers in some winters

Note that more IBAs in this region will be included in the *Important Bird Areas in Asia*, due to be published in early 2004.

Key *IBA name*: NP = National Park.
Status: PA = IBA is a protected area; (PA) = IBA partially protected; — = unprotected; BR = IBA is wholly or partially a Biosphere Reserve (see pp.34–35); R = IBA is wholly or partially a Ramsar Site (see pp.31–32). White-rumped Vulture of region G03 has (or had) populations in several IBAs in this region.

PHOTO: ZULFIKAR ALI/WWF–PAKISTAN

The Salt Range lakes support a large wintering population of White-headed Ducks.

PHOTO: TONY MARTIN/BIRDLIFE

Table 2. Threatened birds of the Indus basin.

Species			Distribution and population
Dalmatian Pelican *Pelecanus crispus*	[bird, outline]	CD	Non-breeding visitor, mostly to southern Sind and the Indus delta
White-headed Duck *Oxyura leucocephala*	[bird, outline]	EN	Several hundred have wintered in Punjab Salt Range, but numbers have recently declined
Marbled Teal *Marmaronetta angustirostris*	[bird, filled]	VU	Winters in the wetlands of Sind, although numbers appear to be declining
Pallas's Fish-eagle *Haliaeetus leucoryphus*	○	VU	A small and declining breeding population is augmented by wintering birds
Sarus Crane *Grus antigone*	○	VU	A tiny breeding population survives in eastern Sind
Indian Skimmer *Rynchops albicollis*	○	VU	A small and declining population, presumably still nests on quieter reaches of the major rivers
Jerdon's Babbler *Chrysomma altirostre*	(?)	VU	Local along the Indus and its tributaries, in tall riverine grasslands

Other threatened waterbirds recorded from this region as rare visitors are: Greater Adjutant *Leptoptilos dubius*, Lesser White-fronted Goose *Anser erythropus* and Siberian Crane *Grus leucogeranus*. In addition to the waterbirds, Greater Spotted Eagle *Aquila clanga* (VU; see F01) and Imperial Eagle *A. heliaca* (VU; see G01) occur in winter.

○ = region estimated to support <10% of global breeding population, (?) = proportion of global breeding population unknown; [bird, filled] = region estimated to support 10–50% of global non-breeding population, [bird, outline] = <10%

irrigation and drainage schemes needs to be assessed through environmental impact assessments, particularly to investigate whether they will affect the hydrology of any key sites for threatened birds; if so, mitigation measures are required to minimise any negative impacts.

WETLAND EXPLOITATION

The vegetation bordering wetlands provides vital feeding, shelter and nesting sites for many waterbirds, but it is commonly cut or burnt, for example to make thatch and other grass products, and to make way for agriculture and settlements. Large trees near lakes and rivers are felled for fuel, timber and fodder, resulting in a shortage of nest sites for Pallas's Fish-eagle. Exploitation of vegetation fringing important wetlands needs to be controlled, to leave sufficient undisturbed habitat for waterbirds. Potential Pallas's Fish-eagle eyries must be protected from cutting and disturbance, possibly through the creation of small village sanctuaries. The planting of nest trees or provision of artificial nesting platforms should be undertaken for this species, carefully sited to avoid disturbance and persecution. Environmental awareness initiatives are needed at key sites, stressing the importance of healthy wetlands (e.g. in preserving water quality and fish stocks) and the need for exploitation of wetland resources to be sustainable.

DAMS AND IRRIGATION

Hydroelectric power schemes and irrigation projects have required the construction of several barrages on the main Indus channel and associated waterways. The diversion of winter flow from rivers seems to have caused significant changes in the distribution of wildfowl, delaying the inundation of lakes and markedly changing their ecology. Barrages reduce the incidence of flooding, which in turn allows the gradual encroachment of temporary cultivation on riverbeds and islands, a potential threat to Indian Skimmer which nests colonially in these habitats. However, large areas of potentially important new wetlands (and wet grassland suitable for Jerdon's Babbler) have been created by irrigation projects, usually by seepage from drainage canals or outfall drains. These new wetlands should be managed for threatened waterbirds wherever possible. Special measures are required to protect all active Indian Skimmer colonies from encroachment and disturbance (once they have been located by surveys), and perhaps also abandoned sites which the species might recolonise.

SILTATION

Sediment run-off and flooding have increased in the Indus watershed because of deforestation in headwater regions and heavy erosion in the densely populated lowlands, causing greater sediment load in the major rivers and more rapid silt deposition in wetlands. As a result, some important sites for waterbirds (e.g. Kushdil Khan) are drying out and becoming overgrown with vegetation. Programmes of forest conservation and reforestation are required in the upper catchment in the Western Himalayas (see F04), possibly coupled with dredging of heavily silted wetlands.

DISTURBANCE

Pakistan's human population is concentrated near sources of water, and rivers carry abundant boat traffic and are almost universally settled by people. The margins of wetlands tend to be heavily grazed and cultivated, and intensively used for fishing, hunting, reed-cutting and recreation. As a consequence, wetland habitats suffer frequent disturbance, which makes them much less attractive to breeding waterbirds. Disturbance needs to be controlled, e.g. at Pallas's Fish-eagle eyries and Indian Skimmer colonies, possibly by designating these sites as local sanctuaries, where human activities can be regulated while they are nesting (e.g. by fences and guards). Effective protection of potential Indian Skimmer breeding sites (i.e. large river islands) might result in its return to previously occupied stretches of river.

Table 3. Conservation issues and strategic solutions for birds of the Indus basin.

Conservation issues	Strategic solutions
Habitat loss and degradation	
■ *CONVERSION TO AGRICULTURE* ■ *WETLAND EXPLOITATION* ■ *DAMS AND IRRIGATION* ■ *SILTATION* ■ *DISTURBANCE* ■ *POLLUTION* ■ *REDUCED FOOD SUPPLY*	➤ Assess the environmental impact of proposed irrigation and drainage schemes ➤ Plant nest trees or erect artificial nest platforms for Pallas's Fish-eagle ➤ Manage wetlands created by dams and irrigation to maximise their value for waterbirds ➤ Minimise disturbance near Pallas's Fish-eagle eyries and Indian Skimmer colonies ➤ Develop programmes to reduce pollution levels in wetlands ➤ Control populations of introduced fish, and avoid further introductions
Protected areas coverage and management	
■ *GAPS IN PROTECTED AREAS SYSTEM* ■ *WEAKNESSES IN RESERVE MANAGEMENT*	➤ Establish new protected areas at key wetlands for threatened birds, and designate Pallas's Fish-eagle eyries and Indian Skimmer colonies (once they have been located) as local sanctuaries ➤ Prepare management plans for wetland reserves, to reconcile the needs of people and wildlife ➤ Strengthen reserve management through improved funding, infrastructure and staff training
Exploitation of birds	
■ *HUNTING*	➤ Strengthen and enforce hunting laws, with education programmes to publicise the legal status and importance of threatened waterbirds
Gaps in knowledge	
■ *INADEQUATE DATA ON THREATENED BIRDS*	➤ Survey breeding Pallas's Fish-eagles and Indian Skimmers, and determine the measures required to protect their nest sites ➤ Monitor the numbers and distribution of wintering White-headed Duck and Marbled Teal ➤ Study the poorly known wetlands in the Indus-irrigated area of Sind

■ *POLLUTION*

Several types of pollution affect wetlands in Pakistan. Agrochemical run-off causes build-up of toxins or fertilisers in wetlands; the accumulation of pesticides through the food chain can reduce the breeding success of raptor populations, and could be affecting Pallas's Fish-eagle, while fertilisers cause wetland eutrophication. Large quantities of domestic sewage and industrial effluent are released into the Indus and its tributaries, which could be a particular problem for Marbled Teal, as the shallow wetlands it prefers appear especially vulnerable to the build-up of pollutants. A coordinated response to the problem of pollution is required, involving monitoring of outflows, water treatment and control of chemicals used or produced by agriculture and industry.

■ *REDUCED FOOD SUPPLY*

Stocking of wetlands in Pakistan (e.g. Khabbaki lake) with herbivorous fish (e.g. *Tilapia*) is thought to have contributed to the reduction in duck populations through direct competition for resources. Efforts may be required to control the populations of these fish at some important wetlands for threatened birds, and further introductions should be avoided.

Protected areas coverage and management

■ *GAPS IN PROTECTED AREAS SYSTEM*

Some of the most important wetlands in the Indus watershed are officially protected, but other sites with significant numbers of threatened waterbirds should become new protected areas, including Sunari lake, Hamal (Katchri), Badam, Mahboub Shah lake, Ghauspur jheel and Mangla reservoir. Pallas's Fish-eagle eyries and Indian Skimmer colonies also need to be protected, possibly by designation as local sanctuaries.

■ *WEAKNESSES IN RESERVE MANAGEMENT*

The management infrastructure is inadequate at many protected areas: managers and guards are scarce, underfunded, poorly equipped and poorly motivated. This situation needs to be rectified through an appropriate injection of funds for salaries, equipment and training. Given the population pressure at many of the wetlands, management plans need to be prepared for the reserves to reconcile the needs of people and wildlife, with important wetlands clearly divided between human use portions and wildlife sanctuaries.

Exploitation of birds

■ *HUNTING*

Hunting of waterbirds is popular in many regions of Pakistan, both for food and sport. In Chitral district alone an estimated 1,700 armed hunters have constructed 770 artificial ponds to attract ducks. In other areas, netting operations continue throughout the winter to supply markets in larger towns with waterbirds. Some sites suffer heavy, uncontrolled shooting by army personnel, and collection of eggs is widespread. Hunting laws need to be strengthened and enforced, particularly in protected areas. Education programmes are needed to publicise the legal status and importance of threatened species to local hunters and communities, including using information boards near key breeding and wintering sites. Outside protected areas, sustainable hunting should be promoted.

Gaps in knowledge

■ *INADEQUATE DATA ON THREATENED BIRDS*

There are some important gaps in knowledge of the threatened waterbirds of Pakistan. Surveys are required, principally along rivers, to locate Pallas's Fish-eagle eyries and Indian Skimmer colonies, and determine what measures are required for their protection. Regular, coordinated counts need to be continued at wintering sites for White-headed Duck and Marbled Teal; both species may change their wintering grounds from year to year, so long-term monitoring is required for their requirements to be fully understood. The ornithologically unknown wetlands within the Indus-irrigated area of Sind, mainly in the Rice Canal Command, need surveying.

NORTH INDIAN WETLANDS

W12

THIS region includes the wetlands on the Gangetic plains, together with the coastal and freshwater wetlands of Rajasthan and Gujarat. It supports more breeding Sarus Cranes (on shallow wetlands and associated agricultural land) and Indian Skimmers (on the vast system of rivers) than any other region, and the scattered lakes and reservoirs also provide habitat for non-breeding flocks of Sarus Crane and significant numbers of breeding and wintering Pallas's Fish-eagles. The 'central population' of Siberian Crane winters in the region, but has declined to near extinction in recent years. The swampy wetlands of northern India, particularly in Bihar, were once the stronghold for Pink-headed Duck, a species which may now be extinct, although it could possibly survive in the more inaccessible parts of its former range. The conservation of grassland birds on the Gangetic plains is covered in G02.

Threatened species	CR	EN	VU	Total
●	1	—	3	4
passage migrant	—	—	—	—
non-breeding visitor[1]	1	1	3	5
Total	2	1	6	9

Key: ● = breeding in this wetland region.
= passage migrant.
= non-breeding visitor.
[1] The Conservation Dependent Dalmatian Pelican is also a non-breeding visitor to this region.

- **Key habitats** Freshwater wetlands on riverine plains, and associated agricultural land; coastal wetlands.
- **Countries and territories** **India** (Punjab, Haryana, Delhi, Rajasthan, Gujarat, Uttar Pradesh, Madhya Pradesh, Bihar); **Nepal**.

Keoladeo National Park in Rajasthan is one of the most actively protected wetlands in Asia, but even here maintenance of inflow during drought is often a major problem. PHOTO: OTTO PFISTER

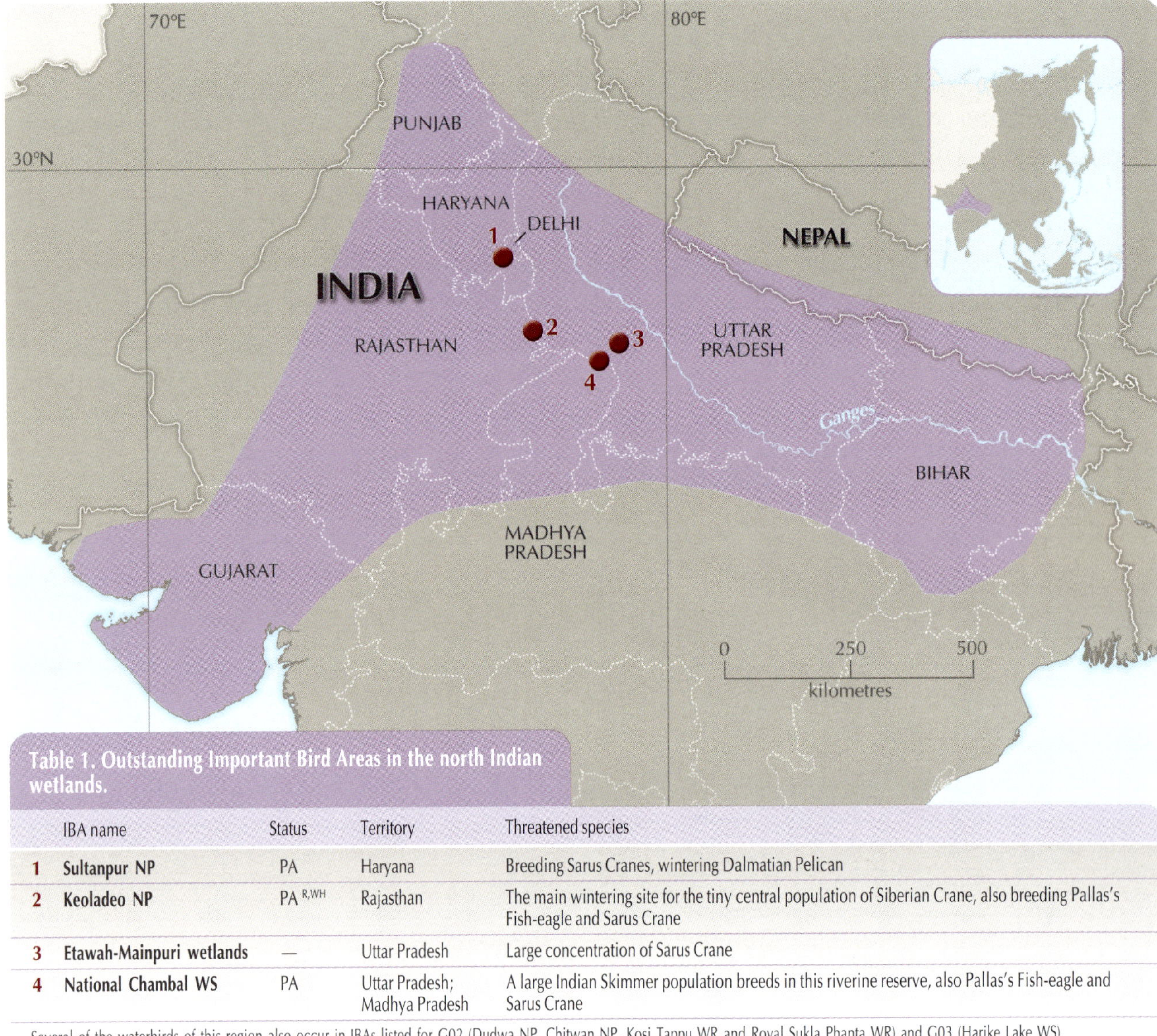

Table 1. Outstanding Important Bird Areas in the north Indian wetlands.

	IBA name	Status	Territory	Threatened species
1	**Sultanpur NP**	PA	Haryana	Breeding Sarus Cranes, wintering Dalmatian Pelican
2	**Keoladeo NP**	PA [R,WH]	Rajasthan	The main wintering site for the tiny central population of Siberian Crane, also breeding Pallas's Fish-eagle and Sarus Crane
3	**Etawah-Mainpuri wetlands**	—	Uttar Pradesh	Large concentration of Sarus Crane
4	**National Chambal WS**	PA	Uttar Pradesh; Madhya Pradesh	A large Indian Skimmer population breeds in this riverine reserve, also Pallas's Fish-eagle and Sarus Crane

Several of the waterbirds of this region also occur in IBAs listed for G02 (Dudwa NP, Chitwan NP, Kosi Tappu WR and Royal Sukla Phanta WR) and G03 (Harike Lake WS).
Note that more IBAs in this region will be included in the *Important Bird Areas in Asia*, due to be published in early 2004.

Key *IBA name*: NP = National Park; WS = Wildlife Sanctuary.
Status: PA = IBA is a protected area; (PA) = IBA partially protected; — = unprotected; R = IBA is wholly or partially a Ramsar Site (see pp.31–32); WH = IBA is wholly or partially a World Heritage Site (see p.34). White-rumped, Indian and/or Slender-billed Vultures of region G03 have (or had) populations in several IBAs in this region.

OUTSTANDING IBAs FOR THREATENED BIRDS (see Table 1)

Four IBAs have been selected, covering important breeding sites of Pallas's Fish-eagle, Sarus Crane and Indian Skimmer, as well as wintering Dalmatian Pelican and Siberian Crane. More sites of value to these birds will be documented during BirdLife's ongoing IBA Project.

CURRENT STATUS OF HABITATS AND THREATENED SPECIES

The vast Gangetic plains are densely populated and utilised throughout. The natural wetlands there have been greatly modified by drainage, irrigation and encroachment for agriculture and development over many centuries. Despite these changes, some extensive, rich wetlands remain, and this region is remarkable for the degree to which wildlife is able to coexist with man in wetland and agricultural areas, linked to the Hindu philosophical tradition. The region also includes extensive coastal wetlands in Gujarat, and freshwater lakes and reservoirs in Gujarat and Rajasthan, vitally important for both wildlife and man in these arid states. Huge numbers of waterbirds occur, including several threatened species, but significant recent declines are linked to wetland drainage, unsustainable exploitation for fuel and fodder, increased agricultural intensification and use of agrochemicals, human disturbance and hunting. Many of the major tributaries flowing into the Ganges have been dammed for hydropower and irrigation schemes, which has created some new wetlands, but is negatively affecting the habitat of riverine birds. The global range of Pink-headed Duck was centred on this region, but it almost certainly became extinct during the twentieth century.

CONSERVATION ISSUES AND STRATEGIC SOLUTIONS (summarised in Table 3)

Habitat loss and degradation

CONVERSION TO AGRICULTURE

Rapid human population growth over recent decades has caused intense competition for land and aquatic resources, and large-scale conversion of wetlands to agriculture. Water is diverted from rivers to supply irrigation projects, causing

The 'central population' of Siberian Crane winters in this region, but has declined to near extinction in recent decades.

PHOTO: JACOB WIJPKEMA

Table 2. Threatened birds of the north Indian wetlands.

Species			Distribution and population
Dalmatian Pelican *Pelecanus crispus*	[bird]	CD	Small numbers of non-breeding birds regular in Gujarat and near Delhi
Spot-billed Pelican *Pelecanus philippensis*	[bird]	VU	Small numbers of non-breeding birds on the larger wetlands
Lesser Adjutant *Leptoptilos javanicus*	[bird]	VU	Rare visitor, occasionally breeds
Greater Adjutant *Leptoptilos dubius*	[bird]	EN	Rare non-breeding visitor
Pink-headed Duck *Rhodonessa caryophyllacea*	EX?	CR	This region was historically a stronghold, but the species is now presumed extinct
Baer's Pochard *Aythya baeri*	[bird]	VU	Small numbers of non-breeding birds regular in Nepal
Pallas's Fish-eagle *Haliaeetus leucoryphus*	○	VU	A significant but declining breeding population, augmented in winter by non-breeding birds
Siberian Crane *Grus leucogeranus*	[bird]	CR	Tiny numbers visit Rajasthan in winter
Sarus Crane *Grus antigone*	◉	VU	This region supports a high proportion of the global population
Indian Skimmer *Rynchops albicollis*	●	VU	A large population breeds along major rivers, but may be in decline

Other threatened waterbirds recorded from this region as rare (or possibly extinct) visitors are: White-headed Duck *Oxyura leucocephala*, Lesser White-fronted Goose *Anser erythropus*, Baikal Teal *Anas formosa* and Marbled Teal *Marmaronetta angustirostris*. In addition to the waterbirds, Greater Spotted Eagle *Aquila clanga* (VU; see F01) and Imperial Eagle *A. heliaca* (VU; see G01) occur in winter.

● = region estimated to support >90% of global breeding population, ◉ = 50–90%, ○ = 10–50%; [bird] = region estimated to support <10% of global non-breeding population; EX? = probably extinct

rivers and associated wetlands to become shallower and more vulnerable to desiccation and drainage. Wetland margins are increasingly being cultivated, including riverbanks and islands. There have been major changes in cropping patterns, including intensification of cultivation of sugarcane in Uttar Pradesh and soybean in Madhya Pradesh. As a result, the number of extensive wetlands with populations of large waterbirds has declined. Moreover, shortage of wetlands during the dry season forces waterbirds to gather in dense concentrations which are highly vulnerable to drought, hunting or other localised threats. In some areas, Sarus Cranes are forced to feed in fields, causing major economic losses and antagonism between farmers and birds.

This loss of natural wetlands needs to be stopped, with the remaining fragments protected as a network of community reserves. Small patches of non-cultivable marshland should be managed for wildlife within agricultural landscapes. Forms of agriculture beneficial to threatened birds should be encouraged, notably wet rice cultivation (used by breeding Sarus Cranes). State and national governments should develop appropriate regulations and policies for wetland conservation, and the Indian Ministry of Rural Development should modify their current land classification to remove wetlands from the 'wasteland' category. Environmental awareness programmes, stressing the traditional relationships between man and wildlife and the role of wetlands in maintaining water quality and commercial fish stocks, will help minimise further wetland conversion and promote the protection of Sarus Crane nest sites in agricultural areas.

INCREASED CULTIVATION OF WATER CHESTNUT

Water chestnut *Trapa* is an aquatic plant which produces edible fruit, and is an important crop in northern India. It forms extensive surface mats that reduce wetland productivity and inhibit feeding by waterbirds. The cultivation of water chestnut has recently increased in this region, devaluing many wetlands as waterbird sites. It is planted during the Sarus Crane breeding season, causing disturbance and egg loss, and is harvested in winter, preventing non-breeding waterbirds from using the wetlands. In addition, large amounts of pesticide are sprayed directly into the wetlands to protect the crop (see below). The cultivation of water chestnut should be strictly

Indian Skimmer colonies are vulnerable to changes in river flow caused by dams and irrigation projects.

PHOTO: OTTO PFISTER

regulated, particularly at key wetlands for threatened birds, e.g. by establishing zones in large wetlands where its cultivation is prohibited.

WETLAND EXPLOITATION

The vegetation bordering wetlands provides vital feeding, shelter and nesting sites for many waterbirds, but it is commonly cut or burnt, for example to make thatch and other grass products. The tall trees used by Pallas's Fish-eagle, storks and other waterbirds as nest sites are sometimes felled for timber, fuel and fodder. Mud removal (for construction) and clearance of aquatic vegetation further reduce habitat availability, although moderate cutting or grazing of vegetation is essential to keep shallow wetlands from becoming overgrown. In general an ample border of natural vegetation, including tall trees, should be maintained around wetlands. Potential eyries for Pallas's Fish-eagle need to be protected from cutting and disturbance. The planting of nest trees or provision of artificial nesting platforms should be undertaken for this species, carefully sited to avoid disturbance and persecution.

DAMS AND IRRIGATION

Several factors affect the natural water supply to wetlands, particularly dams and irrigation projects. With so many demands on water, maintenance of inflow during drought is often a major problem, even at Keoladeo National Park, one of the most actively protected wetlands in Asia. Dams have been constructed for hydropower and irrigation schemes on many of the major tributaries of the Ganges. In some cases dam projects have created reservoirs or other wetlands (often by seepage from irrigation canals) that provide important new habitat for some waterbirds. However, although dams control flooding, they cause problems for Indian Skimmer and other sandbar-nesting birds because: (1) dams block sediment flow and cause intermittent sediment-poor flooding: the first starves sandbars of sediment while the second scours them away; (2) many dams store water during the wet season and release it in the dry season (when skimmers breed), a factor that reduces seasonal changes in water levels such that they may never drop sufficiently low to expose much sand: alternatively, insufficient water release during the dry season may prevent both the formation of river islands and nesting by skimmers, or connect river islands with the bank allowing access by predators; (3) the control of flooding regimes is often not beneficial as it allows agricultural development on banks and islands, further reducing the area of bare sand available for nesting; (4) if the dam is used to generate electricity, sandbars may be flooded on a daily basis because of fluctuations in water release. The impact of dams on river flow is exacerbated by the widespread use of large pumps to remove water from rivers to irrigate the surrounding land.

Changes of water level in rivers and other wetlands, and the impact of these changes on threatened waterbird populations, need to be monitored. There may be a need to control drainage and irrigation activities, even some way upstream, if wetland sanctuaries are being negatively affected. When dams and irrigation projects create new wetlands, they should be managed to maximise their value for waterbirds. Given the impact that dam projects can have on Indian Skimmer and other sandbar-nesting birds, proposed new hydropower and irrigation schemes upstream of important nesting colonies should be carefully considered (including through environmental impact assessments). Where dams are already in place on rivers with important skimmer colonies, for example on the Chambal River, conservationists should work with the dam authorities to minimise the potential problems detailed above. It may also be possible to manage the colonies themselves to reduce the potential problems, e.g. by artificially maintaining the river sandbanks and islands, preventing agricultural encroachment into these habitats and developing methods to protect active nests from flooding.

SILTATION AND FLOODING

Dry-season flooding may destroy colonies of Indian Skimmers situated on low sandbanks. This can result from natural phenomena, but is exacerbated by deforestation. The excessive erosion of deforested headwaters may also lead to heavy siltation and deposition of sediment in wetlands. However, the incidence of flooding and silt deposition is greatly modified by the dams on many of the region's rivers (see above). Programmes of reforestation and forest conservation are required in the catchments of the region's rivers (including in the Himalayas: see F04), possibly coupled with dredging of heavily silted wetlands.

■ *DEVELOPMENT (URBAN, INDUSTRIAL, ETC.)*
Wetland habitat is being lost because of the constant spread of villages and industry. The mining of banks and beds of wetlands and rivers for sand, gravel and stones causes disturbance, lowers food supply and reduces nesting habitat for birds. Intensive dredging regimes destroy river islands, contributing to the decline of the Indian Skimmer. Excavation of sand for nearby lime and brick industries has caused siltation at several sites. These activities need to be controlled when they are damaging wetland habitat, especially inside protected areas.

■ *DISTURBANCE*
Wetlands are intensively used and disturbed by large numbers of fishermen and hunters, for transportation and by local people for bathing, etc. The large number of people and cattle visiting the fringes of wetlands increases the risk of eggs and chicks being trampled. Wetland reserves need to be patrolled to minimise disturbance in the more sensitive areas, particularly during the breeding season. Outside protected areas, human disturbance needs to be controlled at Pallas's Fish-eagle eyries, Indian Skimmer colonies and Sarus Crane nesting and roosting areas, possibly by designating these sites as local sanctuaries where human activities can be regulated (e.g. by guarding during the breeding season).

■ *POLLUTION/PESTICIDES*
Wetlands are affected by air- and water-borne industrial pollutants, agrochemical run-off (causing eutrophication) and municipal waste from adjacent towns, and diminishing water supplies mean that many wetlands are less frequently flushed by pulses of clean water. Pesticides are used in the cultivation of water chestnut (see above) and other crops. Large waterbirds and raptors are usually predators or scavengers, and they are thus particularly vulnerable to build-up of toxic chemicals. The use of pesticides (such as dieldrin and aldrin) or fertilisers should be strictly controlled, especially in areas close to important wetlands, and traditional organic agricultural practices encouraged wherever possible.

■ *REDUCED FOOD SUPPLY*
Many wetlands suffer intensive uncontrolled fishing and this must reduce food supply for piscivorous species such as Pallas's Fish-eagle. Waterways are sometimes deliberately poisoned to kill fish for eating, destroying aquatic fauna including birds. A sustained recovery of fish stocks could be achieved by preventing fishing in parts of wetlands, and perhaps imposing fishing quotas, and fishing with chemicals and dynamite should be banned.

Protected areas coverage and management

■ *GAPS IN PROTECTED AREAS SYSTEM*
The region has many wetland reserves and sanctuaries, which provide some protection for the threatened waterbirds. However, given the dispersed distributions of species such as Pallas's Fish-eagle and Sarus Crane, protected areas alone cannot ensure their survival. To persist in this densely populated and intensively used region, these birds need conservation measures at the landscape level (e.g. control of water levels and promotion of sympathetic farming practices), together with a network of protected areas (including small, locally-managed Community Conservation

Table 3. Conservation issues and strategic solutions for birds of the north Indian wetlands.

Conservation issues	Strategic solutions
Habitat loss and degradation	
■ CONVERSION TO AGRICULTURE ■ INCREASED CULTIVATION OF WATER CHESTNUT ■ WETLAND EXPLOITATION ■ DAMS AND IRRIGATION ■ SILTATION AND FLOODING ■ DEVELOPMENT (URBAN, INDUSTRIAL, ETC.) ■ DISTURBANCE ■ POLLUTION/PESTICIDES ■ REDUCED FOOD SUPPLY	➤ Maintain patches of wetland within agricultural landscapes ➤ Encourage wet rice cultivation to provide additional breeding habitat for Sarus Crane ➤ Develop government regulations and policies for wetland conservation, and remove wetlands from the 'wasteland' category in the current land classification ➤ Regulate the cultivation of water chestnut in important wetlands ➤ Plant nest trees or erect artificial nest platforms for Pallas's Fish-eagles ➤ Manage wetlands created by dams and irrigation to maximise their value for waterbirds ➤ Assess the environmental impact of proposed dam projects, especially on important rivers for threatened waterbirds ➤ Control flow regimes below dams to protect Indian Skimmer colonies ➤ Regulate human activities at key wetlands to minimise disturbance ➤ Limit use of agrochemicals, and encourage traditional organic farming methods ➤ Improve management of fish stocks, and ban fishing with chemicals
Protected areas coverage and management	
■ GAPS IN PROTECTED AREAS SYSTEM ■ WEAKNESSES IN RESERVE MANAGEMENT	➤ Develop networks of small, locally managed community conservation areas to protect waterbird nest and roost sites ➤ Establish new protected areas for Indian Skimmer and Pallas's Fish-eagle, and review the design and management of National Chambal Sanctuary and other riverine reserves ➤ Manage wetland reserves to maximise their value for threatened waterbirds
Exploitation of birds	
■ HUNTING AND PERSECUTION	➤ Improve enforcement of hunting laws, by patrolling protected wetlands and monitor the sale of waterbirds at markets in Bihar and elsewhere ➤ Minimise conflict between farmers and Sarus Cranes through education programmes and an award scheme
Gaps in knowledge	
■ INADEQUATE DATA ON THREATENED BIRDS	➤ Conduct surveys to map Pallas's Fish-eagle eyries and Indian Skimmer colonies, and determine what measures are required for their protection ➤ Study dam management and river flow regimes to help determine what adjustments are required to protect Indian Skimmer colonies ➤ Search for Pink-headed Duck, principally in its former strongholds in Bihar

Areas, a new category of reserve created by a 2002 amendment to the Indian Wildlife Protection Act) to protect some of the most important nesting, roosting and feeding sites. This region is the last stronghold of Indian Skimmer, as well as being important for Pallas's Fish-eagle, and the protection of their habitats is therefore a high priority; new riverine reserves should be established in areas with important populations, for example along the Ganges River in Uttar Pradesh and Bihar. The design and management of existing reserves which include riverine habitats (and skimmer colonies), such as National Chambal Sanctuary, should be reviewed to ensure that they confer the maximum protection to the species (by minimising disturbance and the chances of flooding or drying out).

WEAKNESSES IN RESERVE MANAGEMENT

Some wetland reserves are inappropriately managed. At Keoladeo National Park the spread of aquatic vegetation after grazing control caused a drastic reduction in the amount of open water and hence in the reserve's suitability for some waterbird species. In most wetland reserves, vegetation (e.g. *Paspalum distichum*, *Vetiveria zizanoides*, etc.) needs to be cleared periodically and moderate grazing encouraged. In Gujarat and Uttar Pradesh, some lakebeds require digging or dredging in the dry season to ensure suitable depths for pelicans, overgrazing should be controlled, and encroaching *Prosopis juliflora* annually removed. The educational potential of Keoladeo National Park has not been fully realised; it is ideally situated near major population centres, and visited by large numbers of tourists, so efforts should be made to provide as much information as possible regarding conservation.

Exploitation of birds

HUNTING AND PERSECUTION

Ducks are hunted with shotguns or nets in northern India, either for food or sport, and in some regions storks are also targeted. Even the Sarus Crane, once immune from persecution because of traditional beliefs, is increasingly affected by the removal of its nests or eggs by rice farmers who consider it an agricultural pest. Efforts are needed to control hunting, including by patrolling wetland protected areas and intercepting illegal hunters, and by monitoring and controlling the sale of waterbirds as food in markets, particularly in Bihar and other areas where large-scale hunting is a problem. Measures are required to minimise conflicts between Sarus Cranes and local people, including setting aside small areas of uncultivated wetland and grassland for nesting (see above), and education programmes stressing the threatened status of this species and the traditional relationship between cranes and man. An award scheme to recognise as 'Sarus Protectors' those farmers who have provided suitable habitat on their land has proved successful in maintaining the traditional religious links between farmers and cranes, and in increasing the participation of local people in crane conservation; this approach should be used more widely.

Gaps in knowledge

INADEQUATE DATA ON THREATENED BIRDS

Surveys are required to identify key feeding and roosting areas for Sarus Crane, and along rivers to locate Pallas's Fish-eagle eyries and Indian Skimmer colonies, and to determine what measures are required for the protection of these sites. The effects of dams and irrigation schemes on the flow regimes in the regions' rivers should be studied to determine how to prevent flooding of active Indian Skimmer colonies. Although it is probably extinct, searches should be made for Pink-headed Duck, principally in its former strongholds in Bihar, including analysis of satellite images to locate potentially suitable areas of seasonally flooded swamps on riverine floodplains.

This region is the global stronghold for Sarus Crane, but it is being affected by changing agricultural practices and the loss of natural wetlands.

PHOTO: TIM LOSEBY

SOUTH INDIAN and SRI LANKAN WETLANDS

THE wetlands of this region support a high proportion of the global population of Spot-billed Pelicans, with many colonies associated with water storage reservoirs or 'tanks' on the Deccan plateau in southern India and the dry zone lowlands of Sri Lanka. Lesser Adjutant also occurs, but in relatively low numbers, and small numbers of Spoon-billed Sandpiper have been found wintering in coastal wetlands.

- **Key habitats** Freshwater and coastal wetlands.
- **Countries and territories** **India** (Karnataka, Andhra Pradesh, Kerala, Tamil Nadu, Orissa); **Sri Lanka**.

Threatened species

	CR	EN	VU	Total
●	—	—	2	2
[passage migrant symbol]	—	—	—	—
[non-breeding visitor symbol]	—	—	1	1
Total	—	—	3	3

Key: ● = breeding in this wetland region.
[passage migrant symbol] = passage migrant.
[non-breeding visitor symbol] = non-breeding visitor.

The South Indian and Sri Lankan wetlands region overlaps with part of Conservation International's Western Ghats and Sri Lanka Hotspot (see pp.20–21).

Wetlands in the dry zone of Sri Lanka support large numbers of Spot-billed Pelicans, and a small population of Lesser Adjutants. PHOTO: RAY TIPPER

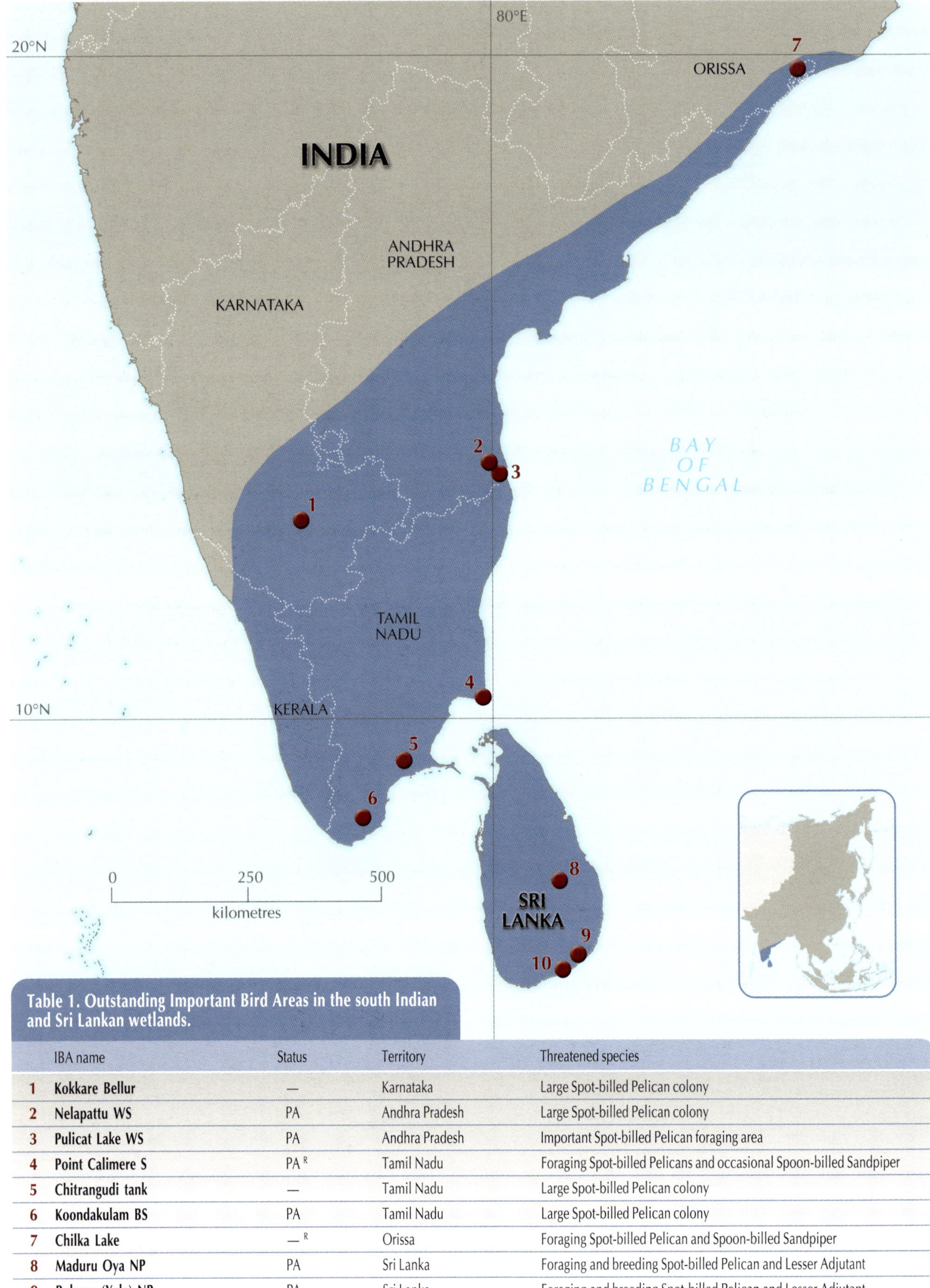

Table 1. Outstanding Important Bird Areas in the south Indian and Sri Lankan wetlands.

	IBA name	Status	Territory	Threatened species
1	**Kokkare Bellur**	—	Karnataka	Large Spot-billed Pelican colony
2	**Nelapattu WS**	PA	Andhra Pradesh	Large Spot-billed Pelican colony
3	**Pulicat Lake WS**	PA	Andhra Pradesh	Important Spot-billed Pelican foraging area
4	**Point Calimere S**	PA R	Tamil Nadu	Foraging Spot-billed Pelicans and occasional Spoon-billed Sandpiper
5	**Chitrangudi tank**	—	Tamil Nadu	Large Spot-billed Pelican colony
6	**Koondakulam BS**	PA	Tamil Nadu	Large Spot-billed Pelican colony
7	**Chilka Lake**	— R	Orissa	Foraging Spot-billed Pelican and Spoon-billed Sandpiper
8	**Maduru Oya NP**	PA	Sri Lanka	Foraging and breeding Spot-billed Pelican and Lesser Adjutant
9	**Ruhuna (Yala) NP**	PA	Sri Lanka	Foraging and breeding Spot-billed Pelican and Lesser Adjutant
10	**Bundala NP**	PA R	Sri Lanka	Foraging and breeding Spot-billed Pelican and Lesser Adjutant

Note that more IBAs in this region will be included in the *Important Bird Areas in Asia*, due to be published in early 2004.

Key *IBA name*: BS= Bird Sanctuary; NP = National Park; S = Sanctuary; WS = Wildlife Sanctuary.
Status: PA = IBA is a protected area; (PA) = IBA partially protected; — = unprotected.; R = IBA is wholly or partially a Ramsar Site (see pp.31–32)

Spot-billed Pelicans nest in trees adjacent to wetlands, but there is now a shortage of potential sites in many areas.

PHOTO: OTTO PFISTER

Table 2. Threatened birds of the south Indian and Sri Lankan wetlands.

Species			Distribution and population
Spot-billed Pelican *Pelecanus philippensis*	●	VU	A high proportion of the global population nests in colonies in this region, with recent estimates of 2,000–2,500 birds in southern India and c.5,000 on Sri Lanka
Lesser Adjutant *Leptoptilos javanicus*	○	VU	Scarce and local breeding bird on Sri Lanka, suspected to breed in small numbers in southern India
Spoon-billed Sandpiper *Eurynorhynchus pygmeus*	(wader symbol)	VU	Small numbers regularly winter on the east coast of India

Other threatened waterbirds recorded from this region as rare (or extinct) visitors are: Pink-headed Duck *Rhodonessa caryophyllacea* (extinct), Spotted Greenshank *Tringa guttifer* and Indian Skimmer *Rynchops albicollis*. In addition to the waterbirds, Greater Spotted Eagle *Aquila clanga* (VU; see F01) occurs in winter in southern India (but not Sri Lanka).

● = region estimated to support 50–90% of global breeding population, ○ = <10%; (wader symbol) = region estimated to support <10% of global non-breeding population

OUTSTANDING IBAs FOR THREATENED BIRDS (see Table 1)

Ten IBAs have been selected in this region, primarily for their importance to Spot-billed Pelican.

CURRENT STATUS OF HABITATS AND THREATENED SPECIES

The water storage reservoirs or 'tanks' favoured by Spot-billed Pelicans have been constructed close to virtually every rural village, and provide a vital source of water and other resources in this relatively arid region. They are therefore fairly secure, although some are being encroached for agriculture or industrial development, disturbed and/or polluted. The estuaries and coastal lagoons visited by non-breeding Spot-billed Pelicans and small numbers of Spoon-billed Sandpiper are being degraded by aquaculture, industrial development and siltation.

CONSERVATION ISSUES AND STRATEGIC SOLUTIONS (summarised in Table 3)

Habitat loss and degradation

■ *CONVERSION TO AGRICULTURE*

Drainage and conversion for cultivation affect many wetlands in this region, both freshwater and coastal. The margins of freshwater wetlands are often lost under cultivation, with an estimated 37% of village tanks being encroached by agriculture in Karnataka. Some wetlands are drying out because of excessive water extraction to supply irrigation projects, and a few, e.g. Bundala Sanctuary on Sri Lanka, are adversely affected by the inflow of excess irrigation water. Important unprotected nesting and feeding areas for Spot-billed Pelicans and Lesser Adjutant need to be designated as new wetland sanctuaries and protected from conversion for agriculture, and their water supplies regulated.

■ *CONVERSION TO AQUACULTURE*

Nearly 400 km^2 of the Andhra Pradesh coastline had been converted to prawn ponds by the mid-1990s, and a further 600 km^2 were earmarked for development, particularly around the Krishna estuary. Some large inland wetlands are also affected, e.g. Kolleru Lake in Andhra Pradesh, once a major Spot-billed Pelican feeding area but now almost completely converted to aquaculture. The margins of Chilka Lake in Orissa are steadily being converted to prawn ponds, and aquaculture projects are replacing wetland habitat in Sri Lanka. Awareness campaigns are required to emphasise the importance of these wetlands for biodiversity (including fish stocks) to local and state governments, to

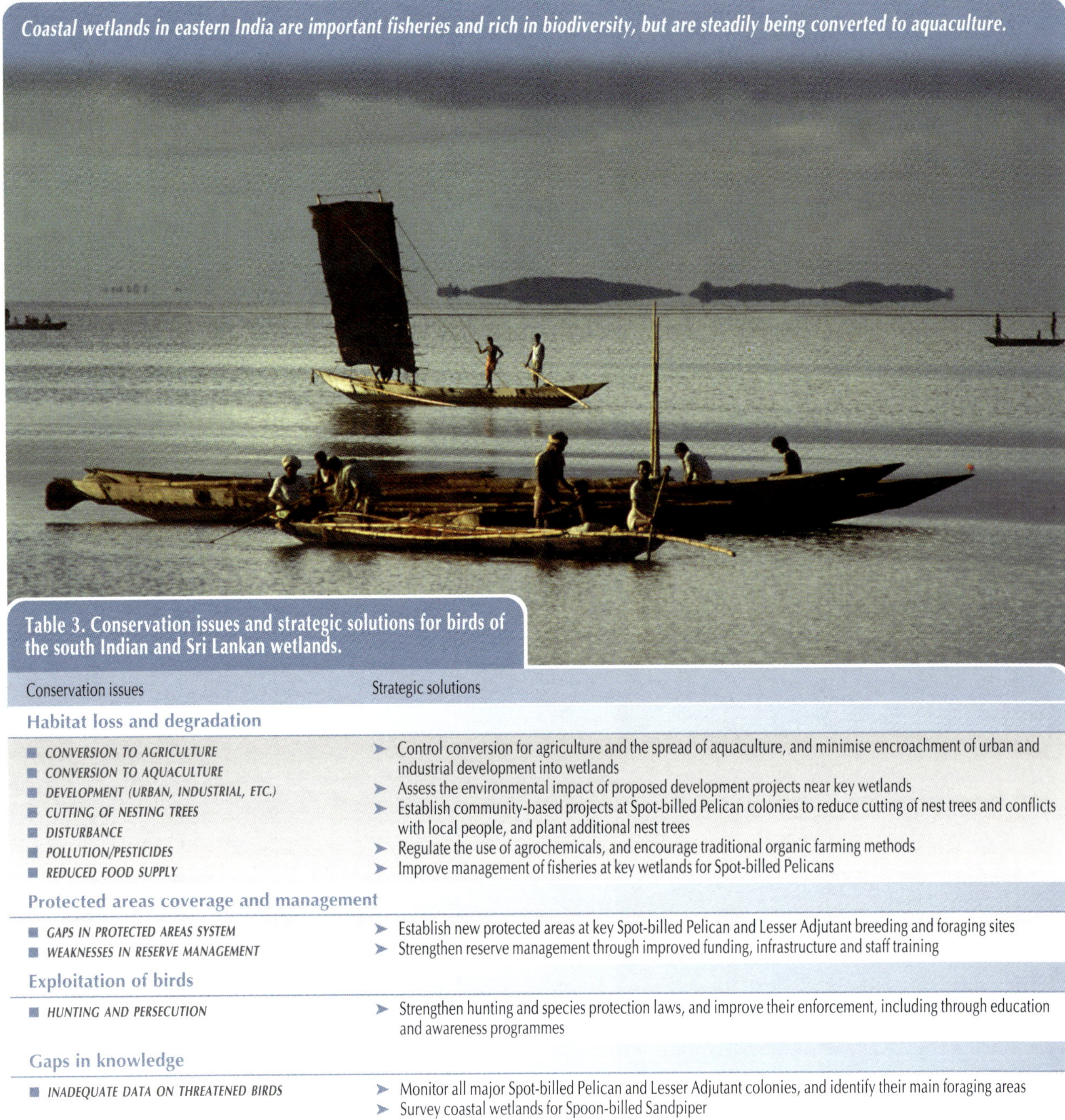

Coastal wetlands in eastern India are important fisheries and rich in biodiversity, but are steadily being converted to aquaculture.

PHOTO: JON HORNBUCKLE

Table 3. Conservation issues and strategic solutions for birds of the south Indian and Sri Lankan wetlands.

Conservation issues	Strategic solutions
Habitat loss and degradation	
■ CONVERSION TO AGRICULTURE ■ CONVERSION TO AQUACULTURE ■ DEVELOPMENT (URBAN, INDUSTRIAL, ETC.) ■ CUTTING OF NESTING TREES ■ DISTURBANCE ■ POLLUTION/PESTICIDES ■ REDUCED FOOD SUPPLY	➤ Control conversion for agriculture and the spread of aquaculture, and minimise encroachment of urban and industrial development into wetlands ➤ Assess the environmental impact of proposed development projects near key wetlands ➤ Establish community-based projects at Spot-billed Pelican colonies to reduce cutting of nest trees and conflicts with local people, and plant additional nest trees ➤ Regulate the use of agrochemicals, and encourage traditional organic farming methods ➤ Improve management of fisheries at key wetlands for Spot-billed Pelicans
Protected areas coverage and management	
■ GAPS IN PROTECTED AREAS SYSTEM ■ WEAKNESSES IN RESERVE MANAGEMENT	➤ Establish new protected areas at key Spot-billed Pelican and Lesser Adjutant breeding and foraging sites ➤ Strengthen reserve management through improved funding, infrastructure and staff training
Exploitation of birds	
■ HUNTING AND PERSECUTION	➤ Strengthen hunting and species protection laws, and improve their enforcement, including through education and awareness programmes
Gaps in knowledge	
■ INADEQUATE DATA ON THREATENED BIRDS	➤ Monitor all major Spot-billed Pelican and Lesser Adjutant colonies, and identify their main foraging areas ➤ Survey coastal wetlands for Spoon-billed Sandpiper

encourage them to control the spread of aquaculture, especially inside protected areas.

■ *DEVELOPMENT (URBAN, INDUSTRIAL, ETC.)*

Many wetlands are being encroached by industrial facilities, roads and settlements, and this needs to be prevented or carefully controlled, especially inside protected areas. Bundala National Park on Sri Lanka is threatened by a number of proposed developments in its immediate vicinity, not least the Hambantota 'mega-city': an oil refinery, harbour, airport, hotels, etc. Development projects should be reviewed through environmental impact assessment and, if necessary, revised in order to reconcile the needs of nature conservation and economic development.

■ *CUTTING OF NESTING TREES*

Both Spot-billed Pelican and Lesser Adjutant nest in trees adjacent to wetlands, but there is now a chronic shortage of potential nest sites in many areas because of the cutting or lopping of trees for timber, fuelwood and fodder, and several large pelican colonies have been lost. In southern India, most pelican colonies are in or close to villages as this is often where the only suitable trees remain, sometimes resulting in conflict with local people (see *Hunting and persecution* below). The trees used by nesting pelicans and storks need protection, including through the establishment of wetland sanctuaries, and trees should be planted periodically to provide additional nesting habitat. Artificial nest platforms could also be experimented with in selected protected areas, but only if this does not encourage local people to cut down nesting trees. Community forestry projects should be established near important colonies to develop alternative sources of timber, fuel and fodder, and thus lessen the pressure on colony trees. Other community-based projects could be developed to minimise conflict between pelicanries and people, e.g. tourism, with awareness campaigns to build local pride in their global importance.

■ *DISTURBANCE*

Human disturbance (particularly as a result of fishing activity) affects most important sites for Spot-billed Pelican. Over 30,000 people depend on Pulicat Lake for their survival and, even though the area is legally protected, current legislation is inadequate to control over-exploitation and disturbance by fishermen. At Chilka Lake, around 9,000 fishing vessels are active throughout the day and night, with detrimental effects on waterbird populations. Human activities need to be managed inside protected areas (and ideally also at unprotected wetlands) so that nesting Spot-billed Pelicans and Lesser Adjutants are not disturbed, and levels of disturbance in important foraging areas are minimised.

■ *POLLUTION/PESTICIDES*

Agrochemicals are widely applied to farmland, and are contaminating wetlands; these are likely to build up in the food chain and affect piscivores and scavengers such as Spot-billed Pelican and Lesser Adjutant, and increased eutrophication and excessive aquatic vegetation reduce the value of wetlands to pelicans. The salinity of wetlands at Point Calimere is being increased by nearby salt-works. The use of pesticides or fertilisers should be controlled, especially in areas close to important wetlands, along with improvements in sewage treatment and management of salt-works. Traditional organic agricultural practices should be encouraged wherever possible.

■ *REDUCED FOOD SUPPLY*

Intensive fishing in lakes frequented by Spot-billed Pelicans presumably reduces their food supply, and may be a particular problem near colonies because large quantities of fish are needed to feed the nestlings. A sustained recovery of fish stocks could be achieved by preventing fishing in parts of wetlands, and perhaps imposing fishing quotas.

Protected areas coverage and management

■ *GAPS IN PROTECTED AREAS SYSTEM*

Several protected areas in southern India and Sri Lanka include breeding colonies and important foraging areas of Spot-billed Pelican. All other nesting sites of this species and Lesser Adjutant should be considered for official protection in both countries, together with a satellite network of secure foraging sites to ensure that sufficient food is available through each breeding season.

■ *WEAKNESSES IN RESERVE MANAGEMENT*

Given the economic importance of wetlands in this densely populated region, wetland protected areas need to be managed with a view to balancing the needs of people and wildlife. Improved control of human activities is necessary in many reserves, for example by zoning sections of lakes where fishing is not allowed, to maintain fish stocks and undisturbed foraging grounds for birds. Protected areas in general are under-funded, under-staffed and poorly equipped, which needs to be addressed through improved finances and better training. A successful project has been conducted at the Kokkare Bellur Spot-billed Pelican colony to captive-rear and then release into the wild the many nestlings which fall from nests; this approach could be followed at other colonies to boost productivity.

Exploitation of birds

■ *HUNTING AND PERSECUTION*

Hunting is a problem at many wetlands in southern India and Sri Lanka, and eggs and young are taken at some Spot-billed Pelican colonies; this was one of the main reasons for the loss of Kolleru and Moonradaippu pelicanries. Hunting

Human disturbance, particularly as a result of fishing activity, affects most important sites for Spot-billed Pelican.

PHOTO: OTTO PFISTER

laws need to be strengthened and strictly enforced at key wetlands, particularly in protected areas. Education programmes should publicise the legal status and importance of threatened species to local hunters and communities. The control of gun ownership might be the best method of addressing this problem in some areas.

At some Spot-billed Pelican colonies in villages local people have tried to discourage nesting in trees close to their houses because of the odours, or in economically important tamarind *Tamarindus indica* trees because the fruit mature while the birds are nesting. Measures to reduce such conflicts could include the planting of new nest trees in less sensitive areas, and the enforcement of species protection laws backed up by conservation awareness programmes and possibly some form of compensation.

Gaps in knowledge

■ *INADEQUATE DATA ON THREATENED BIRDS*

Considerable research has been conducted to locate and monitor the size and breeding success of Spot-billed Pelican colonies in southern India. This work needs to be continued and expanded for this species and Lesser Adjutant in India and Sri Lanka, to help identify the waterbird colonies where conservation action is most urgently required. More data are required on key foraging sites and patterns of wetland use by breeding Spot-billed Pelicans (which travel considerable distances from some colonies to feed) and Lesser Adjutants, using marked birds and radio- and satellite-tracking. Shorebird surveys are needed to determine the size (and hence global significance) of the Spoon-billed Sandpiper population that winters in eastern India.

ASSAM and SYLHET PLAINS

W14

IN India, all wetlands in this region fall within the catchment of the Brahmaputra river, while in Bangladesh it includes inland wetland habitats associated with the Brahmaputra–Jamuna and Meghna–Kalni rivers. This huge lowland area of marshy plains and large lakes is now the global stronghold of Greater Adjutant, following the historical crash of the vast breeding colonies in Myanmar. It also supports important breeding populations of Spot-billed Pelican, Lesser Adjutant and Pallas's Fish-eagle, and a large non-breeding population of Baer's Pochard. The conservation of grassland birds on the Assam and Sylhet plains is covered in G02, and forest wetland birds (such as White-winged Duck) in F06.

- **Key habitats** Freshwater wetlands on riverine plains.
- **Countries and territories** **India** (West Bengal, Arunachal Pradesh, Assam, Meghalaya); **Bangladesh**.

Threatened species

	CR	EN	VU	Total
● (breeding)	1	1	4	6
(passage migrant)	—	—	—	—
(non-breeding visitor)[1]	—	—	3	3
Total	1	1	7	9

Key: ● = breeding in this wetland region.
= passage migrant.
= non-breeding visitor.
[1] The Conservation Dependent Dalmatian Pelican is also a non-breeding visitor to this region.

The Assam and Sylhet plains region is within Conservation International's Indo-Burma Hotspot (see pp.20–21).

Lesser Adjutant is a fairly common breeding bird in Assam.
PHOTO: JACOB WIJPKEMA

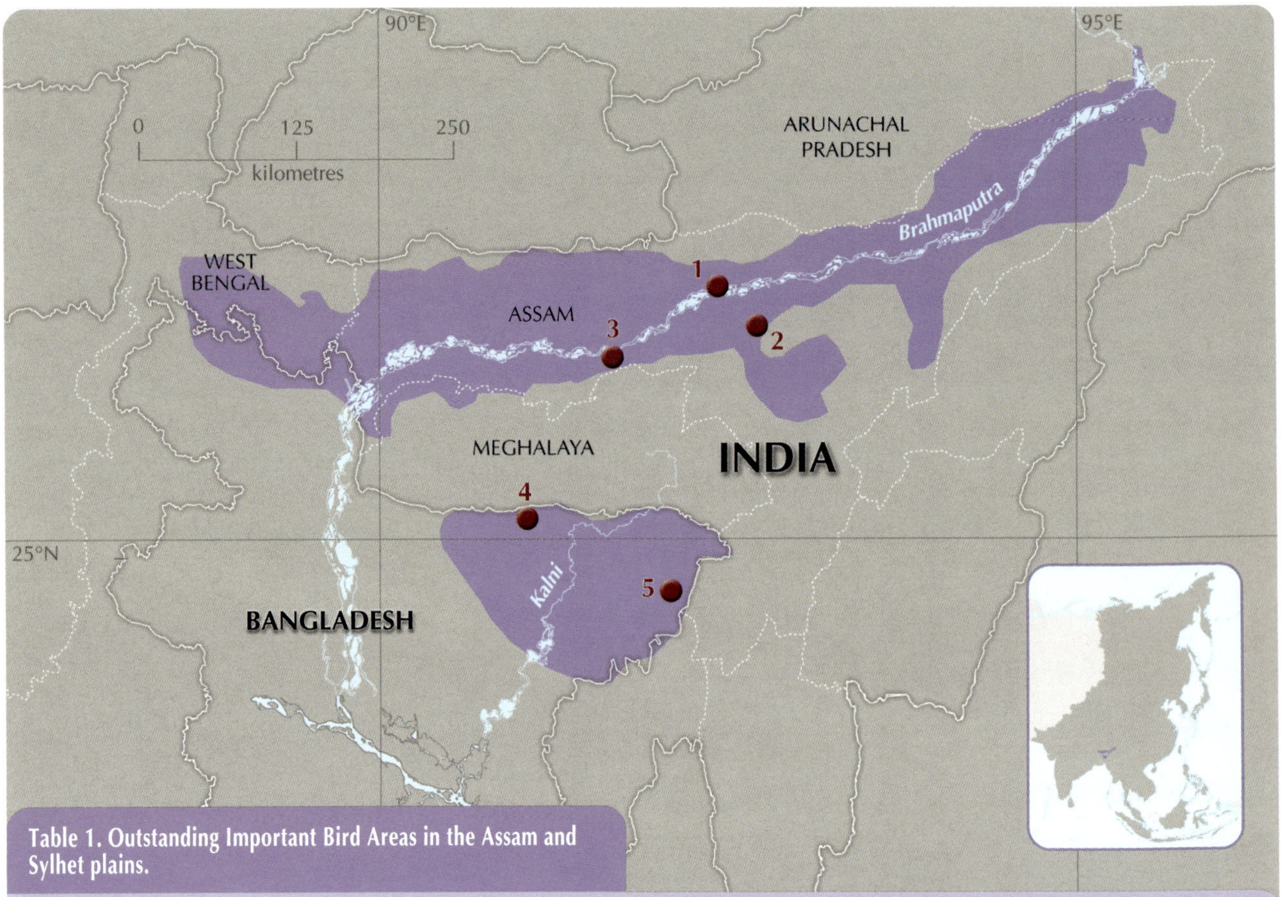

Table 1. Outstanding Important Bird Areas in the Assam and Sylhet plains.

	IBA name	Status	Territory	Threatened species
1	**Orang NP**	PA	Assam	Breeding Spot-billed Pelican, Lesser Adjutant and Pallas's Fish-eagle
2	**Nagaon**	—	Assam	Large breeding colonies of Greater Adjutant
3	**Deepor Beel WS**	PA R	Assam	Non-breeding Lesser Adjutant, Greater Adjutant and Baer's Pochard
4	**Tangua Haor**	PA R	Bangladesh	Breeding Pallas's Fish-eagle, non-breeding Baer's Pochard
5	**Hakaluki Haor**	—	Bangladesh	Breeding Pallas's Fish-eagle, non-breeding Baer's Pochard

Important waterbird populations occur in several of the IBAs listed for region G02 (Jaldapara WS, D'Ering Memorial WS, Manas NP, Kaziranga NP and Dibru-Saikhowa NP). Note that more IBAs in this region will be included in the *Important Bird Areas in Asia*, due to be published in early 2004.

Key *IBA name*: NP = National Park.
Status: PA = IBA is a protected area; (PA) = IBA partially protected; — = unprotected; R = IBA is wholly or partially a Ramsar Site (see pp.31–32). White-rumped and/or Slender-billed Vultures of region G03 have (or had) populations in several IBAs in this region.

The most extensive natural wetlands to survive in India are in Assam, including in Kaziranga National Park.

PHOTO: TIM LOSEBY

Several of the threatened waterbirds that breed in this region require tall trees for nesting, including Greater Adjutant.

PHOTO: JON HORNBUCKLE

OUTSTANDING IBAs FOR THREATENED BIRDS (see Table 1)

Five IBAs have been selected in this region, primarily because of their importance to Greater Adjutant, Baer's Pochard and Pallas's Fish-eagle, as well as Spot-billed Pelican and Lesser Adjutant. Many more sites for these birds will be documented during BirdLife's ongoing IBA Project.

CURRENT STATUS OF HABITATS AND THREATENED SPECIES

A century ago, the lowlands of north-east India and north-east Bangladesh were covered by shallow wetlands, seasonally flooded grasslands and swamp forest, but large areas have now been drained and converted to cultivation and pasture. The plains of Assam have been intensively developed, but there are still some extensive natural wetlands there. Habitat loss has been more severe in West Bengal and northern Bangladesh, where only a few fragments of wetland remain. Many of the surviving wetlands are included in the region's extensive network of protected areas, some of which are very large, having been established for the protection of Indian rhinoceros *Rhinoceros unicornis* and other large mammals. However, some of these reserves are affected by the long-running political unrest in several north-east Indian states, which might also disrupt other conservation initiatives in the region.

CONSERVATION ISSUES AND STRATEGIC SOLUTIONS (summarised in Table 3)

Habitat loss and degradation

■ *CONVERSION TO AGRICULTURE*

Many wetlands are shallow and particularly susceptible to being drained and converted for pasture or cultivation. Throughout much of the Brahmaputra valley and Bangladesh local people have cut canals to drain water from small wetlands, substantially reducing waterbird habitats. Similarly, irrigation projects divert water from rivers in the dry season, thus reducing the inflow to wetlands. Partly owing to deforestation in their upper catchments, the

Table 2. Threatened birds of the Assam and Sylhet plains.

Species			Distribution and population
Dalmatian Pelican *Pelecanus crispus*	(non-breeding <10%)	CD	Small non-breeding population in Assam
Spot-billed Pelican *Pelecanus philippensis*	(breeding 10–50%)	VU	Several hundred pairs breed in Assam
Lesser Adjutant *Leptoptilos javanicus*	(breeding 10–50%)	VU	Fairly common breeding bird in Assam
Greater Adjutant *Leptoptilos dubius*	(breeding 50–90%)	EN	A high proportion of the known global population breeds in Assam
Marbled Teal *Marmaronetta angustirostris*	(non-breeding <10%)	VU	Several recent records of small numbers
Pink-headed Duck *Rhodonessa caryophyllacea*	EX?	CR	Recorded in the past, but the species is now presumed extinct
Baer's Pochard *Aythya baeri*	(non-breeding 10–50%)	VU	Large non-breeding population in Assam and, especially, north-east Bangladesh
Pallas's Fish-eagle *Haliaeetus leucoryphus*	(breeding 10–50%)	VU	Significant breeding population in the *haors* (lake depression) wetlands of north-east Bangladesh and the Brahmaputra floodplain
Sarus Crane *Grus antigone*	(breeding <10%)	VU	A small population survives in Assam
Indian Skimmer *Rynchops albicollis*	(non-breeding <10%)	VU	Scarce non-breeding visitor along major rivers

Other threatened waterbirds recorded from this region as rare visitors are: Oriental Stork *Ciconia boyciana*, Lesser White-fronted Goose *Anser erythropus*, Baikal Teal *Anas formosa*, Hooded Crane *Grus monacha*, Spotted Greenshank *Tringa guttifer* and Spoon-billed Sandpiper *Eurynorhynchus pygmeus*. In addition to the waterbirds, Greater Spotted Eagle *Aquila clanga* (VU; see F01) and Imperial Eagle *A. heliaca* (VU; see G01) occur in winter. Note that three species that occur in forested wetlands in this region, White-bellied Heron *Ardea insignis*, White-winged Duck *Cairina scutulata* and Masked Finfoot *Heliopais personata*, are covered in F06.

● = region estimated to support 50–90% of global breeding population, ◐ = 10–50%, ○ = <10%; (bird symbol, dark) = region estimated to support 10–50% of global non-breeding population, (bird symbol, outline) = <10%; EX? = probably extinct

There is an important breeding population of Pallas's Fish-eagles in the haor wetlands of north-east Bangladesh.

PHOTO: TIM LOSEBY

Brahmaputra and other rivers carry increasing silt loads which they deposit in riverine wetlands, making them shallower, facilitating reclamation and leading to succession to grassland. Important sites need to be protected by controlling or eliminating drainage and conversion to agriculture, careful management of irrigation projects, even some way upstream, and digging or dredging to counter excessive siltation. Awareness campaigns are required, targeted at local governments with jurisdiction over important wetlands and civil society, to emphasise the global importance of these wetlands for large waterbirds and other biodiversity, and the role that clean natural wetlands play in preserving water quality and fish stocks.

■ *DEVELOPMENT (URBAN, INDUSTRIAL, ETC.)*

The region's growing network of roads and expanding industry are reducing wetland habitat. In Bangladesh, a government scheme of oil and gas exploration may cause damage to wetlands. In general, such developments should avoid important wetlands, and major development projects need careful consideration through environmental impact assessments.

■ *CUTTING OF NEST TREES*

Loss of nesting trees is a major threat to the region's important breeding populations of Spot-billed Pelican, Lesser Adjutant, Greater Adjutant and Pallas's Fish-eagle. Trees are cut commercially for match production or to provide firewood to brick factories, and privately for building materials, furniture or fuel. Many of the surviving mature trees are on private land near human habitation, where nesting storks are sometimes deliberately persecuted to remove their unpleasant noise and aroma. Special protection needs to be given to the trees used by nesting pelicans, storks and eagles, including the establishment of small wetland sanctuaries or 'mini-reserves'. As many

Table 3. Conservation issues and strategic solutions for birds of the Assam and Sylhet plains.

Conservation issues	Strategic solutions
Habitat loss and degradation	
■ CONVERSION TO AGRICULTURE ■ DEVELOPMENT (URBAN, INDUSTRIAL, ETC.) ■ CUTTING OF NEST TREES ■ REDUCED FOOD SUPPLY ■ DISTURBANCE ■ POLLUTION/PESTICIDES ■ INTRODUCED WEEDS	➤ Minimise drainage and extraction of water for irrigation near key wetlands for threatened birds, including through awareness campaigns targeted at local governments with jurisdiction over these areas ➤ Assess the environmental impact of proposed development projects, and develop new roads and industry away from key wetlands ➤ Work with landowners to protect pelican, stork and eagle nest trees, and plant trees and erect artificial nest platforms to provide additional habitat ➤ Reduce pressure on waterbird nest trees by promoting community forestry projects and alternatives to fuelwood, and replant native swamp forests ➤ Improve management of fisheries at key wetlands, and ban fishing using chemicals ➤ Continue traditional management at waste disposal sites to ensure a supply of carrion for Greater Adjutants, and/or provide supplementary feeding ➤ Regulate human activities at key wetlands to minimise disturbance ➤ Limit the use of agrochemicals, and encourage traditional organic farming methods ➤ Control water hyacinth infestations using mechanical or biological methods
Protected areas coverage and management	
■ GAPS IN PROTECTED AREAS SYSTEM ■ WEAKNESSES IN RESERVE MANAGEMENT	➤ Protect all extensive areas of natural wetland by establishing new protected areas or extending existing reserves ➤ Create wildlife sanctuaries in the *haor* basin of north-east Bangladesh ➤ Develop a network of sanctuaries to protect wetland fragments where threatened waterbirds nest and feed ➤ Improve the capacity of the government departments responsible for environment and forestry in India and Bangladesh
Exploitation of birds	
■ HUNTING ■ PERSECUTION	➤ Strengthen hunting and species protection laws and improve their enforcement, especially at key wetlands ➤ Control gun ownership
Gaps in knowledge	
■ INADEQUATE DATA ON THREATENED BIRDS	➤ Locate and monitor key waterbird colonies and foraging sites, notably those of Greater Adjutant ➤ Conduct winter surveys for Baer's Pochard ➤ Search for Pink-headed Duck

waterbirds colonies are on private land, notably those of Greater Adjutant, negotiation with landowners is required to ensure that any disagreements over colony protection are settled amicably and in favour of conservation. To provide additional nesting habitat, trees should be planted and artificial nest platforms erected near existing and abandoned waterbird colonies and at other potential nesting sites, carefully sited to avoid disturbance and persecution. Existing efforts in the *haor* basin of north-east Bangladesh to replant native swamp forest on village common lands that were once forested should be promoted (both there and possibly elsewhere). Meanwhile, pressure could be reduced on wood resources, by (e.g.) community forestry projects and the use of gas for brick production and domestic purposes.

REDUCED FOOD SUPPLY

Many wetlands in this region are important fisheries, but the already intense fishing pressure has increased in recent years, reducing the availability of food for pelicans and storks (as well as people). Fishing communities should return to a longer-term perspective in fishery management (with less frequent harvesting of fish), by establishing sanctuaries in parts of the wetlands, and enforcing conservation provisions on the leaseholders of the most valuable fisheries. Some fishermen use pesticides to kill fish, at least in Dibru-Saikhowa National Park, a practice which should be banned. In north-east India, Greater Adjutants frequently feed at municipal dumps or in agricultural areas, but recent changes in agricultural and municipal practices appear to have reduced the quantity of carrion available to them. The continuation of traditional management methods should be considered at key waste disposal sites (e.g. those at Tezpur and Gauhati) to provide a regular food supply to the storks; provision of waste and carrion at designated rural sites might also prove effective.

DISTURBANCE

Many wetlands are heavily used by people, and at least two Spot-billed Pelican colonies in Assam have been abandoned or shifted because of human disturbance. Most large waterbirds have declined dramatically in Bangladesh owing to habitat disturbance and hunting, and the large and rapidly expanding human population (130 million in 2001, 220 million by 2020) will undoubtedly exert huge pressure on the remaining natural wetlands and their resources. Protected areas need to provide waterbirds with secure and undisturbed foraging and nesting habitat: they should be zoned and patrolled to prevent excessive use or disturbance by people in their core areas.

POLLUTION/PESTICIDES

The use of synthetic agrochemicals is largely uncontrolled in India and Bangladesh, and their run-off affects aquatic ecology, especially where cultivation has spread to the fringes of wetlands. Inflow of fertilisers promotes excessive growth of aquatic vegetation, notably of water hyacinth (see below), while toxic pesticides and herbicides build up in the food chain. Other sources of pollution are tea treatment plants, oil fields and sewage, which cause eutrophication and deposit toxins in wetlands near towns and cities. The use of pesticides or fertilisers and the output of industrial effluents need to be monitored and controlled, especially in areas close to important wetlands, along with improvements in sewage treatment. Traditional organic agricultural practices should be encouraged.

INTRODUCED WEEDS

The clogging of the Kaziranga wetlands in Assam with water hyacinth *Eichhornia crassipes* is a major threat to Spot-billed Pelicans, which do not nest when floods fail to flush this weed from the wetlands, thereby reducing the open water that they require for feeding. Mechanical

The haor wetlands of north-east Bangladesh are important fisheries.

PHOTO: PAUL THOMPSON

removal or biological control methods should be used to clear the carpets of weed.

Protected areas coverage and management

■ *GAPS IN PROTECTED AREAS SYSTEM*

This is the only part of the Indian subcontinent where there is an opportunity to protect extensive tracts of natural wetlands and seasonally flooded grasslands. Some large protected areas have already been established, and new reserves and extensions to existing reserves should be considered to increase the coverage of these rapidly dwindling habitats. The *haor* basin of north-east Bangladesh currently receives little protection, and the outstanding IBAs listed above should be considered for declaration as wildlife sanctuaries; Tangua Haor was recently designated as a Ramsar Site, although the district administration has issued fishing permits for the site, a potential conflict of interests that needs to be resolved. Many threatened waterbird colonies and feeding areas are in relatively small, unprotected wetland fragments, and Greater Adjutants nest mainly in urban and suburban areas within a few kilometres of rubbish dumps. A network of small, community-managed wetland sanctuaries or 'mini-reserves' needs to be developed in Assam and elsewhere to protect these sites.

Large numbers of Baer's Pochards winter in Assam and north-east Bangladesh.

PHOTO: TIM LOSEBY

■ *WEAKNESSES IN RESERVE MANAGEMENT*

In both India and Bangladesh, a capacity-building exercise is required to focus on governmental departments responsible for environment and conservation, through the provision of extra funding and training of personnel.

Exploitation of birds

■ *HUNTING*

Waterbirds, especially ducks, are commonly shot or trapped (often using mist-nets at night) in many parts of north-east India and Bangladesh; their meat is sold in local markets, for example in Assam. Hunting laws need to be strengthened and strictly enforced at important wetlands, particularly in protected areas. Education programmes are required to publicise the legal status and importance of threatened species to local hunters and communities, including using information boards at key breeding and wintering sites. Control of gun ownership might be a practical method of addressing this problem in some areas.

■ *PERSECUTION*

Pelicans, storks and other waterbirds are persecuted in Assam (and presumably elsewhere in the region) because fishermen and fish-pond owners believe that are damaging their business by reducing fish stocks. Eggs and chicks may be collected for food or destroyed, incubating adults shot off the nest, or birds deliberately poisoned. As with hunting, threatened species need to be legally protected from these activities, while awareness programmes need to publicise the laws and improve the public's perception of waterbirds.

Gaps in knowledge

■ *INADEQUATE DATA ON THREATENED BIRDS*

This is one of the most important regions of Asia for several species of threatened large waterbirds, notably Greater Adjutant, and further research is needed to locate and monitor their breeding colonies and major foraging sites, and help determine where conservation action is required. This region appears to be important for wintering Baer's Pochards, but the information on its status at some sites is unclear, apparently because of the difficulty of distinguishing it from similar *Aythya* ducks; winter surveys of Baer's Pochards are required, with observers specially trained to identify it. Searches should be made for Pink-headed Duck although it is probably extinct.

BAY OF BENGAL COAST

THIS region includes the extensive coastal wetlands which extend from the Sundarbans of India and Bangladesh along the coasts of Bangladesh and Myanmar to the Irrawaddy delta. The highest known counts of Spotted Greenshank and Spoon-billed Sandpiper are from shifting intertidal mudflats and islands (known as chars) in the outer Ganges–Brahmaputra–Meghna delta in Bangladesh, and it is possible that this area and the poorly known coastal wetlands of Myanmar will prove to be the main wintering grounds of both species. The coast of Bangladesh also supports an important concentration of non-breeding Indian Skimmers.

- **Key habitats** Coastal wetlands.
- **Countries and territories** **India** (West Bengal); **Bangladesh**; **Myanmar**.

Threatened species

	CR	EN	VU	Total
● (breeding)	—	—	2	2
(passage migrant)	—	—	—	—
(non-breeding visitor)	—	1	2	3
Total	—	1	4	5

Key: ● = breeding in this wetland region.
= passage migrant.
= non-breeding visitor.

The Bay of Bengal coast region overlaps with part of Conservation International's Indo-Burma Hotspot (see pp.20–21).

The shifting mudflats and islands in the Ganges–Brahmaputra–Meghna delta in Bangladesh are believed to be one of the main wintering areas for both Spotted Greenshank and Spoon-billed Sandpiper. PHOTO: PAUL THOMPSON

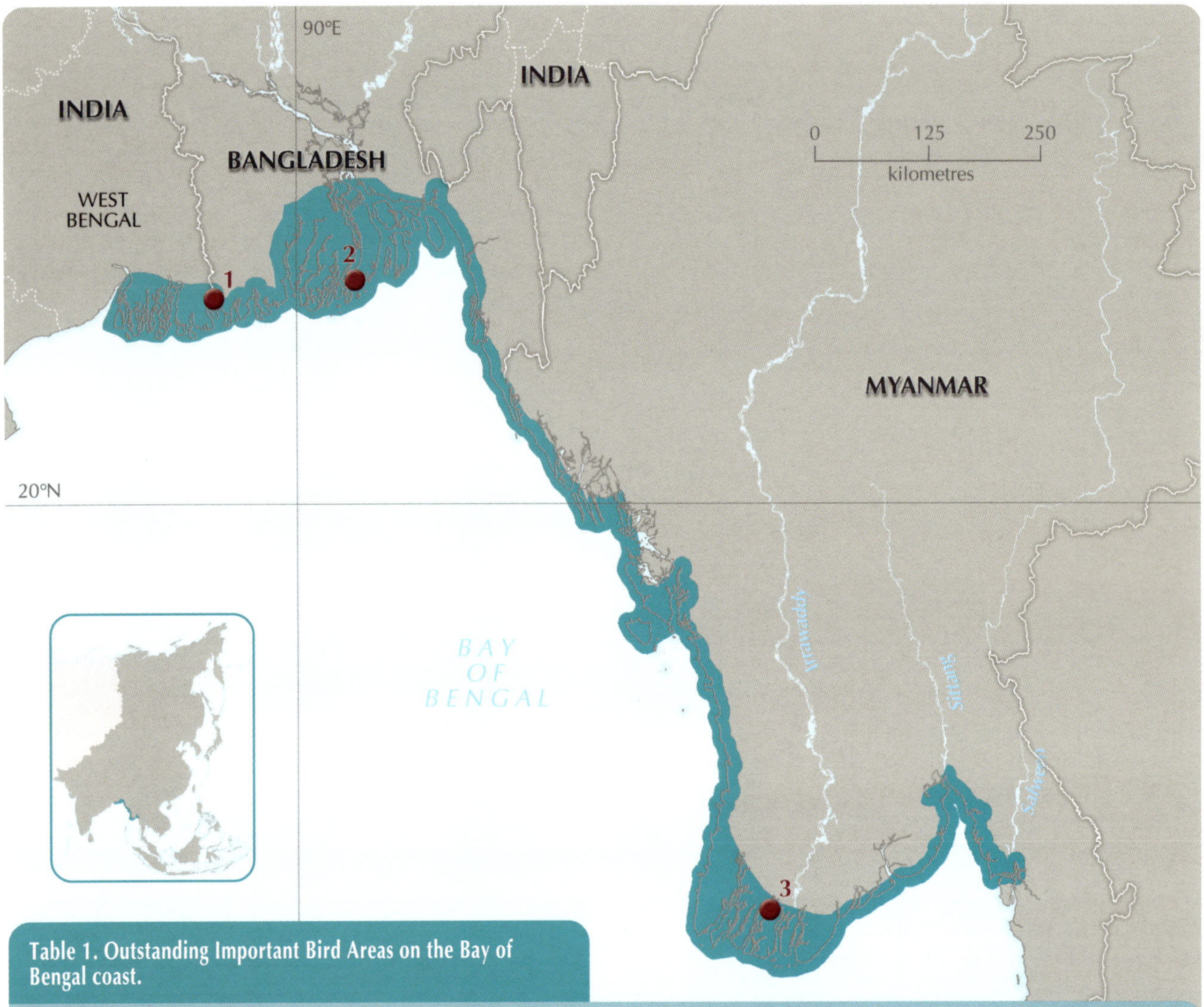

Table 1. Outstanding Important Bird Areas on the Bay of Bengal coast.

	IBAname	Status	Territory	Threatened species
1	**Sundarbans**[F06]	(PA) [BR,R,WH]	India and Bangladesh	Huge area of mangroves supporting breeding Lesser Adjutant and Pallas's Fish-eagle
2	**Ganges–Brahmaputra–Meghna delta**	—	Bangladesh	Non-breeding Spoon-billed Sandpiper, Spotted Greenshank and Indian Skimmer
3	**Irrawaddy delta**	(PA)	Myanmar	Historical records of Lesser Adjutant, Greater Adjutant, Spotted Greenshank and Spoon-billed Sandpiper, but few recent surveys

Note that more IBAs in this region will be included in the *Important Bird Areas in Asia*, due to be published in early 2004.

Key *Status*: PA = IBA is a protected area; (PA) = IBA partially protected; — = unprotected; BR = IBA is wholly or partially a Biosphere Reserve (see pp.34–35); R = IBA is wholly or partially a Ramsar Site (see pp.31–32); WH = IBA is wholly or partially a World Heritage Site (see p.34); F06 = also supports a threatened forest bird of region F06.

The intertidal wetlands on the Bay of Bengal coast may prove to be the main wintering grounds of Spoon-billed Sandpiper.

PHOTO: SHIMPEI WATANABE

The Irrawaddy Delta and other coastal wetlands in Myanmar are probably important for wintering Spotted Greenshank.

PHOTO: RAY TIPPER

OUTSTANDING IBAs FOR THREATENED BIRDS (see Table 1)

Three very large IBAs have been selected, including the intertidal flats of the Ganges–Brahmaputra–Meghna delta, which support some of the largest known concentrations of Spoon-billed Sandpiper, Spotted Greenshank and Indian Skimmer. It is likely that the Irrawaddy delta and possibly other coastal wetlands in Myanmar will also prove to be important for these species.

CURRENT STATUS OF HABITATS AND THREATENED SPECIES

The coasts of this region are fringed with intertidal mudflats and mangrove swamps, with the most extensive areas of these habitats in the Ganges–Brahmaputra–Meghna and Irrawaddy deltas. Some of these wetlands have been converted for agriculture and aquaculture, but large areas of natural habitat remain. However, the mangroves in both deltas are heavily exploited and degraded, and the waterbirds there are also under pressure from hunting and human disturbance.

CONSERVATION ISSUES AND STRATEGIC SOLUTIONS (summarised in Table 3)

Habitat loss and degradation

■ *COASTAL RECLAMATION*

Large areas of coastal wetland in Myanmar are reported to have been converted to agriculture, principally wet rice cultivation in lower-lying areas. Reclamation of intertidal mudflats in the Ganges–Brahmaputra–Meghna delta in Bangladesh is likely to proceed, but the impacts on shorebirds and skimmers are unknown, largely because these birds do not occur at fixed sites but shift their distribution as estuarine conditions change. However, sea-level rise associated with global warming is likely to reduce mudflat areas, as embankments limit the scope for settled or cultivated areas to revert to intertidal habitat. Reclamation of coastal wetlands for agricultural development needs to be better controlled, and prevented inside protected areas.

■ *CONVERSION TO AQUACULTURE*

Shrimp ponds are proliferating along the coasts of Bangladesh and Myanmar, seriously encroaching on natural wetlands. Protected areas should be kept free of aquacultural developments, and elsewhere traditional, extensive aquacultural practices should be promoted, to maximise the value of shrimp- and fish-ponds for waterbirds.

■ *CUTTING OF MANGROVES*

The mangrove forests of the Sundarbans are heavily exploited for timber, pulpwood and fuelwood, and those in the Irrawaddy delta are now mainly degraded, presumably because of excessive extraction of forest products. Logging (presumably of mangroves) apparently eliminated a colony of Greater Adjutant in the Sundarbans, and the species no longer breeds in the area. The sustainable use of mangroves

Table 2. Threatened birds of the Bay of Bengal coast.

Species			Distribution and population
Lesser Adjutant *Leptoptilos javanicus*	○	VU	Small breeding population in the Sundarbans of India and Bangladesh
Pallas's Fish-eagle *Haliaeetus leucoryphus*	○	VU	Small and declining breeding population
Spotted Greenshank *Tringa guttifer*	(dark bird symbol)	EN	Significant numbers winter in the Ganges–Brahmaputra–Meghna delta, and possibly in Myanmar
Spoon-billed Sandpiper *Eurynorhynchus pygmeus*	(dark bird symbol)	VU	Significant numbers winter in the Ganges–Brahmaputra–Meghna delta, and possibly in Myanmar
Indian Skimmer *Rynchops albicollis*	(light bird symbol)	VU	Large non-breeding congregation in Ganges–Brahmaputra–Meghna delta

Other threatened waterbirds recorded from this region as rare visitors are: Spot-billed Pelican *Pelecanus philippensis*, Greater Adjutant *Leptoptilos dubius* and Sarus Crane *Grus antigone*. In addition to the waterbirds, Greater Spotted Eagle *Aquila clanga* (VU; see F01) occurs in winter. Note that Masked Finfoot *Heliopais personata*, which occur in forested wetlands in the Sundarbans (and possibly elsewhere in this region), is covered in F06.

○ = region estimated to support <10% of global breeding population; (dark bird symbol) = region estimated to support 50–90% of global non-breeding population, (light bird symbol) = 10–50%

Large areas of wetland in the Irrawaddy Delta have been converted to rice paddies.

PHOTO: TAKASHI KUROSAKI

by local communities and commercial companies should be promoted, including the maintenance of areas of mature growth (to provide nesting trees), and further mangrove protected areas established in suitable areas in Myanmar. Where suitable nest trees are not available, artificial nest platforms could be considered for adjutant storks and Pallas's Fish-eagle. An education and awareness programme is needed to inform decision-making bodies and local communities about the importance of healthy, well-managed mangrove forests as spawning grounds for fish.

■ *DEVELOPMENT (URBAN, INDUSTRIAL, ETC.)*
Estuarine areas are likely to be considered for industrial and urban expansion as Bangladesh and Myanmar develop. Enhancement of accretion has been used over the last two decades to create new land for human settlement and agriculture in the Ganges–Brahmaputra–Meghna delta in Bangladesh, contributing to constant changes in the intertidal areas. Development projects affecting coastal wetlands need to be carefully considered and regulated, with environmental impact assessments conducted and reclamation projects cancelled if they would cause large-scale environmental damage.

■ *REDUCED FOOD SUPPLY*
There is intense over-fishing of shrimp fry and fish larvae in some coastal areas of Bangladesh, which may be reducing the food supply of some threatened species. For their own sustainability and the benefit of wildlife, fishing communities should be helped to implement conservation measures such as local fish sanctuaries, closed seasons, and reduced use of fishing gear that target fry or have a high bycatch.

■ *DISTURBANCE*
The large and rapidly increasing human population exerts huge pressure on wetlands and aquatic resources. The Sundarbans are disturbed day and night by large numbers of wood-cutters, fishermen and honey collectors. Any increased use of mudflats by people presumably threatens shorebirds and skimmers through disturbance. Human use of important wetlands (particularly within protected areas) therefore needs to be managed, keeping some areas undisturbed so that waterbirds are able to roost, forage and breed successfully.

Protected areas coverage and management

■ *GAPS IN PROTECTED AREAS SYSTEM*
Parts of the Sundarbans and Irrawaddy delta are nature reserves, but the areas of intertidal mudflats and mangroves under protection in both should be expanded. In the Ganges–Brahmaputra–Meghna delta of Bangladesh, key areas of tidal creek and high-tide waterbird roosts should be officially protected, possibly as a network of seasonal wildlife sanctuaries. Surveys will be required in the poorly known wetlands of Myanmar to identify the most appropriate sites for official protection.

Exploitation of birds

■ *HUNTING*
Hunting of waterbirds is widespread, usually using mist-nets in Myanmar. Hunting of threatened bird species should be controlled, by patrolling wetland protected areas to intercept hunters, and reducing gun and net ownership at important sites. Education and awareness programmes are needed within communities throughout the region, to improve understanding of the effects of hunting on the threatened birds and the relevant laws.

Gaps in knowledge

■ *INADEQUATE DATA ON THREATENED BIRDS*
In Bangladesh, continued surveillance of coastal districts is necessary to monitor the numbers, distribution and management requirements of wintering Spotted Greenshank, Spoon-billed Sandpiper and Indian Skimmer. Extensive surveys are required in the coastal wetlands of Myanmar, principally the massive Irrawaddy delta, to improve understanding of the current status of threatened waterbirds, to identify priority wetlands for conservation (including the establishment of new protected areas), and to clarify the most important issues affecting these wetlands and waterbirds.

Table 3. Conservation issues and strategic solutions for birds of the Bay of Bengal coast.

Conservation issues	Strategic solutions
Habitat loss and degradation	
■ COASTAL RECLAMATION ■ CONVERSION TO AQUACULTURE ■ CUTTING OF MANGROVES ■ DEVELOPMENT (URBAN, INDUSTRIAL, ETC.) ■ REDUCED FOOD SUPPLY ■ DISTURBANCE	➤ Control reclamation of coastal wetlands for agriculture, especially in protected areas ➤ Prevent conversion of wetlands for aquaculture in protected areas, and promote traditional, extensive aquacultural practices ➤ Promote sustainable use of mangroves by local communities and commercial companies, with mature trees retained for nesting waterbirds ➤ Assess the environmental impact of development projects affecting key wetlands ➤ Help local communities to improve management of their fisheries, for the benefit of people and waterbirds ➤ Regulate human activities at key wetlands to minimise disturbance
Protected areas coverage and management	
■ GAPS IN PROTECTED AREAS SYSTEM	➤ Expand the existing protected areas in the Sundarbans and the Irrawaddy delta ➤ Protect key areas of tidal creek and waterbird roosts in the Ganges–Brahmaputra–Meghna delta, possibly as seasonal wildlife sanctuaries
Exploitation of birds	
■ HUNTING	➤ Control hunting and trapping of threatened birds, including by patrolling protected areas, and reducing gun and net ownership at key sites
Gaps in knowledge	
■ INADEQUATE DATA ON THREATENED BIRDS	➤ Monitor the numbers and distribution of wintering Spotted Greenshank, Spoon-billed Sandpiper and Indian Skimmer in Bangladesh ➤ Survey the coastal wetlands of Myanmar, principally the Irrawaddy delta, to identify priority wetlands for conservation action

MYANMAR PLAINS

THE Irrawaddy, Chindwin and Sittang valleys were once hugely important breeding grounds for large waterbirds such as Spot-billed Pelican and Greater Adjutant, but their populations crashed in the early twentieth century because of the wholesale destruction of the forests where they used to nest. There is little recent information on the status of these and other threatened waterbird species from Myanmar, but it is possible that the sandbars on these rivers still support breeding populations of Indian Skimmers and that significant breeding populations of birds such as Pallas's Fish-eagle and Sarus Crane occur around large lakes and swamps, while Baer's Pochard may winter in substantial numbers. Pink-headed Duck, a species which may now be extinct, could possibly survive in the poorly known wetlands of Myanmar. This region also supports populations of White-rumped and Slender-billed Vultures, which may prove to be important, given the recent rapid declines in their numbers in the Indian subcontinent (see G03).

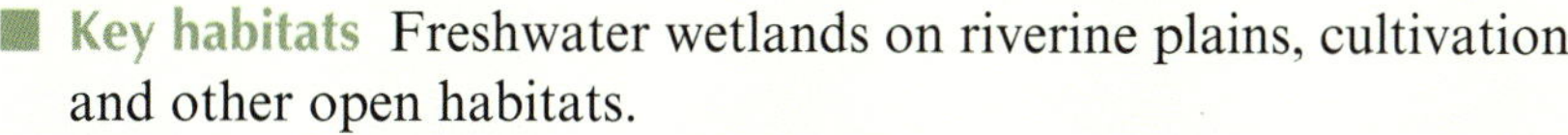

- **Key habitats** Freshwater wetlands on riverine plains, cultivation and other open habitats.
- **Countries and territories** **Myanmar**.

Threatened species	CR	EN	VU	Total
● (breeding)	3	—	5	8
passage migrant	—	—	—	—
non-breeding visitor	—	1	2	3
Total	3	1	7	11

Key: ● = breeding in this wetland region.
= passage migrant.
= non-breeding visitor.

The Myanmar plains region is within Conservation International's Indo-Burma Hotspot (see pp.20–21).

Small populations of Sarus Cranes have been found in several parts of Myanmar during recent surveys. PHOTO: ELEANOR BRIGGS

OUTSTANDING IBAs FOR THREATENED BIRDS (see Table 1)

Five IBAs have been selected, which are known to support some of the threatened waterbirds (and possibly have populations of several more), but future fieldwork is likely to locate many other important wetland IBAs in the Irrawaddy, Chindwin and Sittang river basins.

CURRENT STATUS OF HABITATS AND THREATENED SPECIES

Extensive lakes and marshes associated with the Irrawaddy, Chindwin and Sittang rivers once provided foraging grounds for very large numbers of waterbirds, while tall forests nearby provided secure breeding grounds. Over the past century, many wetlands in Myanmar were bunded and converted to

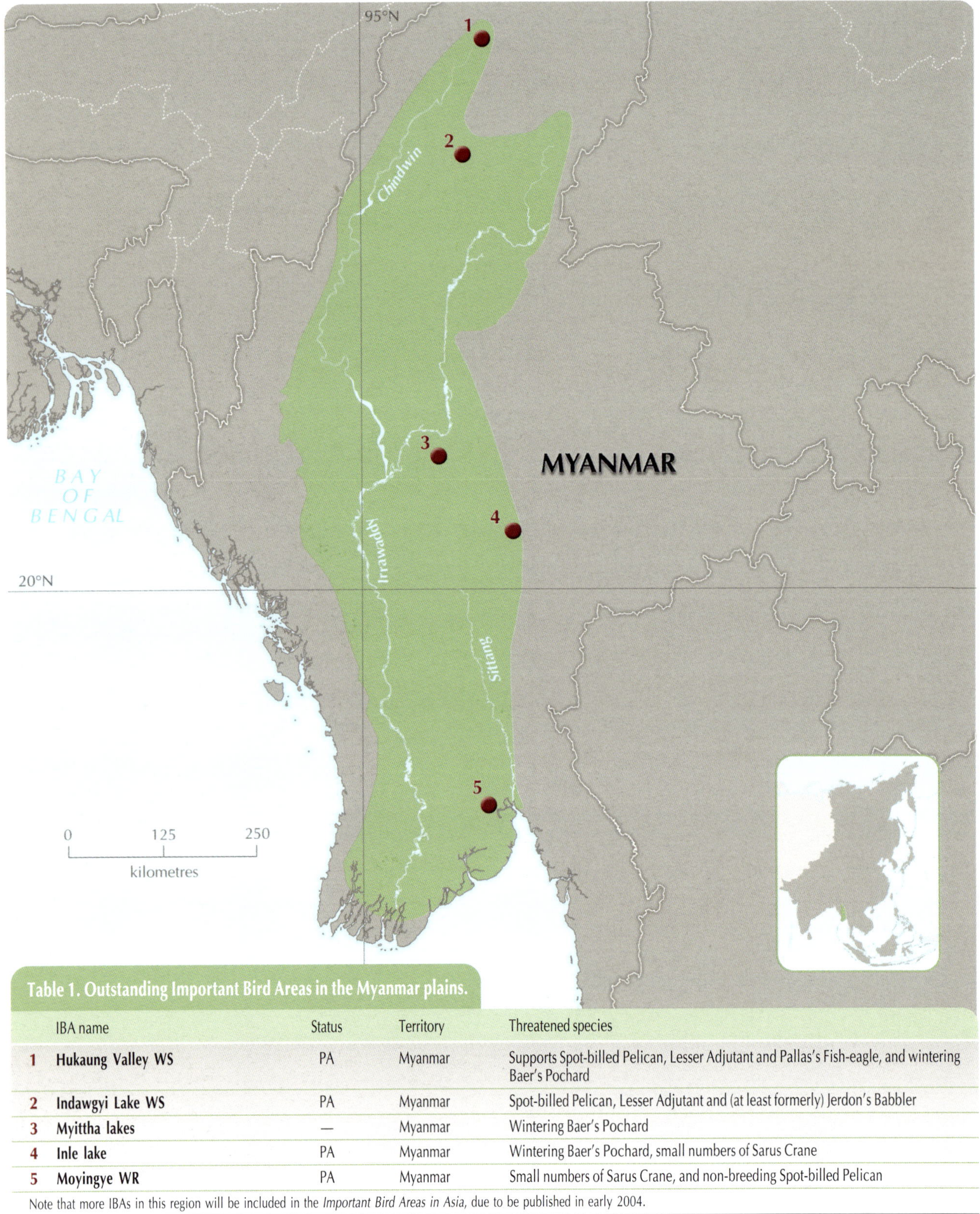

Table 1. Outstanding Important Bird Areas in the Myanmar plains.

	IBA name	Status	Territory	Threatened species
1	**Hukaung Valley WS**	PA	Myanmar	Supports Spot-billed Pelican, Lesser Adjutant and Pallas's Fish-eagle, and wintering Baer's Pochard
2	**Indawgyi Lake WS**	PA	Myanmar	Spot-billed Pelican, Lesser Adjutant and (at least formerly) Jerdon's Babbler
3	**Myittha lakes**	—	Myanmar	Wintering Baer's Pochard
4	**Inle lake**	PA	Myanmar	Wintering Baer's Pochard, small numbers of Sarus Crane
5	**Moyingye WR**	PA	Myanmar	Small numbers of Sarus Crane, and non-breeding Spot-billed Pelican

Note that more IBAs in this region will be included in the *Important Bird Areas in Asia*, due to be published in early 2004.

Key *IBA name*: WR = Wetland Reserve.
Status: PA = IBA is a protected area; (PA) = IBA partially protected; — = unprotected.

There used to be huge breeding colonies of Spot-billed Pelican and other waterbirds in Myanmar, but numbers crashed following the clearance of the tall forests used for nesting.

PHOTO: TIM LOSEBY

Table 2. Threatened birds of the Myanmar plains.

Species			Distribution and population
Spot-billed Pelican *Pelecanus philippensis*	🐦	VU	Now rare and thought not to breed
Lesser Adjutant *Leptoptilos javanicus*	○	VU	Probably breeds in small numbers
Greater Adjutant *Leptoptilos dubius*	🐦	EN	Now rare and thought not to breed
Pink-headed Duck *Rhodonessa caryophyllacea*	EX?	CR	Recorded historically in this region, but the species may be extinct
Baer's Pochard *Aythya baeri*	🐦	VU	Several hundred recently found wintering on scattered wetlands
Pallas's Fish-eagle *Haliaeetus leucoryphus*	○	VU	Breeding population of unknown size on the upper Irrawaddy
White-rumped Vulture *Gyps bengalensis*	(?)	CR	Urban and cultivated areas, light woodland and open habitats in the lowlands
Slender-billed Vulture *Gyps tenuirostris*	(?)	CR	Cultivated areas, light woodland and open habitats in the lowlands
Sarus Crane *Grus antigone*	○	VU	Small populations breed in several areas
Indian Skimmer *Rynchops albicollis*	○	VU	Several recent sightings, probably breeds on the major rivers
Jerdon's Babbler *Chrysomma altirostre*	(?)	VU	Formerly widespread in tall riverine grasslands, but no recent records

In addition to the waterbirds, Greater Spotted Eagle *Aquila clanga* (VU; see F01) occurs in winter. Note that three species which occur in forested wetlands in Myanmar, White-bellied Heron *Ardea insignis*, White-winged Duck *Cairina scutulata* and Masked Finfoot *Heliopais personata*, are covered in region F06.

○ = region estimated to support <10% of global breeding population, (?) = proportion of global breeding population unknown; 🐦 = region estimated to support <10% of global non-breeding population, EX? = probably extinct

permanent agriculture, and the forests were felled, such that waterbird numbers have plummeted. Some parts of the plains are still seasonally inundated, and then provide feeding areas for waterfowl, but they are used for rice cultivation during the dry season (except perhaps along some larger rivers in remote northern regions). Myanmar has some of the most extensive and least disturbed stretches of waterway in South-East Asia, and may retain relatively healthy populations of riverine birds such as Indian Skimmer. However, there is little recent information on the status of threatened waterbirds and their habitats, and surveys are urgently needed to identify priority areas for conservation and the threats to wetlands and waterbirds that need to be addressed.

CONSERVATION ISSUES AND STRATEGIC SOLUTIONS (summarised in Table 3)

Habitat loss and degradation

■ *CONVERSION TO AGRICULTURE*

Many floodplain wetlands in Myanmar have been drained and converted for permanent agriculture, and there is still pressure to convert wetlands to rice paddies. All important wetlands for threatened birds must be safeguarded from any

Fishing provides a vital source of protein to the large human population in the lowlands of Myanmar.

PHOTO: SIMBA CHAN

further reclamation. An education and awareness programme is needed to inform decision-making bodies and local communities about the importance of healthy wetlands in maintaining water quality and preserving fish stocks.

■ *CUTTING OF NESTING TREES*
The loss of secure breeding sites in tall lowland forest appears to have been the main factor in the rapid disappearance of the vast Spot-billed Pelican and Greater Adjutant colonies in Myanmar. Any surviving waterbird colonies should be protected by working with landowners and managers, and developing new protected areas and/or community conservation projects. These and former nesting sites should be developed for the future by planting suitable tree species for nesting (or even by constructing artificial nest platforms).

■ *DISTURBANCE*
The plains of Myanmar have a large rural population, and many wetlands are doubtless badly affected by human disturbance. Rivers are used for transport, and the disturbance caused is likely to put serious pressure on riverine birds such as Indian Skimmer which nest on sandbars and islands. Human use of important wetlands and stretches of river (particularly within protected areas) therefore needs to be managed, keeping some areas undisturbed to maintain populations of wildlife and aquatic resources.

Wetlands provide many resources to local people, such as fodder for their livestock, but human disturbance is likely to be affecting threatened waterbirds in many areas.

PHOTO: SIMBA CHAN

■ *POLLUTION*
There is evidence that pollution is a problem at some wetlands in Myanmar, for example at Inle lake where discharge of effluents is affecting fish and waterbird populations and causing eutrophication. Dynamite and chemical fishing are reported to be common. Gold mining in the upper Irrawaddy river is affecting sedimentation and water turbidity, as well as causing disturbance to sandbanks. The levels and sources of pollution need to be investigated, and existing (and if necessary new) laws used to control the release of toxic chemicals into the environment.

Protected areas coverage and management

■ *GAPS IN PROTECTED AREAS SYSTEM*
Only a few of Myanmar's wetlands are legally protected within reserves. The government is currently reviewing the country's protected areas system, and there is a need to identify and protect further important wetland sites. Particular priorities are the protection of sites with nesting colonies of large waterbirds (or the potential for former nesting colonies to be restored), and of undisturbed stretches of riverine habitat and representative examples of other types of wetland.

Exploitation of birds

■ *HUNTING*
Levels of persecution and poaching are reported to be high in Myanmar. Migratory ducks are apparently netted by hunters and brought to markets in large quantities in winter. Hunting and trapping of all threatened birds should be banned, wetland protected areas more intensively patrolled, and levels of trade in markets monitored. Gun and net ownership should be controlled at important sites. The plight of large waterbirds and the laws protecting them need to be widely and clearly communicated, with a view to reducing persecution and disturbance.

Gaps in knowledge

■ *INADEQUATE DATA ON THREATENED BIRDS*
There is little recent information on threatened waterbirds and vultures and their habitats in Myanmar, and surveys are required to improve understanding of their current status, identify priority areas for conservation, and clarify the most important issues affecting these species and sites. Although it is possibly extinct, surveys should be conducted for Pink-headed Duck, including analysis of satellite images to locate potentially suitable areas.

Table 3. Conservation issues and strategic solutions for birds of the Myanmar plains.

Conservation issues	Strategic solutions
Habitat loss and degradation	
■ CONVERSION TO AGRICULTURE ■ CUTTING OF NESTING TREES ■ DISTURBANCE ■ POLLUTION	➤ Safeguard key wetlands for threatened birds from reclamation for agriculture ➤ Protect active waterbird colonies, and plant nest trees or erect artificial nest platforms for threatened waterbirds ➤ Regulate human activities at key wetlands to minimise disturbance ➤ Enforce laws to control wetland pollution
Protected areas coverage and management	
■ GAPS IN PROTECTED AREAS SYSTEM	➤ Establish a network of wetland protected areas in Myanmar, including all colonies of large waterbirds and representative examples of all types of wetland
Exploitation of birds	
■ HUNTING	➤ Ban hunting and trapping of all threatened bird species ➤ Patrol protected areas, and control gun and net ownership at key sites
Gaps in knowledge	
■ INADEQUATE DATA ON THREATENED BIRDS	➤ Conduct surveys of wetlands throughout Myanmar, to locate the key sites for the conservation of threatened waterbirds ➤ Search for Pink-headed Duck

THAILAND WETLANDS

THIS region includes the wetlands on the floodplains of the Chao Phraya river and its tributaries, and on the Gulf of Thailand coast. It once supported populations of large waterbirds such as Spot-billed Pelican, Giant Ibis and Sarus Crane, but these now occur as vagrants only, or not at all, mainly because their habitat has been lost. The enigmatic White-eyed River-martin is known from a single site in this region, Bung Boraphet, but it is possible that the species survives in riverine habitats in Thailand or a neighbouring country.

- **Key habitats** Freshwater wetlands on riverine plains, coastal wetlands.
- **Countries and territories** **Thailand**.

Threatened species	CR	EN	VU	Total
● (breeding)	1	—	—	1
(passage migrant)	—	—	—	—
(non-breeding visitor)	—	2	5	7
Total	1	2	5	8

Key: ● = breeding in this wetland region.
= passage migrant.
= non-breeding visitor.

The Thailand wetlands region is within Conservation International's Indo-Burma Hotspot (see pp.20–21).

The coastal wetlands in Khao Sam Roi Yot National Park face many pressures, including conversion to aquaculture and plantations. PHOTO: MIKE CROSBY/BIRDLIFE

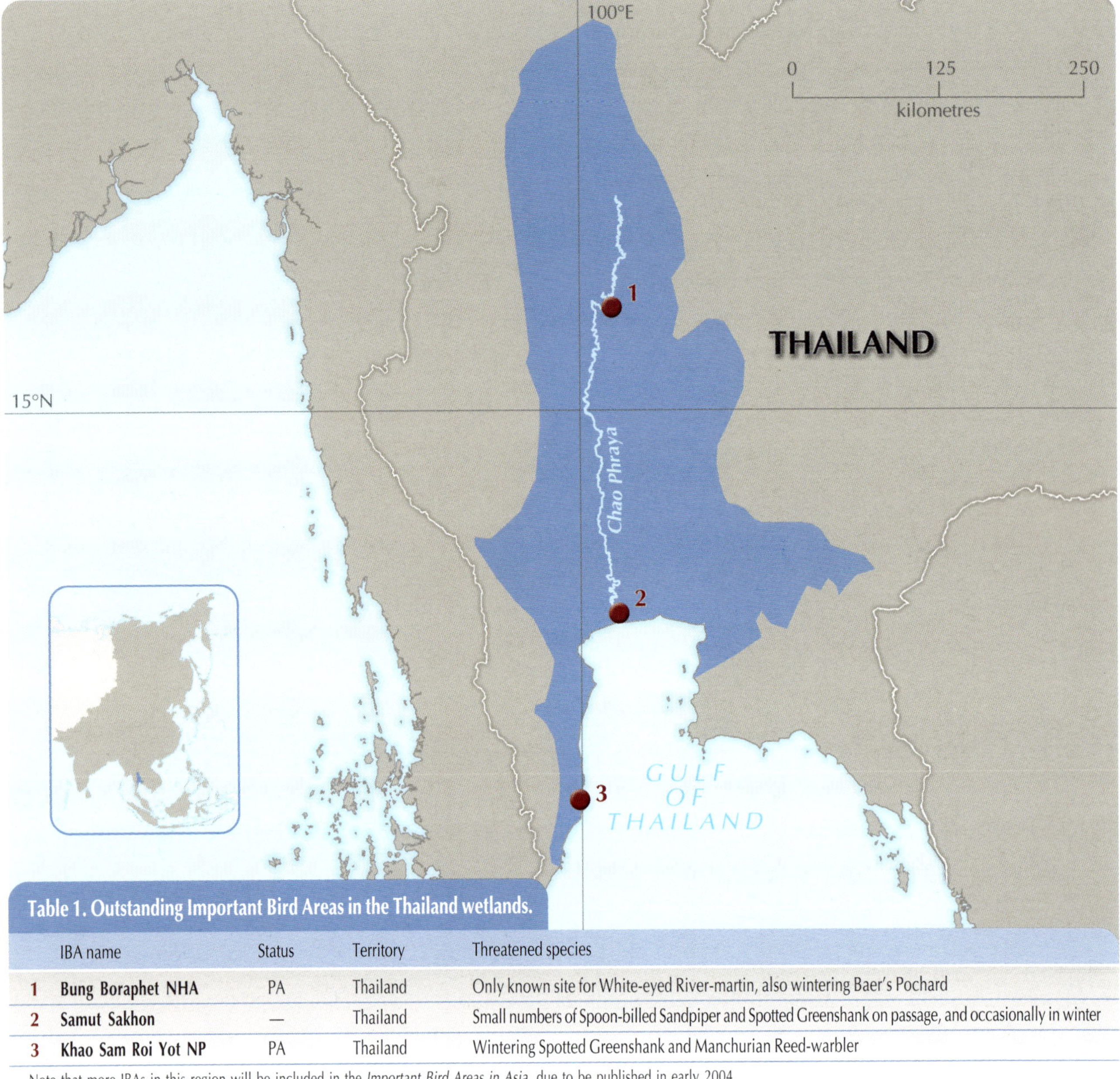

Table 1. Outstanding Important Bird Areas in the Thailand wetlands.

	IBA name	Status	Territory	Threatened species
1	**Bung Boraphet NHA**	PA	Thailand	Only known site for White-eyed River-martin, also wintering Baer's Pochard
2	**Samut Sakhon**	—	Thailand	Small numbers of Spoon-billed Sandpiper and Spotted Greenshank on passage, and occasionally in winter
3	**Khao Sam Roi Yot NP**	PA	Thailand	Wintering Spotted Greenshank and Manchurian Reed-warbler

Note that more IBAs in this region will be included in the *Important Bird Areas in Asia*, due to be published in early 2004.

Key *IBA name*: NHA = non-hunting area; NP = National Park.
Status: PA = IBA is a protected area; (PA) = IBA partially protected; — = unprotected.

Table 2. Threatened birds of the Thailand wetlands.

Species			Distribution and population
Spot-billed Pelican *Pelecanus philippensis*	<10% non-breeding	VU	Scarce non-breeding visitor
Lesser Adjutant *Leptoptilos javanicus*	<10% non-breeding	VU	Scarce non-breeding visitor
Greater Adjutant *Leptoptilos dubius*	<10% non-breeding	EN	Rare but almost annual visitor
Baer's Pochard *Aythya baeri*	<10% non-breeding	VU	Winters at several lakes, usually in small numbers
Spotted Greenshank *Tringa guttifer*	<10% non-breeding	EN	Scarce on passage, small numbers overwinter
Spoon-billed Sandpiper *Eurynorhynchus pygmeus*	<10% non-breeding	VU	Scarce on passage and in winter
White-eyed River-martin *Eurychelidon sirintarae*	● ?	CR	Discovered at Bung Boraphet in central Thailand in 1968, but no confirmed records since the 1970s and perhaps extinct
Manchurian Reed-warbler *Acrocephalus tangorum*	? non-breeding	VU	Small numbers winter in reed-swamps at Khao Sam Roi Yot National Park

Other threatened waterbirds recorded from this region as rare visitors are: Chinese Egret *Egretta eulophotes*, Milky Stork *Mycteria cinerea*, Black-faced Spoonbill *Platalea minor*, Baikal Teal *Anas formosa*, Sarus Crane *Grus antigone* and Indian Skimmer *Rynchops albicollis*. Giant Ibis *Thaumatibis gigantea* formerly occurred in Thailand, but is now extinct there. In addition to the waterbirds, Greater Spotted Eagle *Aquila clanga* (VU; see F01) and Imperial Eagle *A. heliaca* (VU; see G01) occur in winter. Note that two species which occur in forested wetlands in Thailand, White-winged Duck *Cairina scutulata* and Masked Finfoot *Heliopais personata*, are covered in region F06.

● = region estimated to support >90% of global breeding population; (bird symbol) = region estimated to support <10% of global non-breeding population; ?(bird symbol) = proportion of global non-breeding population unknown

OUTSTANDING IBAs FOR THREATENED BIRDS (see Table 1)

Three IBAs have been selected in the region, including the only known site for White-eyed River-martin, and important non-breeding localities for several other threatened waterbirds.

CURRENT STATUS OF HABITATS AND THREATENED SPECIES

Huge areas of Thailand would formerly have been covered by wetlands, notably in the central plains, but most freshwater wetlands were converted for agriculture during the nineteenth and twentieth centuries and few natural areas remain. This process has led to the extinction of the national breeding populations of most large waterbirds, including threatened species such as Giant Ibis and Sarus Crane. Several threatened species occur on the remaining freshwater wetlands, including on reservoirs, and in the coastal wetlands on the Gulf of Thailand. However, they are under pressure from habitat loss and degradation, disturbance, pesticide use and other threats.

CONSERVATION ISSUES AND STRATEGIC SOLUTIONS (summarised in Table 3)

Habitat loss and degradation

CONVERSION TO AGRICULTURE

Most natural wetlands in Thailand have already been converted to agricultural land, and many remaining sites are under pressure. The numerous irrigation projects divert river water away from wetlands, leaving them shallower, overgrown and more easily drained. Reedbeds are encroached by settlements and crops, such that very few are left in the country; the most important reedbed in Thailand (which supports a large wintering population of Manchurian Reed-warbler), within Khao Sam Roi Yot National Park, is being replaced by plantations of casuarinas, eucalyptus and coconut palms. The changes in land use in Khao Sam Roi Yot National Park have proved difficult to oppose, and the laws protecting habitat within reserves therefore need to be tightened and strictly enforced.

Healthy wetlands are vital in maintaining water quality and preserving fish stocks.

PHOTO: PHIL BENSTEAD

CONVERSION TO AQUACULTURE AND SALT-PRODUCTION

Prawn-ponds and saltpans are proliferating along the Gulf of Thailand coast, greatly reducing the available area of foraging habitat for shorebirds. At Khao Sam Roi Yot, prawn farms take up large areas of land and pump saline water into previously freshwater areas. Inland wetlands are converted to lotus production and fish farms, with associated vegetation clearance, infrastructural development and disturbance. Aquacultural development should be prohibited within existing protected areas, and more tightly regulated in other important wetlands. An education and awareness programme—managed by the Wildlife Conservation Division—is needed to inform decision-making bodies and local communities about the importance of healthy wetlands in maintaining water quality and preserving fish stocks.

LOSS OF NESTING HABITAT

A major constraint on the use of wetlands by large waterbirds appears to be the lack of undisturbed nesting habitat near suitable swampy feeding sites. Existing waterbird colonies should be protected, and these and former nesting sites enhanced by planting suitable tree species for nesting (or even by constructing artificial nest platforms) and by reducing human disturbance. Given that several threatened species that formerly bred in Thailand still visit, breeding colonies could re-establish themselves should suitable conditions become available.

DEVELOPMENT (URBAN, INDUSTRIAL, ETC.)

Coastal areas are threatened by industrial and urban growth. At Khao Sam Roi Yot, coastal plots are being sold to property developers; if this continues, it will destroy shorebird habitat inside the national park. Shorebird wintering areas in the Inner Gulf, such as Bang Poo and Samut Sakhon, were reduced and degraded by industrial development during the 1990s. Inland, lakes and rivers are affected by new roads, dams and tourist facilities. Development projects affecting wetlands need tighter control through the application of land-use laws and policies, with environmental impact assessments conducted and mitigation enforced.

DISTURBANCE

Coastal wetlands, such as Pattani bay and Ko Libong, are disturbed by large numbers of fishermen and collectors of bivalves, anemones, sea cucumbers and crabs, reducing their value to migrant shorebirds. Lakes and reservoirs are heavily used by fishermen, hunters, reed-cutters, etc. In the 1980s, around 30,000 people lived around the margins of Bung Boraphet, causing considerable disturbance. However, it is rivers that are perhaps the most disturbed aquatic habitat in Thailand, with whole communities of nesting riverine birds having vanished from large segments of their ranges because of constant use and navigation of rivers, and associated habitat modification and hunting. Human use of important wetlands (particularly within protected areas) therefore needs to be managed, keeping some areas undisturbed to maintain populations of wildlife and aquatic resources.

POLLUTION/PESTICIDES

Pesticides use and pollution are widespread in Thailand, and are probably affecting threatened waterbirds populations. Although persistent pesticides are apparently little used in rice-growing areas, organochlorines used extensively on the nearby upland cotton crops could contaminate the Bung Boraphet basin. Intertidal areas at Pattani bay and Ko Libong are being polluted by heavy metals and chlorinated hydrocarbons, mostly as a direct result of local industrial activities, agricultural run-off and local dumping of waste. Existing laws to control the release of toxic chemicals into the environment need to be strictly enforced, and new legislation introduced if necessary.

Protected areas coverage and management

■ *GAPS IN PROTECTED AREAS SYSTEM*

Several protected areas in Thailand include important wetland habitats, but some wetland types are not well represented, such as freshwater swamp. A number of IBAs which include these wetland habitats should be considered for establishment as new nature reserves. Wetlands and swamps are relatively easy to rehabilitate, developing in only a few years to attractive habitats for waterbirds, so even degraded wetlands could be designated as protected areas if this can be followed by appropriate habitat management.

■ *WEAKNESSES IN RESERVE MANAGEMENT*

There are serious management problems in the wetland areas in and around Khao Sam Roi Yot National Park, with rapid encroachment, and community support for the park suppressed by the activities of wealthy landlords. The other wetland reserves in the country are also understaffed, underfunded and poorly equipped to deal with the intense competition for the aquatic resources they are designed to protect. It is necessary to strengthen management in important wetland reserves through better training, pay and equipment for reserve staff, and by the preparation of management plans which incorporate ecological management with the needs of local people. Measures to improve reserve management may include more intensive patrolling and stricter enforcement of environmental laws.

White-eyed River-martin is known only from Bung Boraphet in Thailand.

PHOTO: H. E. McCLURE/BIRDLIFE

Exploitation of birds

■ *HUNTING*

In Thailand, large birds, including ducks, storks, cranes and pelicans, are frequently shot or trapped for food, pets or sale to zoos and markets. Eggs and young are liable to be taken from any nests that are found. It is unlikely that these waterbirds could return in any numbers given such persecution. In coastal regions, shorebirds are netted in large numbers for food. Even the White-eyed River-martin may have been hunted to extinction by swallow-trappers working with nets at Bung Boraphet to supply local food markets. Hunting and trapping of all threatened bird species should be banned, and wetland protected areas more intensively patrolled to intercept hunters. Gun and net ownership should be controlled at important sites. The plight of large waterbirds and the laws protecting them also need to be widely and clearly communicated, with a view to reducing persecution and disturbance.

Gaps in knowledge

■ *INADEQUATE DATA ON THREATENED BIRDS*

White-eyed River-martin is one of the most enigmatic birds in the world, known only from Bung Boraphet in Thailand, at which it apparently no longer occurs. It is likely to breed along rivers, like the related African River-martin *Pseudochelidon eurystomina*, but might conceivably utilise some entirely different habitat, or even be semi-nocturnal. Apart from continued vigilance at Bung Boraphet, searches in its putative breeding range in Thailand and adjacent countries should be continued with a broader outlook.

Table 3. Conservation issues and strategic solutions for birds of the Thailand wetlands.

Conservation issues	Strategic solutions
Habitat loss and degradation	
■ CONVERSION TO AGRICULTURE ■ CONVERSION TO AQUACULTURE AND SALT PRODUCTION ■ LOSS OF NESTING HABITAT ■ DEVELOPMENT (URBAN, INDUSTRIAL, ETC.) ■ DISTURBANCE ■ POLLUTION/PESTICIDES	➤ Strengthen and enforce laws to prevent conversion for agriculture and aquaculture within protected areas ➤ Protect active waterbird colonies, and plant nest trees or erect artificial nest platforms to encourage recolonisation of former sites ➤ Assess the impact of development projects that could affect important wetlands ➤ Regulate human activities at key wetlands to minimise disturbance ➤ Enforce laws to control wetland pollution
Protected areas coverage and management	
■ GAPS IN PROTECTED AREAS SYSTEM ■ WEAKNESSES IN RESERVE MANAGEMENT	➤ Improve coverage of all wetland habitat types in Thailand by establishing new protected areas ➤ Strengthen reserve management through improved funding, infrastructure and staff training ➤ Formulate and implement management plans for Khao Sam Roi Yot and other wetland protected areas
Exploitation of birds	
■ HUNTING	➤ Ban hunting and trapping of all threatened bird species ➤ Patrol protected areas, and control gun and net ownership at key sites
Gaps in knowledge	
■ INADEQUATE DATA ON THREATENED BIRDS	➤ Search for White-eyed River-martin in Thailand and adjacent countries

LOWER MEKONG BASIN

THE plains of the lower Mekong still retain some areas of near-primary habitat, with mosaics of open deciduous dipterocarp forest, seasonally inundated wetlands and grasslands, and riverine habitats. The region supports the entire world population of Giant Ibis, together with a high proportion of the global population of White-shouldered Ibis and large numbers of several other threatened birds. The swamp forests around Tonle Sap, the world's largest floodplain lake, have the most important waterbird colonies remaining in mainland South-East Asia, with globally significant breeding populations of Spot-billed Pelican, Lesser Adjutant and Greater Adjutant. The seasonally inundated grasslands around the lake are probably the global stronghold for Bengal Florican, and grasslands in the Mekong floodplain and delta support several hundred pairs of Sarus Crane. In addition to the waterbirds, this region has important populations of White-rumped and Slender-billed Vultures (particularly given their rapid declines in South Asia: see G03) and several threatened birds found in forests and associated wetlands (White-winged Duck, Masked Finfoot and Green Peafowl: see F06).

- **Key habitats** Freshwater wetlands on riverine plains, seasonally inundated grassland, open deciduous forest, swamp forest.
- **Countries and territories** **Laos**; **Cambodia**; **Vietnam**.

Threatened species	CR	EN	VU	Total
Breeding	4	2	4	10
Passage migrant	—	—	—	—
Non-breeding visitor	—	—	1	1
Total	4	2	5	11

Key: ● = breeding in this wetland region. = passage migrant. = non-breeding visitor.

The Lower Mekong basin region is within Conservation International's Indo-Burma Hotspot (see pp.20–21).

The Mekong flood-plain was originally a mosaic of open forest, small seasonal pools and wet grasslands. PHOTO: ELEANOR BRIGGS

W18

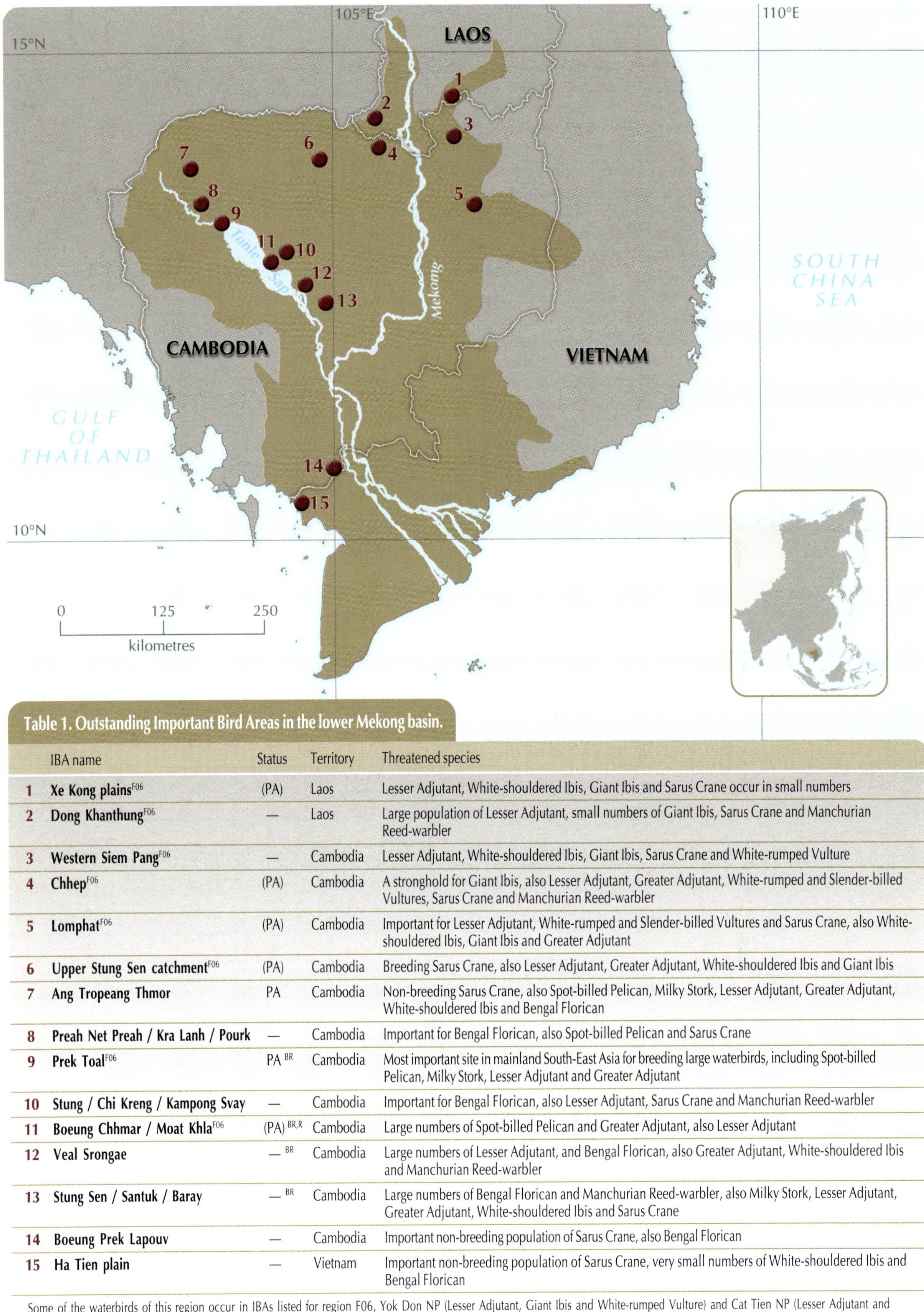

Table 1. Outstanding Important Bird Areas in the lower Mekong basin.

	IBA name	Status	Territory	Threatened species
1	**Xe Kong plains**[F06]	(PA)	Laos	Lesser Adjutant, White-shouldered Ibis, Giant Ibis and Sarus Crane occur in small numbers
2	**Dong Khanthung**[F06]	—	Laos	Large population of Lesser Adjutant, small numbers of Giant Ibis, Sarus Crane and Manchurian Reed-warbler
3	**Western Siem Pang**[F06]	—	Cambodia	Lesser Adjutant, White-shouldered Ibis, Giant Ibis, Sarus Crane and White-rumped Vulture
4	**Chhep**[F06]	(PA)	Cambodia	A stronghold for Giant Ibis, also Lesser Adjutant, Greater Adjutant, White-rumped and Slender-billed Vultures, Sarus Crane and Manchurian Reed-warbler
5	**Lomphat**[F06]	(PA)	Cambodia	Important for Lesser Adjutant, White-rumped and Slender-billed Vultures and Sarus Crane, also White-shouldered Ibis, Giant Ibis and Greater Adjutant
6	**Upper Stung Sen catchment**[F06]	(PA)	Cambodia	Breeding Sarus Crane, also Lesser Adjutant, Greater Adjutant, White-shouldered Ibis and Giant Ibis
7	**Ang Tropeang Thmor**	PA	Cambodia	Non-breeding Sarus Crane, also Spot-billed Pelican, Milky Stork, Lesser Adjutant, Greater Adjutant, White-shouldered Ibis and Bengal Florican
8	**Preah Net Preah / Kra Lanh / Pourk**	—	Cambodia	Important for Bengal Florican, also Spot-billed Pelican and Sarus Crane
9	**Prek Toal**[F06]	PA [BR]	Cambodia	Most important site in mainland South-East Asia for breeding large waterbirds, including Spot-billed Pelican, Milky Stork, Lesser Adjutant and Greater Adjutant
10	**Stung / Chi Kreng / Kampong Svay**	—	Cambodia	Important for Bengal Florican, also Lesser Adjutant, Sarus Crane and Manchurian Reed-warbler
11	**Boeung Chhmar / Moat Khla**[F06]	(PA) [BR,R]	Cambodia	Large numbers of Spot-billed Pelican and Greater Adjutant, also Lesser Adjutant
12	**Veal Srongae**	— [BR]	Cambodia	Large numbers of Lesser Adjutant, and Bengal Florican, also Greater Adjutant, White-shouldered Ibis and Manchurian Reed-warbler
13	**Stung Sen / Santuk / Baray**	— [BR]	Cambodia	Large numbers of Bengal Florican and Manchurian Reed-warbler, also Milky Stork, Lesser Adjutant, Greater Adjutant, White-shouldered Ibis and Sarus Crane
14	**Boeung Prek Lapouv**	—	Cambodia	Important non-breeding population of Sarus Crane, also Bengal Florican
15	**Ha Tien plain**	—	Vietnam	Important non-breeding population of Sarus Crane, very small numbers of White-shouldered Ibis and Bengal Florican

Some of the waterbirds of this region occur in IBAs listed for region F06, Yok Don NP (Lesser Adjutant, Giant Ibis and White-rumped Vulture) and Cat Tien NP (Lesser Adjutant and White-shouldered Ibis).
Note that more IBAs in this region will be included in the *Important Bird Areas in Asia*, due to be published in early 2004.

Key *Status*: PA = IBA is a protected area; (PA) = IBA partially protected; — = unprotected; BR = IBA is wholly or partially inside Tonle Sap Biosphere Reserve (see pp.34–35); R = IBA is wholly or partially a Ramsar Site (see pp.31–32); F06 = also supports threatened forest birds of region F06.

OUTSTANDING IBAs FOR THREATENED BIRDS (see Table 1)

This region supports by far the largest waterbird populations remaining in South-East Asia, and a total of fifteen IBAs have been selected, chiefly for their importance to Spot-billed Pelican, Lesser Adjutant, Greater Adjutant, White-shouldered Ibis, Giant Ibis, White-rumped Vulture, Slender-billed Vulture and Bengal Florican.

CURRENT STATUS OF HABITATS AND THREATENED SPECIES

The lower Mekong and its major feeder rivers, the Sesan, Sekong and Srepok, retain the most intact riverine floodplain habitats in the Asia region. There are still extensive areas of open deciduous forest studded with small seasonal pools (*trapaengs*) and seasonally wet meadows (*veals*), most notably on the plains of northern and eastern Cambodia, which are the last stronghold for Giant Ibis and breeding grounds for White-shouldered Ibis and Sarus Crane. In comparison to the plains of Myanmar (W16) and Thailand (W17), human population density in most of the lower Mekong basin is low, but the region's wetlands have nevertheless been subject to professional hunting and seasonal subsistence fishing over a long period, and these pressures are growing. The Mekong and its tributaries support the only remaining examples of near-intact riverine habitats and bird communities in South-East Asia. However, islands in these rivers are used increasingly by itinerant fishermen and their families for camps, and there is regular motorised boat traffic, which may account for the apparent extinction of the Mekong's Indian Skimmer population. Various proposed hydropower schemes in the Mekong catchment threaten river flow rates and seasonal water levels (see below). In Vietnam and parts of southern Cambodia, the Mekong floodplain and delta has been almost totally converted to rice paddy, while much of the remainder is intensively used. Large areas of mangrove and *Melaleuca* wetlands in the delta were destroyed during the Vietnam War, then replanted, but later cleared for aquaculture. The inundation zone of Tonle Sap lake still has extensive tracts of swamp forest, and grasslands which are subject to a complex and ancient Khmer agricultural ecology. The swamp forests support the only remaining large colonies of communally nesting waterbirds in South-East Asia, and the traditionally managed grasslands have a high proportion of the global population of Bengal Florican.

The large undeveloped tracts of flood-plain habitats in northern and eastern Cambodia are the last stronghold for Giant Ibis.

PHOTO: ALLAN MICHAUD

CONSERVATION ISSUES AND STRATEGIC SOLUTIONS (summarised in Table 3)

Habitat loss and degradation

■ *CONVERSION TO AGRICULTURE*

Wetlands and grasslands in this region are threatened by the large-scale intensification of agriculture, in particular the

Table 2. Threatened birds of the lower Mekong basin.

Species			Distribution and population
Spot-billed Pelican *Pelecanus philippensis*	◐ (10–50%)	VU	Large breeding population in the swamp forest around Tonle Sap lake
Milky Stork *Mycteria cinerea*	○ (<10%)	VU	Very small numbers breed in the swamp forest around Tonle Sap lake
Lesser Adjutant *Leptoptilos javanicus*	◐ (10–50%)	VU	Significant numbers breed at Tonle Sap lake, and at lower densities throughout the forested plains
Greater Adjutant *Leptoptilos dubius*	◐ (10–50%)	EN	Significant numbers breed in the swamp forest around Tonle Sap lake
White-shouldered Ibis *Pseudibis davisoni*	● (50–90%)	CR	Significant numbers breed in Cambodia; close to extinction in Laos and southern Vietnam
Giant Ibis *Thaumatibis gigantea*	● (>90%)	CR	Entire global population now confined to this region, mainly in northern and eastern Cambodia, but also very small numbers in adjacent parts of Laos and Vietnam
White-rumped Vulture *Gyps bengalensis*	?	CR	Widely distributed on the plains, in deciduous forest and open habitats
Slender-billed Vulture *Gyps tenuirostris*	?	CR	Widely distributed on the plains, in deciduous forest and open habitats
Sarus Crane *Grus antigone*	○ (<10%)	VU	Several hundred pairs breed in northern Cambodia and southern Laos; outside breeding season moves to north-western and southern Cambodia and the upper Mekong delta in Vietnam
Bengal Florican *Houbaropsis bengalensis*	● (50–90%) ?	EN	Significant population around Tonle Sap lake, dispersing to adjacent areas in the wet season; close to extinction in the Mekong delta in Vietnam
Manchurian Reed-warbler *Acrocephalus tangorum*	? (non-breeding)	VU	Recently found wintering in damp grasslands in Cambodia

Other threatened waterbirds recorded from this region as rare (or perhaps extinct) visitors are: Black-faced Spoonbill *Platalea minor*, Pallas's Fish-eagle *Haliaeetus leucoryphus* and Indian Skimmer *Rynchops albicollis*. In addition to the waterbirds, Greater Spotted Eagle *Aquila clanga* (VU; see F01) and Imperial Eagle *A. heliaca* (VU; see G01) occur in winter. Note that three species which occur in forests and forested wetlands in the lower Mekong basin, White-winged Duck *Cairina scutulata*, Masked Finfoot *Heliopais personata* and Green Peafowl *Pavo muticus*, are also covered in region F06.

● = region estimated to support >90% of global breeding population, ● = 50–90%, ◐ = 10–50%; ○ = <10%, ? = proportion of global breeding population unknown; ? (bird symbol) = region's proportion of global non-breeding population unknown

expansion of irrigated wet rice cultivation. This continuing conversion of seasonally inundated grassland to cultivation causes loss of natural habitat, falling water levels, acidification of soils and increased disturbance. Even in the undeveloped tracts of northern and eastern Cambodia (the last hope for species such as Giant Ibis), seasonally flooded meadows and pools are the first places to be cultivated as villages gradually spread into unsettled areas. In many places, temporary settlements are occupied in the dry season along margins of rivers and wetlands, reducing habitat for large waterbirds. For example, in Kompong Thom province in Cambodia around 30% of wetlands were thought to be seasonally settled by people who plant crops such as pulses, melons, maize and pumpkin. The spread of agriculture needs to be controlled by setting aside land in northern and eastern Cambodia and southern Laos as natural wetlands, involving the establishment of new protected areas, and by protecting a network of permanent pools and seasonal meadows as 'safe havens' (see *Protected areas* below). In the inundation zone of Tonle Sap lake, the traditional Khmer agricultural ecology should be promoted and any intensification of rice cultivation must be carefully managed, to leave short-grass areas suitable for breeding Bengal Florican. The remaining seasonally inundated grasslands in the Mekong Delta regions of Cambodia and Vietnam are extremely important for Sarus Crane and other waterbirds, and should be protected from conversion.

Protected areas have been established at some key wetlands for threatened birds in Cambodia.

PHOTO: ELEANOR BRIGGS

■ *CONFLICT WITH FISHERIES*

The conservation of Tonle Sap lake and the surrounding wetlands involves several complex issues, the entire area being a private fishing concession leased piecemeal to fishing businesses and cooperatives. Annual profits from fishing in 'Lot 2' alone (the area that contains the bulk of the waterbird colonies) run into hundreds of thousands of US dollars, and conservation interests have little chance of competing against such an important economic activity. Negotiations are needed to resolve current and potential conflicts between fishermen and conservationists, with the aim of continuing to manage the whole area as a low-impact fishery (which would maintain fish stocks and benefit wildlife), with conservation management in the areas around waterbird breeding and feeding sites.

Table 3. Conservation issues and strategic solutions for birds of the lower Mekong basin.

Conservation issues	Strategic solutions
Habitat loss and degradation	
■ *CONVERSION TO AGRICULTURE* ■ *CONFLICT WITH FISHERIES* ■ *CUTTING OF NESTING TREES* ■ *DEVELOPMENT (URBAN, INDUSTRIAL, ETC.)* ■ *REDUCED FOOD SUPPLY (FOR VULTURES)* ■ *DISTURBANCE* ■ *POLLUTION*	➤ Promote the traditional Khmer agricultural ecology around Tonle Sap lake ➤ Manage any intensification of rice cultivation in the inundation zone of Tonle Sap lake to leave short-grass areas for breeding Bengal Floricans ➤ Prevent any further conversion of seasonally inundated grasslands in the Mekong Delta ➤ Continue to manage Tonle Sap as a low-impact fishery, to maintain fish stocks and protect threatened birds ➤ Manage the forests at Tonle Sap and other wetlands sustainably, and maintain patches of old growth to provide nesting habitat for waterbirds ➤ Critically review all dam projects in the Mekong watershed, to minimise their impact on ecological processes and the region's biodiversity ➤ Assess the environmental impact of development projects, with mitigation if they damage wetland habitats ➤ Maintain and restore vulture food supplies from traditional extensive livestock management, with supplementary feeding if required ➤ Introduce community conservation programmes to reduce disturbance in the inundation zone of Tonle Sap lake ➤ Ban the use of chemicals or dynamite for fishing ➤ Control the use of pesticides and herbicides, especially near key wetlands
Protected areas coverage and management	
■ *GAPS IN PROTECTED AREAS SYSTEM* ■ *WEAKNESSES IN RESERVE MANAGEMENT*	➤ Establish new wetland protected areas, including 'safe havens' to prevent human disturbance at key waterbird sites in southern Laos and Cambodia ➤ Define and establish protected areas at Tonle Sap, seeking compromise between its management as a fishery and as a sanctuary for wildlife ➤ Develop zonation of large protected areas to minimise conflicts between biodiversity conservation and human usage ➤ Resolve problems with protected area legislation in Laos and Cambodia ➤ Strengthen reserve management through improved funding, infrastructure and staff training
Exploitation of birds	
■ *HUNTING* ■ *EGG AND CHICK COLLECTION*	➤ Improve, publicise and enforce hunting laws, including through monitoring trade in threatened waterbirds at Cambodian food-stalls ➤ Control gun ownership, particularly near important wetlands ➤ Continue and expand the programme to control the exploitation of eggs and nestlings at Tonle Sap waterbird colonies
Gaps in knowledge	
■ *INADEQUATE DATA ON THREATENED BIRDS*	➤ Survey White-shouldered Ibis, Giant Ibis and other large waterbirds in northern and eastern Cambodia ➤ Identify key non-breeding areas for Bengal Florican, and investigate the role of agricultural practices in maintaining its breeding habitat around Tonle Sap lake ➤ Study the ecology and monitor populations of threatened waterbirds, to help improve management of protected areas

There are several large waterbird colonies in the swamp forests around Tonle Sap lake, but many eggs and chicks are taken by professional collectors.

PHOTO: ELEANOR BRIGGS

■ *CUTTING OF NESTING TREES*

In many parts of Asia large waterbirds have lost potential breeding habitat through the felling of tall trees near wetlands. The swamp forests around Tonle Sap lake still provide habitat for several large colonies, although large tracts have apparently already been logged or converted to agriculture; forest clearance is currently proceeding at a slow rate there, but the pressure to exploit the remainder is likely to intensify as timber resources become depleted elsewhere. In the Mekong delta's *Melaleuca* swamp forests, forest fires are a constant risk owing to excessive use of ground-water, which is lowering the water table and drying out the underlying peat. The forests around Tonle Sap and other important wetlands need to be managed sustainably, with patches of old growth maintained near wetlands to provide sites for waterbird colonies.

■ *DEVELOPMENT (URBAN, INDUSTRIAL, ETC.)*

Various hydropower projects are proposed in the Mekong catchment, especially in Yunnan, Laos and Vietnam, and threaten its ecological balance. This is a particular threat to the Mekong's highly diverse (and often migratory) fish fauna that is so important as food for waterbirds and people. Near towns, waterbird foraging areas are being lost to increasing urban and industrial development. This places a demand on timber, which is collected from swamp forests and mangroves for domestic use and construction. Grasslands and wetlands at the Ang Tropeang Thmor Reserve are threatened by plans to develop a village. Proposals to dam the Mekong and its tributaries should be critically reviewed, to minimise damage to ecological processes and globally outstanding biodiversity; some individual projects may cause extreme ecological disruption, and their cancellation may be appropriate. Detailed environmental impact assessments should be mandatory for development proposals, and appropriate mitigation enforced (e.g. re-creation of habitats).

■ *REDUCED FOOD SUPPLY (FOR VULTURES)*

A key factor in vulture declines in South-East Asia appears to be the collapse in food supply, as a result of crashes of large ungulate populations and a transition from extensive to intensive livestock management in many areas (meaning that many fewer livestock carcasses are left for the vultures). Supplementary feeding could be considered as a short-term measure to help the remaining vulture populations, but ultimately food supplies need to be restored through localised changes back to traditional livestock management, and ideally the protection and recovery of large ungulate populations.

■ *DISTURBANCE*

Throughout the Mekong basin human populations are gradually spreading along the rivers into the most remote regions. Local people consume large quantities of fish and other aquatic products, so most wetlands are badly affected by (e.g.) new or seasonal settlements (with ubiquitous dogs), buffalo grazing and trampling, and hunters and fishermen. Waterbirds are very wary of hunters, and easily flushed from their foraging sites or nests. A network of 'safe havens' (see *Protected areas* below) is required for large waterbirds, a scheme involving the exclusion of human activity around important nesting, foraging and roosting sites for storks, ibises and cranes (both within and outside protected areas), particularly in the dry season when habitat availability is most limited and birds most vulnerable to disturbance and hunting. As local people require wetlands for various purposes, zoning is needed, with a few important wetlands classed as strict no-entry zones for a few months during the dry season. This idea should be expanded to cover as many sites as possible in northern and eastern Cambodia and southern Laos, although enforcement in remote regions will be very difficult. Community conservation programmes could be introduced in the inundation zone of Tonle Sap lake, under which Bengal Florican nests would be marked to avoid accidental disturbance.

■ *POLLUTION*

Levels of agricultural or industrial pollution are currently relatively low, although some waterways are deliberately poisoned or dynamited to kill fish for eating: this destroys much aquatic fauna including birds. Large waterbirds and raptors are usually predators or scavengers, and thus vulnerable to build-up of toxic chemicals in the food chain. The use of chemicals or dynamite to kill fish should be banned. Existing and new laws should be enforced in all countries to control the use of pesticides (such as dieldrin and aldrin) and herbicides, especially near important wetlands.

Protected areas coverage and management

■ *GAPS IN PROTECTED AREAS SYSTEM*

Some of the most important wetlands in the lower Mekong basin are officially protected, but several new protected areas should be established, including Dong Khantung in Laos. In Vietnam, a nature reserve is needed for key habitats in the Ha Tien Plain, possibly established under national protected areas legislation, or a private reserve involving the corporate sector. In Cambodia, a national conservation area should be established at Boeung Prek Lapouv, and in western Siem Pang district a system of small seasonally closed sites, or 'safe havens', might be developed to prevent human activity around important nesting, foraging and roosting sites for storks, ibises and cranes. If successful, this model should be extended to elsewhere in northern and eastern Cambodia, and southern Laos. Alternatively, if government support were forthcoming, it may be more appropriate to extend protected area status to part of western Siem Pang district. The protected status of key waterbird sites at Tonle Sap lake needs to be more clearly defined, with the aim of finding a compromise between its management as a fishery and as a sanctuary for wildlife (see *Conflict with fisheries* above). Further core areas should be designated within Tonle Sap Biosphere Reserve, to improve coverage of the swamp forest, including the remaining unprotected large waterbird colonies, and the seasonally inundated grasslands.

■ WEAKNESSES IN RESERVE MANAGEMENT

Many gazetted and proposed protected areas are large, for example many National Protected Areas in Laos, and require careful zonation to minimise conflicts between long-term conservation and legitimate human uses. Throughout the region, there are problems with the current protected areas legislation, as well as inadequate funding, equipment and manpower. A re-evaluation and redefinition of legal terms is therefore required, particularly those applying to protected sites and species in Laos and Cambodia, and increased funding and training is needed to strengthen governmental conservation and forestry departments.

Exploitation of birds

■ HUNTING

Hunting is a widespread practice for many cultural and economic reasons in Indochina. Wildlife is seen as a culinary delicacy and income earner, and populations of all large species of open habitats have declined as a result. Waterbirds are poisoned, shot, or caught on hooked lines. Until the recent establishment of a conservation area, the Sarus Cranes at Ang Tropeang Thmor were hunted for food, or captured for trade to Thailand. In the dry season, the problem intensifies as permanent wetlands attract concentrations of waterbirds that are especially susceptible to intense hunting pressure. Laws need to be strengthened in all countries, and then well publicised (including the environmental justification for this legislation) and enforced, especially at important wetlands. However, Laos, Cambodia and Vietnam have policies and active programmes to control gun ownership, and these may be a more effective and immediate way of reducing hunting pressure than legislation or new protected areas. This gun control should be continued in southern Laos and Vietnam, and if possible also implemented in northern Cambodia, particularly near key wetlands. Enforcement officials should routinely visit food stalls and markets throughout the region to monitor bird trade, confiscate endangered species and fine offenders.

■ EGG AND CHICK COLLECTION

Professional collectors work the waterbird colonies at Tonle Sap, taking many thousands of clutches of eggs and chicks from pelican and stork nests each year. Many of the young birds are reared and fattened before sale, often to middlemen who sell them for food in Battambang and Siem Reap. Surveillance by Wildlife Protection Office staff was thought to have improved the situation, but recent news indicates that the colonies are still poached from their unguarded landward sides. In northern Cambodia and southern Laos, the eggs and chicks of the ground-nesting Sarus Crane are easily accessible, and are taken annually from known nesting sites for food or sale.

An ongoing long-term conservation programme at Tonle Sap needs to be continued and expanded, involving the following components: (1) permanent presence of Wildlife Protection Office staff during the breeding season at colonies; (2) registration of egg and chick collectors and monitoring of their activities; (3) a blanket ban on collection in designated core areas, these encompassing all or part of colonies; (4) guarding of lakeward and landward access to colonies to prevent smuggling of eggs and chicks; (5) development of alternative livelihoods for professional egg and chick collectors. Conservation awareness programmes are also required, targeted at officials, local people and schools around Tonle Sap, to emphasise the importance of protecting wetlands and large waterbirds, and old and new laws relating to activities at waterbird colonies. In the Upper Stung Sen catchment, a pilot Site Support Group (see p.21) has been established, which involves local stakeholders in the protection of nesting waterbirds; this could be a model for many other areas.

Gaps in knowledge

■ INADEQUATE DATA ON THREATENED BIRDS

Further surveys are needed to clarify the distribution, seasonal movements and ecology of large waterbirds and vultures in Cambodia, particularly White-shouldered Ibis and Giant Ibis, and to identify the most important sites for their conservation. Further research is also required to identify key non-breeding sites for Bengal Florican, and to investigate whether existing agricultural systems help in maintaining suitable breeding habitat. The establishment of new protected areas, and the improved management of existing reserves, requires research to determine the most appropriate boundaries and management regimes. For example, the measures required to maintain waterbird numbers at Prek Toal includes the following research elements: (1) evaluation and monitoring of threats (e.g. conversion of forest to farmland, exploitation of waterbirds) from nearby communities; (2) annual monitoring of waterbird numbers and levels of exploitation, with efforts to assess the impact on their populations and to evaluate the potential for sustainable management; (3) continued aerial surveys of the Prek Toal area to locate colonies and estimate waterbird populations; (4) research into the viability of alternative livelihoods and ecotourism.

Surveys are needed to clarify the distribution, seasonal movements and ecology of White-shouldered Ibis and other large waterbirds in Cambodia.

PHOTO: PETE MORRIS/BIRDQUEST

PHILIPPINE WETLANDS

W19

FOUR threatened waterbirds occur in the Philippine archipelago. The freshwater wetlands support the endemic Philippine Duck and the only known non-breeding population of Streaked Reed-warbler. The Philippines also appears to be the main non-breeding area of Chinese Egret, which inhabits coastal beaches, mangrove swamps and estuaries. There were formerly breeding populations of Spot-billed Pelican and Sarus Crane, but these both appear to have declined to extinction.

- **Key habitats** Coastal and freshwater wetlands.
- **Countries and territories** **Philippines**.

Threatened species

	CR	EN	VU	Total
● (breeding)	—	—	1	1
(passage migrant)	—	—	—	—
(non-breeding visitor)	—	1	2	3
Total	—	1	3	4

Key: ● = breeding in this wetland region.
= passage migrant.
= non-breeding visitor.

The Philippine wetlands region is within Conservation International's Philippines Hotspot (see pp.20–21).

Large areas of mangroves have been cleared in the Philippines, but mangrove plantations on open mudflats actually cause further harm by reducing the feeding habitat available to coastal waterbirds. PHOTO: SIMBA CHAN

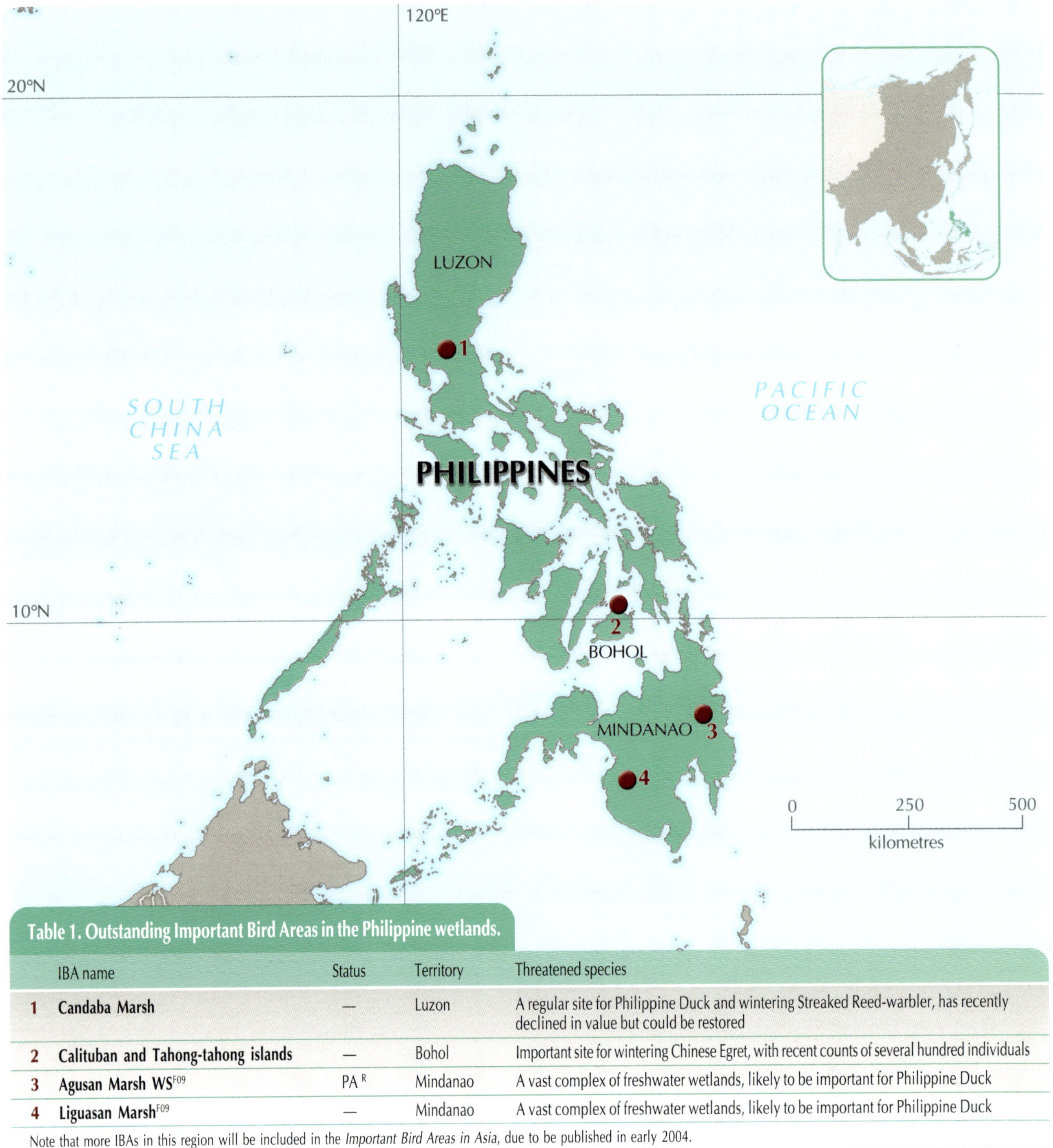

Table 1. Outstanding Important Bird Areas in the Philippine wetlands.

	IBA name	Status	Territory	Threatened species
1	**Candaba Marsh**	—	Luzon	A regular site for Philippine Duck and wintering Streaked Reed-warbler, has recently declined in value but could be restored
2	**Calituban and Tahong-tahong islands**	—	Bohol	Important site for wintering Chinese Egret, with recent counts of several hundred individuals
3	**Agusan Marsh WS**[F09]	PA [R]	Mindanao	A vast complex of freshwater wetlands, likely to be important for Philippine Duck
4	**Liguasan Marsh**[F09]	—	Mindanao	A vast complex of freshwater wetlands, likely to be important for Philippine Duck

Note that more IBAs in this region will be included in the *Important Bird Areas in Asia*, due to be published in early 2004.

Key *IBA name*: WS = Wildlife Sanctuary.
Status: PA = IBA is a protected area; (PA) = IBA partially protected; — = unprotected; R = IBA is wholly or partially a Ramsar Site (see pp.31–32); F09 = also supports threatened forest birds of region F09.

Table 2. Threatened birds of the Philippine wetlands.

Species			Distribution and population
Chinese Egret *Egretta eulophotes*	(50–90% non-breeding symbol)	VU	The coastal wetlands of the Philippines appear to be the most important non-breeding grounds of this species
Philippine Duck *Anas luzonica*	●	VU	Scarce and declining Philippine endemic; breeds in freshwater wetlands but also occurs in coastal habitats
Spotted Greenshank *Tringa guttifer*	(<10% symbol)	EN	Rare non-breeding visitor
Streaked Reed-warbler *Acrocephalus sorghophilus*	(>90% non-breeding symbol) ?	VU	Its only known wintering grounds are in freshwater wetlands in the Philippines

Other threatened waterbirds recorded from this region are: Spot-billed Pelican *Pelecanus philippensis* (probably extinct), Oriental Stork *Ciconia boyciana* (rare visitor), Black-faced Spoonbill *Platalea minor* (rare visitor), Baer's Pochard *Aythya baeri* (rare visitor), Sarus Crane *Grus antigone* (probably extinct), Bristle-thighed Curlew *Numenius tahitiensis* (rare visitor), and Spoon-billed Sandpiper *Eurynorhynchus pygmeus* (rare visitor).

● = region estimated to support >90% of global breeding population; (symbol) = region estimated to support >90% of global non-breeding population, (symbol) = 50–90%, (symbol) = <10%

OUTSTANDING IBAs FOR THREATENED BIRDS (see Table 1)

Four IBAs have been selected, which together support important populations of Chinese Egret, Philippine Duck and Streaked Reed-warbler, the species most reliant on Philippine wetlands for their survival.

CURRENT STATUS OF HABITATS AND THREATENED SPECIES

Large areas of natural freshwater wetland in the Philippines have been converted for cultivation, either through drainage or adoption of wet agriculture. The coastal wetlands have been greatly affected by conversion for aquaculture and cutting of mangroves for firewood, with the area of mangroves estimated to have declined by 67% in the past 60 years. This loss of habitat is one of the main reasons for the declines and extinctions of waterbird populations, but hunting and human disturbance are also likely to have been important factors.

CONSERVATION ISSUES AND STRATEGIC SOLUTIONS (summarised in Table 3)

Habitat loss and degradation

■ *CONVERSION TO AGRICULTURE*

Freshwater wetlands continue to be drained or converted to wet agriculture (rice paddies), severely limiting the amount of habitat available in the dry season. The clearest example is the drainage of Candaba Marsh, once the most important wintering site for Philippine Duck but now too dry to support large numbers. Moreover, the cultivation of rice instead of watermelon at Candaba entails draining the marshes in December or January instead of March or April. Even marsh vegetation is now patchy at the site (to the detriment of Streaked Reed-warbler). Wetland protected areas need to be established and their water levels managed.

■ *CONVERSION TO AQUACULTURE*

Many coastal wetlands are being converted to shrimp- or fish-ponds, as are freshwater wetlands such as Candaba Marsh. The ongoing loss of coastal mangroves is of particular concern, as this must reduce the wintering habitat available to Chinese Egret, and also affect Philippine Duck, which is commonly found in this habitat. Awareness campaigns are required, targeted at local governments with jurisdiction over important wetlands, to inform them of the importance of these wetlands, and persuade them to control wetland conversion so that sufficient natural habitat remains for threatened birds. The existing laws that ban the conversion of mangroves must be more strictly enforced.

■ *CUTTING OF MANGROVES*

In addition to the clearance of mangroves for aquaculture, large areas are affected by cutting for fuelwood. The sustainable use of mangroves by local communities should be promoted (including the maintenance of some areas of mature growth where cutting is not allowed), especially near the critical habitats identified under the National Wildlife Act (which include National Integrated Protected Areas System [NIPAS] sites and the 117 IBAs identified by the Haribon Foundation), through education and awareness campaigns.

■ *POLLUTION*

Many Philippine wetlands are affected by pollution, including from untreated domestic sewage, industrial effluent, toxic chemicals used in aquaculture and agrochemical run-off from farmland. The existing laws to control pollution need to be more strictly enforced by the Department of Environment and National Resources (DENR).

■ *DEVELOPMENT (URBAN, INDUSTRIAL, ETC.)*

Wetland habitat in many lowland areas has been converted to industrial use, or encroached by housing, tourist infra-structure or roads. A site selection process should be followed

The endemic Philippine Duck is under pressure from both habitat loss and hunting for food and sport.

PHOTO: JOE BLOSSOM/WWT

for new roads and other proposed development projects, with the aim of avoiding new development in important wetlands, notably the critical habitats identified under the National Wildlife Act (including NIPAS sites and IBAs).

Protected areas coverage and management

GAPS IN PROTECTED AREAS SYSTEM

The protected areas system in the Philippines is currently being redeveloped through the National Integrated Protected Area System (NIPAS) process, and the recent Local Government Code legislation is likely to be a vital mechanism for the conservation of many sites (see F09 for details). There are currently few wetland protected areas in the Philippines, and more need to be proposed under NIPAS or designated and managed as local protected areas under the Local Government Code. The Haribon Foundation's IBA analysis identifies several unprotected wetlands that are important for threatened birds, notably the outstanding IBAs listed in Table 1. However, given the general lack of data on waterbirds, a new survey initiative is required to identify further sites of importance for threatened waterbirds; the results could be used to produce a new national wetlands directory, to form the basis for the designation of more wetland reserves.

WEAKNESSES IN RESERVE MANAGEMENT

Most sites being designated under NIPAS do not yet have large-scale funding, and a major challenge is to provide the resources and infrastructure necessary to ensure that they are adequately protected. The capacity and resources of the Protected Areas and Wildlife Bureau (PAWB) need to be increased at all levels, especially so that local offices have the capacity to protect NIPAS sites under their jurisdiction.

Exploitation of birds

HUNTING

Hunting for food and sport is a major problem in the Philippines, and may have been the main reason for the decline in the Philippine Duck and the national extinction of Spot-billed Pelican and Sarus Crane. Hunting of all bird species is illegal, but the mechanisms for real enforcement are lacking, and local people in many areas are likely to resist any attempts at strict enforcement of hunting laws. Concerted programmes of education and awareness are needed in and around important wetlands, to improve understanding of the effects of hunting on threatened birds and the relevant laws.

Gaps in knowledge

INADEQUATE DATA ON THREATENED BIRDS

Relatively little research has been conducted on waterbirds in the Philippines, and there are important gaps in knowledge of the distribution and ecology of the threatened species. Much more information is required about the Philippine Duck, including the identification (and subsequent protection) of the most important breeding and non-breeding areas, and detailed ecological and life-history studies to help determine the most appropriate management strategies. In particular (when the security situation allows), surveys are urgently required of the vast Agusan and Liguasan marshes on Mindanao. Studies are also required to identify the most important sites for non-breeding Chinese Egret, Streaked Reed-warbler and shorebirds, including work on the current status and habitat requirements of the warbler at Candaba Marsh. A small flock of Black-faced Spoonbill was located on the Batanes islands in the northern Philippines in winter 2001/2002, and monitoring is required to determine whether this is a regular site for the species.

The Philippines appears to be the main wintering area of Chinese Egret, but further research is required there to clarify its numbers and key sites.

PHOTO: MARTIN HALE

Table 3. Conservation issues and strategic solutions for birds of the Philippine wetlands.

Conservation issues	Strategic solutions
Habitat loss and degradation	
CONVERSION TO AGRICULTURE	Manage water levels in key wetlands to improve conditions for threatened waterbirds
CONVERSION TO AQUACULTURE	Minimise wetland conversion, and strictly enforce laws banning the conversion of mangroves
CUTTING OF MANGROVES	Promote the sustainable use of mangroves by local communities
POLLUTION	Improve enforcement of laws to control pollution
DEVELOPMENT (URBAN, INDUSTRIAL, ETC.)	Avoid siting new roads and other developments in key wetlands
Protected areas coverage and management	
GAPS IN PROTECTED AREAS SYSTEM	Establish new wetland reserves under NIPAS or the Local Government Code
WEAKNESSES IN RESERVE MANAGEMENT	Produce a new directory of Philippine wetlands
	Increase the capacity of the Protected Areas and Wildlife Bureau, especially local offices
Exploitation of birds	
HUNTING	Conduct education and awareness programmes around key wetlands, to reduce hunting of threatened species
Gaps in knowledge	
INADEQUATE DATA ON THREATENED BIRDS	Survey Philippine Duck and wintering Chinese Egret, Streaked Reed-warbler and shorebirds, with ecological studies to help determine the most appropriate management strategies

SUNDALAND WETLANDS

THIS region includes the coasts of southern peninsular Thailand and Peninsular Malaysia, Sumatra, Borneo, Java and associated small islands. The large areas of mangrove swamp (the most extensive in the world) and intertidal mudflat throughout the region provide important habitat for Milky Stork and Lesser Adjutant, both of which have their highest numbers here, and there are non-breeding populations of Chinese Egret, Spotted Greenshank and possibly Spoon-billed Sandpiper. Two waterbirds are endemic to wetlands on the coastal plains of Java, the Sunda Coucal, which survives in mangroves and associated swamps, and the Javanese Lapwing, which was last recorded in marshy grassland in 1940 and may now be extinct.

- **Key habitats** Coastal wetlands.
- **Countries and territories** **Thailand**; **Malaysia** (Peninsular, Sabah, Sarawak); **Singapore**; **Brunei**; **Indonesia** (Sumatra, Kalimantan, Java).

Threatened species

	CR	EN	VU	Total
Breeding	1	—	4	5
Passage migrant	—	—	—	—
Non-breeding visitor	—	1	2	3
Total	1	1	6	8

Key: ● = breeding in this wetland region.
(passage migrant symbol) = passage migrant.
(non-breeding visitor symbol) = non-breeding visitor.

The Sundaland wetlands region is within Conservation International's Sundaland Hotspot (see pp.20–21).

Most of the global population of Milky Stork breeds in the Sundaland wetlands.
PHOTO: JON HORNBUCKLE

Some important sites for Sunda Coucal on Java are surrounded by settlements or encroached by industrial complexes, and several former localities are now covered by cities.

PHOTO: MIKE CROSBY/BIRDLIFE

OUTSTANDING IBAs FOR THREATENED BIRDS (see Table 1)

Eight IBAs have been selected to cover the largest known colonies of Milky Stork and Lesser Adjutant, plus sites for Spot-billed Pelican, Chinese Egret, Spotted Greenshank and the endemic Sunda Coucal.

CURRENT STATUS OF HABITATS AND THREATENED SPECIES

Large areas of coastal wetland have already been lost or degraded in this region. In peninsular Thailand, mangroves have been lost because of cutting for charcoal and timber and clearance for shrimp ponds, with a 22% reduction between 1961 and 1979, and even more rapid loss subsequently. In Peninsular Malaysia, c.25% of mangrove forest has been lost to agriculture and aquaculture, and cutting for wood-chips and timber. The coastal wetlands of

Table 1. Outstanding Important Bird Areas in the Sundaland wetlands.

	IBA name	Status	Territory	Threatened species
1	**Krabi bay**	— R	Thailand	Non-breeding Spotted Greenshank and Chinese Egret
2	**Matang Mangrove Forest R**	PA	Peninsular Malaysia	Colonies of Milky Stork and Lesser Adjutant, although their numbers appear to be in decline
3	**Sembilang** F07	(PA)	Sumatra	Large colony of Milky Stork, also Spot-billed Pelican and Lesser Adjutant
4	**Tanjung Koyan**	—	Sumatra	Large colonies of Milky Stork and Lesser Adjutant
5	**Way Kambas NP** F07	PA	Sumatra	Milky Stork and Lesser Adjutant probably nest, also Spot-billed Pelican
6	**Muara Gembong-Tanjung Sedari**	(PA)	Java	Population of Sunda Coucal, also Milky Stork and Lesser Adjutant
7	**Muara Cimanuk**	—	Java	Population of Sunda Coucal, also Milky Stork
8	**Solo delta**	—	Java	Population of Sunda Coucal, also Milky Stork and Lesser Adjutant

Some waterbirds of this region occur in several of the IBAs listed for region F07, notably Berbak National Park in Sumatra and Danau Sentarum National Park in Kalimantan. Note that more IBAs in this region will be included in the *Important Bird Areas in Asia*, due to be published in early 2004.

Key *IBA name*: NP = National Park; R = Reserve.
Status: PA = IBA is a protected area; (PA) = IBA partially protected; — = unprotected; R = IBA is wholly or partially a Ramsar Site (see pp.31–32); F07 = supports some threatened forest birds of region F07.

Table 2. Threatened birds of the Sundaland wetlands.

Species			Distribution and population
Spot-billed Pelican *Pelecanus philippensis*	○	VU	A small population is believed to breed on Sumatra
Chinese Egret *Egretta eulophotes*	10–50% non-breeding	VU	Significant non-breeding populations on the Thai-Malay peninsula and on Borneo
Milky Stork *Mycteria cinerea*	●	VU	Most of the global population breeds in this region, in Peninsular Malaysia, Sumatra and Java
Lesser Adjutant *Leptoptilos javanicus*	10–50%	VU	Significant numbers breed in Peninsular Malaysia, Sumatra and Borneo
Javanese Lapwing *Vanellus macropterus*	EX?	CR	Known from two areas of marshy grassland on the coasts of east and west Java, but no records since 1940
Spotted Greenshank *Tringa guttifer*	<10% non-breeding	EN	Small but significant non-breeding populations in peninsular Thailand, Peninsular Malaysia and Sumatra
Spoon-billed Sandpiper *Eurynorhynchus pygmeus*	<10% non-breeding	VU	Small numbers recorded in peninsular Thailand, Peninsular Malaysia and Singapore
Sunda Coucal *Centropus nigrorufus*	●	VU	Known only from Java, where it is a scarce inhabitant of mangroves and associated swamps

Other threatened waterbirds recorded from this region are: Black-faced Spoonbill *Platalea minor* and Sarus Crane *Grus antigone*. In addition to the waterbirds, Greater Spotted Eagle *Aquila clanga* (VU; see F01) and Imperial Eagle *A. heliaca* (VU; see G01) winter on the Thai-Malay peninsula, and the former also on Sumatra. Note that two species which occur in forested wetlands in the Thai-Malay peninsula, Sumatra and West Java (at least formerly), White-winged Duck *Cairina scutulata* and Masked Finfoot *Heliopais personata*, are covered in region F07.

● = region estimated to support >90% of global breeding population, (grey) = 10–50%, ○ = <10%; (dark bird) = region estimated to support 10–50% of global non-breeding population, (outline bird) = <10%; EX? = probably extinct

Brunei are almost pristine, but those of Kalimantan, Sumatra and Java have been extensively converted to cultivation (principally rice) and aquaculture. The process is most advanced on Java, where a high proportion of the mangrove and adjacent swamp areas has been destroyed. Despite these pressures, some extensive areas of biodiversity-rich intertidal wetland remain in the region, notably on the east coast of Sumatra.

CONSERVATION ISSUES AND STRATEGIC SOLUTIONS (summarised in Table 3)

Habitat loss and degradation

■ *CONVERSION TO AGRICULTURE*

Conversion of coastal wetlands for cultivation is continuing in many parts of this region, through bunding, installation of sluices, and canalisation. River and coastal sedimentation, exacerbated by deforestation, is affecting estuarine and other coastal areas and making many wetlands shallower and easier to drain. This sedimentation must be minimised, mainly through forest protection (see F07). Awareness initiatives should better inform local authorities and communities about the ecological services that coastal wetlands provide (e.g. mitigation of flooding, control of coastal erosion, fisheries), and of their importance for biodiversity.

■ *CONVERSION TO AQUACULTURE*

Coastal wetlands, mainly mangrove swamps but also some intertidal mudflats, are being converted to shrimp- and fish-ponds throughout this region. Major new plans include large-scale brackish-water fish-farms in southern Sumatra. The extent and impact of conversion for aquaculture must be reduced through application of existing land-use plans, laws and policies, establishment of new nature reserves (see below) and development of alternative options for employment and livelihood improvement. Malaysia has a national policy on coastal resources management, and aims to formulate integrated coastal zone management plans for all states in the country; these plans need to take into account the need for biodiversity conservation, and prevent conversion of key wetlands for threatened birds.

■ *DEVELOPMENT (URBAN, INDUSTRIAL, ETC.)*

This region is experiencing rapid industrial development and human population growth, and many coastal wetlands have been lost to mining, land-filling and coastal reclamation for housing, industry, tourist facilities and roads. For example, some important sites for Sunda Coucal on Java are surrounded by housing estates or encroached by industrial complexes, and several former sites are now built over. The reclamation of wetlands for development needs to be controlled through existing land-use laws and policies (including environmental impact assessments), and new nature reserves (see below).

■ *CUTTING OF MANGROVES*

In addition to the clearance of mangroves for aquaculture, large areas are affected by exploitation. Cutting tall mangroves removes the main nesting habitat of Milky Stork and an important one for Lesser Adjutant, but small-scale exploitation for fuelwood, charcoal, housing and fish-traps appears to have little impact. However, commercial mangrove exploitation poses a real threat to the sustainability of this ecosystem; for example, cutting by the woodchip industry has greatly reduced mangrove cover and quality in Peninsular Malaysia and on Borneo. In many areas illegal logging has destroyed mangrove habitat by clear-felling. Even where mangroves are managed as production forests, the logging tends to leave little old growth suitable for nesting storks. In these production forests, core blocks of habitat should be left undisturbed and unharvested, to provide nesting habitat for storks. There is scope for greater community participation in

Many of the coastal wetlands on Java have already been converted to agriculture and aquaculture.

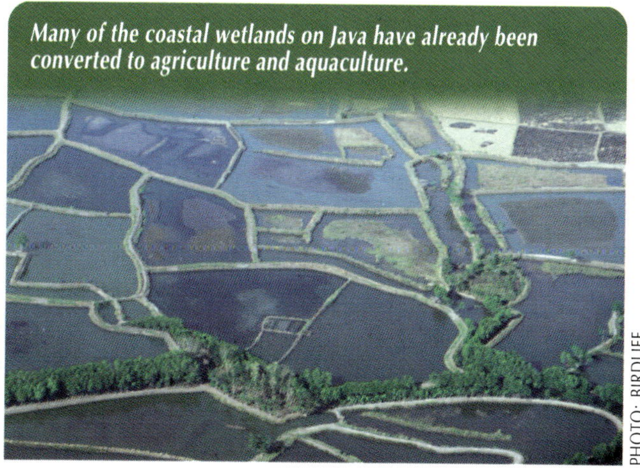

PHOTO: BIRDLIFE

mangrove forest management and conservation. In Indonesia, local governments are developing mangrove and coastal area rehabilitation programmes, which are an important initiative towards restoring lost habitats; however, it is important that mangroves are planted only in areas which were forested in the past, as planting on open mudflats destroys vital waterbird feeding areas.

■ *DISTURBANCE*

Population growth and economic development has caused an increase in human disturbance in wetlands, e.g. by crab-catchers, fishermen, motorboats, etc. Some formerly remote wetlands in Sumatra and Kalimantan have been affected by transmigrants from Java. In many areas of both Malaysia and Indonesia, riverbanks are eroded by speedboat and barge traffic. Disturbance leads to reductions in waterbird numbers and abandonment of breeding sites, e.g. the colony of Milky Storks at Pulau Dua, Java, was lost after it became accessible to people. Human access to important wetland sites must be controlled, particularly at active stork colonies.

■ *POLLUTION*

Pollution threatens waterbirds in Peninsular Malaysia, and is presumably a problem elsewhere in the region. Oil pollution from the Kelang estuary conurbation and the busy shipping lanes in the Straits of Melaka is a major threat to coastal wetlands. Heavy industries along the west coast of Peninsular Malaysia are contaminating coasts and wetlands with high levels of lead, manganese, iron and mercury. Toxic chemicals are sometimes used by prawn farmers to control pests, which could cause problems for species such as Chinese Egret which forage near prawn-ponds. Existing laws to restrict toxic chemical usage need strict enforcement.

Protected areas coverage and management

■ *GAPS IN PROTECTED AREAS SYSTEM*

There are several protected areas in this region which include substantial areas of coastal wetlands, but many important sites remain unprotected. The wetlands of Java are under particular pressure, and new reserves are needed to protect the habitat of Sunda Coucal and other species, including the proposed nature reserves at two of the outstanding IBAs on Java, Muara Gembong-Tanjung Sedari and Muara Cimanuk. Another priority is to protect the large nesting colonies of Milky Stork and Lesser Adjutant on Sumatra.

■ *WEAKNESSES IN RESERVE MANAGEMENT*

Responsibility for reserve management in Indonesia lies with the Directorate General of Forest Protection and Nature Conservation (PKA), but its effectiveness is constrained by shortages of staff, expertise and money. It is necessary to strengthen the PKA through training, improved terms and conditions and equipment for reserve staff, and to improve reserve management through more intensive patrolling, clear demarcation of boundaries, and stricter enforcement of environmental laws.

Exploitation of birds

■ *HUNTING*

Waterbirds are hunted in many parts of this region, and stork nests are robbed of eggs and chicks. In Indonesia, individual storks or pelicans are sometimes seen for sale at local markets, either for food or as pets. Wetland protected areas need to be more intensively patrolled, with campaigns to make hunters more aware of threatened species and existing hunting laws. Efforts are needed to protect nesting colonies of storks from exploitation, particularly on Sumatra.

Gaps in knowledge

■ *INADEQUATE DATA ON THREATENED BIRDS*

The distribution and numbers of the threatened waterbirds are poorly known in some parts of this region, and surveys are required to help identify new areas for conservation action. Searches should continue for Javanese Lapwing until all potential sites on Java have been visited, and surveys (and assessments of habitat extent and quality) at known and potential sites for Sunda Coucal would also be valuable. Surveys are also required to locate Spot-billed Pelican, Milky Stork and Lesser Adjutant colonies on Sumatra (and on Borneo for Lesser Adjutant, e.g. in the Mahakam lakes region), and the numbers and breeding success of storks should be monitored at the larger colonies.

Table 3. Conservation issues and strategic solutions for birds of the Sundaland wetlands.

Conservation issues	Strategic solutions
Habitat loss and degradation	
■ CONVERSION TO AGRICULTURE ■ CONVERSION TO AQUACULTURE ■ DEVELOPMENT (URBAN, INDUSTRIAL, ETC.) ■ CUTTING OF MANGROVES ■ DISTURBANCE ■ POLLUTION	➤ Limit wetland conversion for aquaculture and reclamation for development, based on existing land-use plans, laws and policies ➤ Protect core blocks of undisturbed habitat in mangrove production forests, to provide nesting habitat for storks ➤ Develop community participation in mangrove forest management and conservation, and mangrove and coastal area rehabilitation programmes ➤ Control human access at key wetland sites, particularly stork colonies ➤ Enforce laws to control pollution
Protected areas coverage and management	
■ GAPS IN PROTECTED AREAS SYSTEM ■ WEAKNESSES IN RESERVE MANAGEMENT	➤ Establish new nature reserves to protect coastal wetlands on Java and large stork colonies on Sumatra ➤ Strengthen the PKA through training, improved terms and conditions and equipment for staff ➤ Improve reserve management through more intensive patrolling, boundary demarcation and stricter law enforcement
Exploitation of birds	
■ HUNTING	➤ Strengthen enforcement of existing hunting laws, including through patrolling of nature reserves and protection of nesting colonies of storks
Gaps in knowledge	
■ INADEQUATE DATA ON THREATENED BIRDS	➤ Search for Javanese Lapwing and survey Sunda Coucal on Java ➤ Survey and monitor large waterbird colonies on Sumatra and Borneo

SEABIRDS

THREE threatened seabird species (Short-tailed Albatross, Chinese Crested-tern and Japanese Murrelet) breed only in the Asia region, two breed in eastern Asia and in the U.S.A. (Black-footed Albatross and Red-legged Kittiwake), and two are non-breeding visitors to the Asian region from Christmas Island (Australia) (Abbott's Booby and Christmas Island Frigatebird). These threatened seabirds are mainly concentrated in north-east Asia, particularly on islands in Japan.

	Threatened species			
	CR	EN	VU	Total
●	1	—	2	3
○	—	—	2	2
✈	2	—	—	2
Total	3	—	4	7

Key: ● = breeds only in this region.
○ = also breeds in other region(s).
✈ = non-breeding visitor.

■ **Countries and territories** **Russia** (Chukotka; Koryakia; Kamchatka; Khabarovsk; Primorye; Sakhalin); **Japan** (Hokkaido; Honshu; Izu Islands; Ogasawara Islands; Iwo Islands; Shikoku; Kyushu; Nansei Shoto; Daito Islands; Senkaku Islands); **South Korea**; **China** (*Mainland:* Liaoning; Hebei; Shandong; Jiangsu; Fujian; Guangdong; *Hong Kong*; *Taiwan*); **India** (Andaman Islands); **Sri Lanka**; **Thailand**; **Philippines**; **Malaysia** (Peninsular Malaysia; Sabah; Sarawak); **Singapore**; **Brunei**; **Indonesia** (Kalimantan; Sumatra; Java and Bali; Nusa Tenggara; Maluku).

Japanese Murrelet is virtually confined to Japan, where it is affected by disturbance and predation at its nesting colonies, and pollution and mortality in fishing nets while at sea.
PHOTO: KOJI ONO

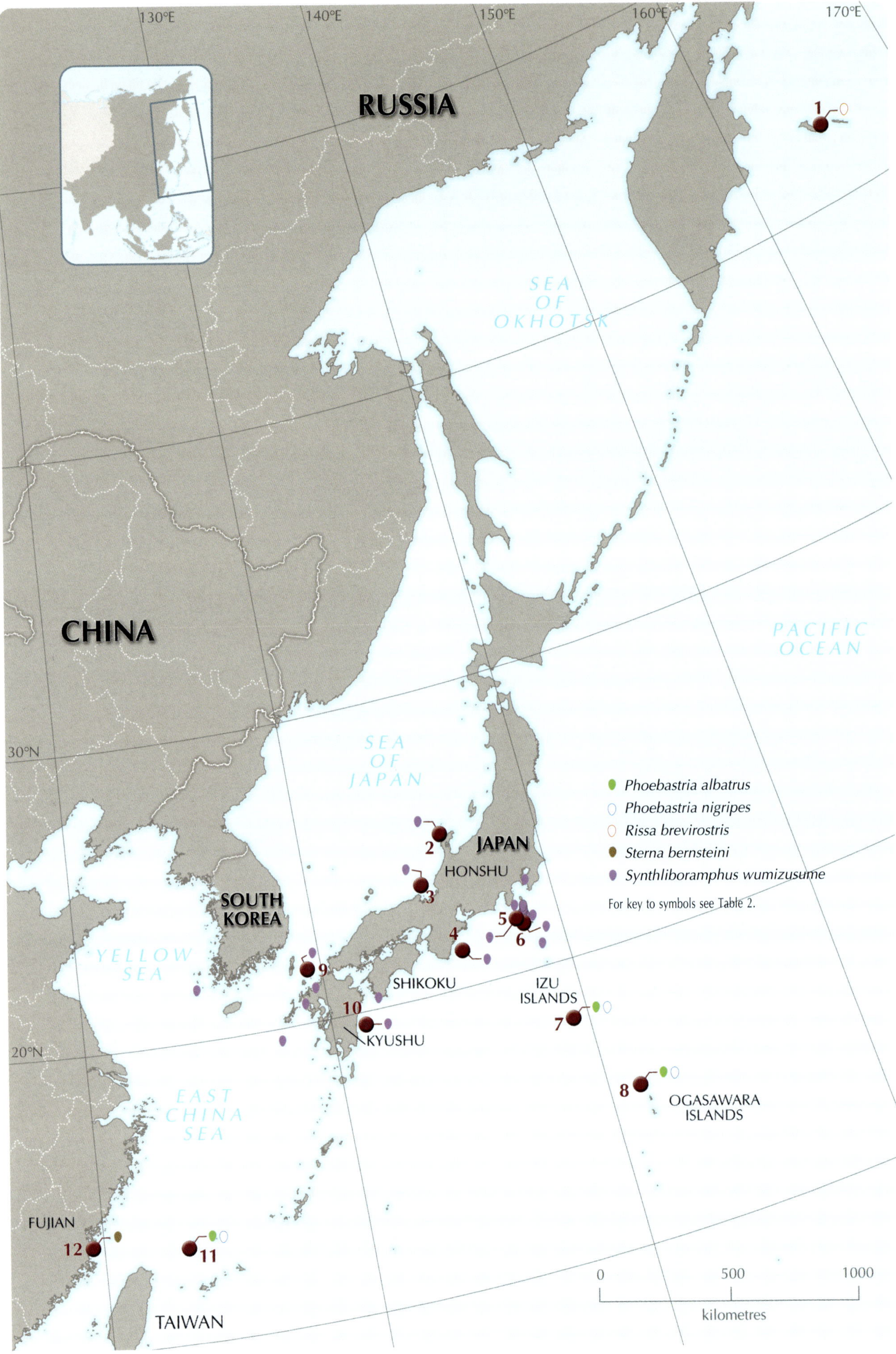

130°E
140°E
150°E
160°E
170°E
RUSSIA
CHINA
SEA OF OKHOTSK
PACIFIC OCEAN
SEA OF JAPAN
30°N
20°N
JAPAN
HONSHU
SOUTH KOREA
YELLOW SEA
SHIKOKU
KYUSHU
IZU ISLANDS
OGASAWARA ISLANDS
EAST CHINA SEA
FUJIAN
TAIWAN
1
2
3
4
5
6
7
8
9
10
11
12
Phoebastria albatrus
Phoebastria nigripes
Rissa brevirostris
Sterna bernsteini
Synthliboramphus wumizusume
For key to symbols see Table 2.
0
500
1000
kilometres

Table 1. Outstanding Important Bird Areas for seabirds.

	IBA name	Status	Territory/Island group	Notes
1	**Commander islands**	—	Kamchatka	Breeding colonies of Red-legged Kittiwake
2	**Nanatsu-jima island**	PA	Honshu	Important breeding colony of Japanese Murrelet
3	**Kutsu-jima island**	—	Honshu	Important breeding colony of Japanese Murrelet
4	**Kiinaga-shima islands**	—	Honshu	Important breeding colony of Japanese Murrelet on Mimiana-jima
5	**Kozu-jima islands**	—	Izu islands	Important breeding colony of Japanese Murrelet on Tadanae-jima
6	**Mitake-jima islands**	—	Izu islands	Important breeding colony of Japanese Murrelet on Onohara-jima
7	**Tori-shima island**	PA	Izu islands	Breeding colonies of Short-tailed Albatross and Black-footed Albatross
8	**Muko-jima**	—	Ogasawara islands	Breeding colony of Black-footed Albatross, with a recent breeding record of Short-tailed Albatross on nearby Yome-jima
9	**Okino-shima island**	PA	Kyushu	Important breeding colony of Japanese Murrelet
10	**Biro-jima island**	—	Kyushu	The largest known breeding colony of Japanese Murrelet
11	**Senkaku islands**	—	Senkaku islands	Breeding colonies of Short-tailed Albatross and Black-footed Albatross
12	**Mazu (Matzu) Dao islands**	PA	Mazu (Matzu) Dao islands	The only know Chinese Crested-tern breeding colony

Note that more IBAs in this region will be included in the *Important Bird Areas in Asia*, due to be published in early 2004.

Key *Status*: PA = IBA is a protected area; (PA) = IBA partially protected; — = unprotected.

Table 2. Threatened Asian seabirds.

Species			Distribution
Short-tailed Albatross *Phoebastria albatrus*	●	VU	Currently known to nest on three islands in southern Japan; non-breeding birds disperse widely in the northern Pacific Ocean
Black-footed Albatross *Phoebastria nigripes*	○	VU	Nests on the Hawaiian Islands (U.S.A.) and on three islands in Japan; non-breeding birds disperse widely in the northern Pacific Ocean
Abbott's Booby *Papasula abbotti*	✈	CR	Breeds on Christmas Island (Australia); non-breeding birds disperse in the Indian and Pacific Oceans (including Indonesia)
Christmas Island Frigatebird *Fregata andrewsi*	✈	CR	Breeds on Christmas Island (Australia); regular non-breeding visitor to Indonesia, Malaysia and Thailand, with records from China, India, Sri Lanka, Singapore and Brunei
Red-legged Kittiwake *Rissa brevirostris*	○	VU	Breeds on the Commander Islands (Russia), and on the Pribilof, Bogoslof and Buldir islands (U.S.A.)
Chinese Crested-tern *Sterna bernsteini*	●	CR	Breeds on islets off eastern China; recent non-breeding records from Taiwan, and old records from Thailand, the Philippines, East Malaysia and Indonesia
Japanese Murrelet *Synthliboramphus wumizusume*	●	VU	Breeds on islands off central and southern Japan and southern South Korea, with records from the Russian Far East (where it may breed) and Taiwan

The Data Deficient Matsudaira's Storm-petrel *Oceanodroma matsudairae* breeds on the Iwo islands in southern Japan, and non-breeding birds disperse west into the Indian Ocean.

● = breeds only in this region; ○ = also breeds in other region(s); ✈ = non-breeding visitor. Colour denotes species as indicated on map.

OUTSTANDING IBAs FOR THREATENED BIRDS (see Table 1)

Twelve IBAs have been selected, covering the most important known colonies of the five threatened seabirds which breed in the region.

CURRENT STATUS OF HABITATS AND THREATENED SPECIES

Short-tailed Albatross was formerly abundant, but declined dramatically in the late nineteenth and early twentieth centuries because of unsustainable exploitation, primarily for feathers to make quilts. The species was assumed to be extinct until its rediscovery on Tori shima in the 1950s; it has subsequently slowly increased in numbers (to c.1,200 individuals), and has started to nest on two other islands. Rapid declines in Black-footed Albatross have recently been noted, caused until 1992 by high mortality from interactions with squid fishing gear and driftnets in the north Pacific, and currently by longline fisheries. Red-legged Kittiwake has also suffered a recent population

Chinese Crested-tern was feared extinct until the discovery of a tiny breeding colony on islands off the coast of eastern China in summer 2000.

PHOTO: CHANG SHOU-HUA

S01

decline, possibly caused by overfishing of its food supply. Japanese Murrelet is a scarce species (estimated at <10,000 individuals) that is thought to be declining for a variety of reasons (see below). Chinese Crested-tern is an exceptionally poorly known bird, historically recorded from a handful of sites in eastern China (where it was assumed to breed) and South-East Asia, feared extinct until four nesting pairs were located in a tern colony on a small island off the coast of Fujian in summer 2000. Abbott's Booby and Christmas Island Frigatebird are both facing rapid decline because of major problems on their breeding grounds on Christmas Island (outside the Asian region in Australia: see BirdLife International 2000).

Short-tailed Albatross was formerly abundant, but declined to near extinction because of exploitation for its feathers.

PHOTO: TAKAO BABA

CONSERVATION ISSUES AND STRATEGIC SOLUTIONS (summarised in Table 3)

Habitat loss and degradation

■ *LOSS OF BREEDING GROUNDS*

The breeding colony of Short-tailed Albatross on Tori-shima has been affected by soil erosion, probably caused by trampling by the albatrosses themselves, but this was addressed in the 1980s by the transplantation of native vegetation into the nesting colony in order to stabilise the substrate and even the nest structures. There is a constant danger that the nesting grounds of this species and Black-footed Albatross could be destroyed by volcanic eruptions on Tori-shima, so sound lures and decoys have been successfully used to attract birds to nest at other locations on the island. The management of the colonies on Tori-shima should be continued and, if feasible, comparable measures should be taken in the Senkaku islands (although this is difficult because they are the subject of territorial claims involving Japan, mainland China and Taiwan), and at any newly discovered nesting colonies.

■ *DISTURBANCE AND INCREASED PREDATION*

Sport fishing off isolated offshore reefs and islets has become very popular in Japan, and direct disturbance and damage to habitats by this activity is believed to be causing declines in Japanese Murrelet at several nesting colonies. The fishermen discard small (dead) fish and use others as lures, thereby attracting Large-billed Crows *Corvus macrorhynchos* and Black-tailed Gulls *Larus crassirostris* from the nearby main islands; predation by the crows (usually of eggs) and gulls (usually of chicks) is putting immense pressure on the murrelets. For example, on Biro-jima island predation by crows was estimated to account for 40% of breeding failure in 1994. Education boards have been constructed near Biro-jima and leaflets produced to inform fishermen about the status of Japanese Murrelet and the importance of this breeding colony. Similar initiatives should be developed at other major colonies, to inform fishermen that this species is

Table 3. Conservation issues and strategic solutions for Asian seabirds.

Conservation issues	Strategic solutions
Habitat loss and degradation	
■ *LOSS OF BREEDING GROUNDS* ■ *DISTURBANCE AND INCREASED PREDATION* ■ *POLLUTION*	➤ Continue management of the albatross colonies on Tori-shima, and consider similar measures at other nesting sites ➤ Encourage sports fishermen in Japan to minimise disturbance and the use of dead fish as lures near Japanese Murrelet colonies ➤ Restrict human access to some islands with breeding Japanese Murrelets ➤ Inform relevant authorities about key areas for threatened seabirds, to reduce the risk of oil spillage in these areas
Protected areas coverage and management	
■ *GAPS IN PROTECTED AREAS SYSTEM*	➤ Establish new protected areas to improve coverage of threatened seabird colonies, notably at the outstanding IBAs listed in Table 1
Exploitation of birds	
■ *HUNTING*	➤ Prevent hunting and disturbance at the Chinese Crested-tern colony on the Mazu (Matzu) Dao islands
Gaps in knowledge	
■ *INADEQUATE DATA ON THREATENED SEABIRDS*	➤ Investigate the conservation needs of Short-tailed Albatross in the Senkaku islands, and locate and protect any new colonies ➤ Monitor Red-legged Kittiwake numbers on the Commander islands, and investigate the possible link between its decline and commercial fishing ➤ Survey islands off eastern China for nesting Chinese Crested-terns ➤ Search for nesting Japanese Murrelets in south-east Russia, and monitor its colonies in Japan
Other conservation issues	
■ *MORTALITY CAUSED BY FISHERIES* ■ *INTRODUCED PREDATORS*	➤ Promote best-practice measures to mitigate seabird by-catch in all longline fisheries ➤ Encourage Japan, South Korea and China to develop and implement National Plans of Action to reduce seabird bycatch by longline fisheries, and join the Agreement on the Conservation of Albatrosses and Petrels ➤ Introduce regulations and measures to reduce by-catch of Japanese Murrelets in drift-nets ➤ Evaluate the feasibility of rat elimination at important Japanese Murrelet colonies

virtually endemic to Japan (and that its conservation is therefore a national responsibility), and that disturbance of murrelets and the practice of using lures should be reduced or stopped near colonies. It may prove necessary to introduce restrictions on human access to some islands with breeding murrelets. New controls on dumping garbage should be introduced on the Izu islands to reduce the population of Large-billed Crow, although more direct methods of control may also be necessary (see F02).

■ *POLLUTION*

Oil spillage at sea is a potential cause of mortality in the seabirds of this region; three Japanese Murrelet were among the 1,315 seabirds recorded killed when a Russian tanker was wrecked in the Sea of Japan in 1997, and an incubating Short-tailed Albatross has been seen with oil on its breast. The possibility of oil exploration around the Senkaku islands (a nesting ground for Short-tailed and Black-footed Albatross) has been discussed in the past. Floating debris, such as fishing-lines, pieces of polystyrene and plastic bags, could be a significant cause of mortality in the surface-feeding Short-tailed and Black-footed Albatross, and has appeared in the food delivered to nestlings and been regurgitated by young. Negotiations are needed with appropriate authorities to inform them of the vulnerability of seabirds to oil spills, the location of islands with nesting colonies, and areas of sea which support wintering concentrations of threatened seabirds. They should be encouraged to adjust oil tanker shipping lanes and oil exploration activities to avoid these areas if possible, and to take measures to minimise the risk of oil spillage near these sensitive areas. Awareness campaigns are required in North-East Asia and globally, aimed at reducing the dumping of rubbish at sea.

The risk of oil spillage needs to be minimised in areas of sea with large wintering concentrations of Japanese Murrelet and other threatened seabirds.

PHOTO: KOJI ONO

Protected areas coverage and management

■ *GAPS IN PROTECTED AREAS SYSTEM*

Some of the most important sites for threatened seabirds are officially protected, including the Short-tailed and Black-footed Albatross colonies on Tori-shima, the breeding grounds of Chinese Crested-tern on the Mazu (Matzu) Dao islands (which are geographically part of Fujian province, but are under the administration of Taipei), and several Japanese Murrelet colonies. New protected areas should be established at some of the outstanding IBAs and at other threatened seabird colonies.

Exploitation of birds

■ *HUNTING*

Seabirds have long been exploited in China, for example Roseate Terns *Sterna dougallii* have been taken for food, and this type of hunting (and the association disturbance at nesting colonies) could be one of the reasons for the apparent decline in Chinese Crested-tern. Indeed, the main threat to the newly discovered colony on the Mazu (Matzu) Dao islands is from fishermen from mainland China visiting the nesting islets to collect seashells or birds' eggs. This site and any other colonies of the species need to managed to prevent hunting and disturbance by fishermen or other visitors (e.g. photographers).

Sports fishermen often visit the small islands used by nesting Japanese Murrelet.

PHOTO: KOJI ONO

S01

Gaps in knowledge

■ *INADEQUATE DATA ON THREATENED SEABIRDS*

If possible, research should be conducted on the status and conservation needs of Short-tailed Albatross at the breeding colony in the Senkaku islands. As the population of this species recovers, it is likely that new colonies will be established, and efforts should be made to locate and protect them; there is a strong possibility that it will return to the islands where it nested in the past, where conditions must be potentially suitable, and searches should be conducted in the appropriate season. The population of Red-legged Kittiwake should be monitored on the Commander islands, together with studies of commercial fisheries in this part of Russia, to investigate whether declines in its numbers could be related to a reduction in food supply caused by unsustainable fishing. If this is found to be the case, measures for more sustainable commercial fishing will need to be developed and promoted.

Surveys are needed in eastern China to try to locate more nesting colonies of Chinese Crested-tern, with immediate conservation measures to safeguard any sites found. The most obvious targets for surveys are the islets off Shandong where the largest series of specimens of the species was collected in the 1930s, but there must be many other potentially suitable islets for nesting off the coast of China between Shandong (and Liaoning) and Fujian; surveys could be conducted for this species together with several other birds that nest on offshore islets, notably Chinese Egret *Egretta eulophotes* and Black-faced Spoonbill *Platalea minor* (see W06). Structured interviews with fishermen in coastal ports could help locate offshore islands and remote coastal localities with breeding colonies of terns and other birds. A record of a juvenile Japanese Murrelet in Peter the Great Bay suggests that the species may nest in this part of Russia, and surveys are required at potential nesting areas, as a forerunner to the development and implementation of appropriate conservation measures.

East Asian fishing fleets use longline fishing techniques in the southern oceans, and their adoption of measures to reduce the bycatch of seabirds is a high global priority.

PHOTO: GRAHAM ROBERTSON/AUSTRALIAN ANTARCTIC DIVISION

Surveys and monitoring are also required for the species in Japan, to improve understanding of its distribution, movements and numbers, and of the impacts of the numerous threats that it faces.

Other conservation issues

■ *MORTALITY CAUSED BY FISHERIES*

Small numbers of Short-tailed Albatross are known to have been killed by longline fishing hooks and fishing-nets, and Black-footed Albatross is currently being badly affected by longline fisheries for tuna, billfish and groundfish, with several thousand birds being killed per year by US-based fisheries alone. Ongoing satellite tracking studies of Black-footed Albatross to assess temporal and spatial overlap with longline fisheries should be continued, with best-practice mitigating measures adopted in all longline fisheries within the species's range. It is important to note that the fishing fleets based in Japan, South Korea and China use longline fishing techniques in the southern oceans, and interact with many threatened seabirds in that region; the adoption of best-practice mitigating measures by these fleets is therefore a high priority for the global conservation of seabirds. It has been proposed that these countries should develop and implement National Plans of Action (NPOAs) to reduce the bycatch of seabirds in longline fishing operations, and join the Agreement on the Conservation of Albatrosses and Petrels (ACAP).

Large numbers of Japanese Murrelet were estimated to have been incidentally caught and killed in drift-nets during the 1990s, possibly involving 10% or more of the total breeding population of the species. Japan, South Korea and Taiwan agreed to a United Nations Resolution to cease large-scale drift-net fisheries in international waters of the North Pacific by the end of 1992, which should have reduced the by-catch of Japanese Murrelet, although drift-nets and coastal gill-nets within the 320-km Exclusive Economic Zone of Japan have probably continued to kill the species in much of its breeding and wintering ranges. Several methods have been devised for reducing (by c.60-70%) by-catch of seabirds (especially alcids) in gill-net fisheries, with little or no reduction in fish catch, including using opaque nylon, 'pingers' that emit sound, fishing during the day rather than at night, and avoiding areas of obvious high bird concentrations. Statutory regulations need to be developed, in consultation with commercial fishery organisations, to reduce by-catch of this species (and other seabirds) using these measures, particularly in areas known to have important concentrations of Japanese Murrelet.

■ *INTRODUCED PREDATORS*

The introduced black rat *Rattus rattus* is a potential predator of seabirds, and carcasses of Japanese Murrelet are frequently found at colonies, suggesting that the predation of adult birds by rats is a widespread problem. For example, the remains of 145 Japanese Murrelet were found on Koya-jima island in 1987, apparently killed by rats (probably brought there accidentally during visits for sports fishing), and the estimated total mortality was 414 birds; few breeding murrelets are currently found there, at what was once a very large colony. Black rats are widespread on Tori-shima, including on the nesting slope used by the albatrosses, and it was feared that they might prey on eggs or hatched young, but there is now evidence that predation by rats is not as serious a problem as was feared. The feasibility of eliminating rats from the more important Japanese Murrelet colonies should be investigated.

APPENDIX: THREATENED BIRD SPECIES COVERED IN THIS ANALYSIS

The following list includes the 303 globally threatened bird species covered in the analysis in this book. It gives their IUCN Red List Categories, the page number of the relevant species account in the *Threatened birds of Asia: the BirdLife International Red Data Book*, and the habitat regions where they occur (with coding to indicate the occurrence status of the species in each region).

Species	Category	RDB	Region	Status
Short-tailed Albatross *Phoebastria albatrus* [4] CMS I; CITES I	VU	46	S01	
Black-footed Albatross *Phoebastria nigripes* [4] CMS II; CITES II	VU	2424	S01	
Spot-billed Pelican *Pelecanus philippensis*	VU	68	W10	
			W12	
			W13	
			W14	
			W15	v
			W16	
			W17	
			W18	
			W19	v
			W20	
Abbott's Booby *Papasula abbotti* [1] CITES I	CR	2425	S01	
Christmas Island Frigatebird *Fregata andrewsi* [1] CITES I	CR	104	S01	
Chinese Egret *Egretta eulophotes* CMS I; CITES I	VU	111	W02	
			W04	
			W06	
			W08	v
			W10	
			W17	v
			W19	
			W20	
White-bellied Heron *Ardea insignis*	EN	137	F06	
White-eared Night-heron *Gorsachius magnificus*	EN	146	F03	
			F06	
Japanese Night-heron *Gorsachius goisagi* CMS I; CITES I	EN	153	F02	
			F08	
			F09	
Milky Stork *Mycteria cinerea* CITES I	VU	169	W17	v
			W18	
			W20	
Storm's Stork *Ciconia stormi*	EN	181	F07	
Oriental Stork *Ciconia boyciana* CMS I; CITES I	EN	194	W02	
			W03	
			W04	
			W05	v
			W06	
			W07	
			W08	
			W10	
			W14	v
			W19	v
Lesser Adjutant *Leptoptilos javanicus*	VU	223	W12	
			W13	
			W14	
			W15	
			W16	
			W17	
			W18	
			W20	
Greater Adjutant *Leptoptilos dubius*	EN	267	W11	v
			W12	
			W14	
			W15	v
			W16	
			W17	
			W18	
White-shouldered Ibis *Pseudibis davisoni*	CR	293	F07	
			W18	
Giant Ibis *Thaumatibis gigantea*	CR	307	W18	
Crested Ibis *Nipponia nippon* CITES I	EN	315	W07	
Black-faced Spoonbill *Platalea minor* CMS I; CITES I	EN	330	W02	v
			W04	
			W06	
			W08	v
			W09	v
			W10	
			W17	v
			W18	v
			W19	v
			W20	v
White-headed Duck *Oxyura leucocephala* [3] CMS I,II; CITES I	EN	354	W05	
			W11	
			W12	v
Swan Goose *Anser cygnoides* [2] CMS I,II; CITES I	EN	363	W02	
			W03	
			W04	
			W05	
			W06	
			W07	
			W08	
			W10	
Lesser White-fronted Goose *Anser erythropus* [3] CMS I,II; CITES I	VU	383	W01	
			W02	
			W03	
			W04	
			W06	
			W07	
			W08	
			W10	v
			W11	v
			W12	v
			W14	v
Crested Shelduck *Tadorna cristata*	CR	399	W02	EX?
White-winged Duck *Cairina scutulata* CITES I	EN	403	F06	
			F07	
Philippine Duck *Anas luzonica* CITES II	VU	441	W19	
Baikal Teal *Anas formosa* CMS I,II; CITES I	VU	449	F01	
			W01	
			W02	
			W03	
			W04	
			W05	
			W06	
			W07	
			W08	
			W10	
			W12	v
			W14	v
			W17	v
Marbled Teal *Marmaronetta angustirostris* [3] CMS I,II; CITES I	VU	479	W11	
			W12	v
			W14	
Pink-headed Duck *Rhodonessa caryophyllacea* CITES I	CR	489	W12	EX?
			W14	EX?
			W16	EX?
Baer's Pochard *Aythya baeri* CMS II; CITES II	VU	502	W02	
			W03	
			W04	
			W05	
			W06	
			W07	
			W08	
			W09	
			W10	

Species	Category	RDB	Region	
Baer's Pochard (continued)			W12	
			W14	
			W16	
			W17	
			W19	v
Scaly-sided Merganser *Mergus squamatus* CMS II; CITES II	EN	523	F01	
			W02	
			W03	
			W04	
			W06	
			W07	v
			W08	
			W09	v
			W10	
Pallas's Fish-eagle *Haliaeetus leucoryphus* [2] CMS I,II; CITES I	VU	542	W05	
			W07	v
			W09	
			W11	
			W12	
			W14	
			W15	
			W16	
			W18	v
Steller's Sea-eagle *Haliaeetus pelagicus* CMS I,II; CITES I	VU	572	W02	
White-rumped Vulture *Gyps bengalensis* CITES II	CR	588	G03	
			W16	
			W18	
Indian Vulture *Gyps indicus* CITES II	CR	614	G03	
Slender-billed Vulture *Gyps tenuirostris* CITES II	CR	621	G03	
			W16	
			W18	
Mountain Serpent-eagle *Spilornis kinabaluensis* CITES II	VU	627	F07	
Nicobar Sparrowhawk *Accipiter butleri* CITES II	VU	630	F06	
Philippine Eagle *Pithecophaga jefferyi* CITES I	CR	633	F09	
Greater Spotted Eagle *Aquila clanga* [3] CMS I,II; CITES I	VU	678	F01	
			W04	v
			W06	
			W07	v
			W08	v
			W09	v
			W10	
			W11	
			W12	
			W13	
			W14	
			W15	
			W16	
			W17	
			W18	
			W20	
Imperial Eagle *Aquila heliaca* [3] CMS I,II; CITES I	VU	712	G01	
			W02	v
			W04	v
			W06	
			W07	v
			W08	v
			W09	v
			W10	
			W11	
			W12	
			W13	v
			W14	
			W15	v
			W17	
			W18	
			W20	
Javan Hawk-eagle *Spizaetus bartelsi* CITES II	EN	736	F07	
Philippine Hawk-eagle *Spizaetus philippensis* CITES II	VU	748	F09	
Wallace's Hawk-eagle *Spizaetus nanus* CITES II	VU	753	F07	
Lesser Kestrel *Falco naumanni* [3] CMS I,II; CITES I	VU	759	G01	
Maleo *Macrocephalon maleo* CITES I	EN	772	F08	

Species	Category	RDB	Region	
Nicobar Megapode *Megapodius nicobariensis*	VU	793	F06	
Moluccan Megapode *Eulipoa wallacei*	VU	800	F08	
Swamp Francolin *Francolinus gularis*	VU	810	G02	
Black Partridge *Melanoperdix nigra*	VU	821	F07	
Manipur Bush-quail *Perdicula manipurensis*	VU	826	G02	
Chestnut-breasted Partridge *Arborophila mandellii*	VU	831	F04	
Sichuan Partridge Arborophila rufipectus	EN	836	F03	
White-faced Hill-partridge *Arborophila orientalis*	VU	844	F07	
White-necklaced Partridge *Arborophila gingica*	VU	847	F03	
Orange-necked Partridge *Arborophila davidi*	EN	852	F06	
Chestnut-headed Partridge *Arborophila cambodiana*	EN	856	F06	
Hainan Partridge *Arborophila ardens*	VU	860	F06	
Himalayan Quail *Ophrysia superciliosa*	CR	865	F04	
Western Tragopan *Tragopan melanocephalus* CITES I	VU	870	F04	
Blyth's Tragopan *Tragopan blythii* CITES I	VU	888	F04	
Cabot's Tragopan *Tragopan caboti* CITES I	VU	898	F03	
Sclater's Monal *Lophophorus sclateri* CITES I	VU	906	F04	
Chinese Monal *Lophophorus lhuysii* CITES I	VU	914	F04	
Edwards's Pheasant *Lophura edwardsi* CITES I	EN	921	F06	
Vietnamese Pheasant *Lophura hatinhensis*	EN	926	F06	
Aceh Pheasant *Lophura hoogerwerfi*	VU	932	F07	
Salvadori's Pheasant *Lophura inornata*	VU	935	F07	
Crestless Fireback *Lophura erythrophthalma*	VU	938	F07	
Wattled Pheasant *Lobiophasis bulweri*	VU	952	F07	
Brown Eared-pheasant *Crossoptilon mantchuricum* CITES I	VU	959	F04	
Cheer Pheasant *Catreus wallichi* CITES I	VU	966	F04	
Elliot's Pheasant *Syrmaticus ellioti* CITES I	VU	981	F03	
Hume's Pheasant *Syrmaticus humiae* CITES I	VU	989	F04	
Reeves's Pheasant *Syrmaticus reevesii*	VU	1001	F03	
Mountain Peacock-pheasant *Polyplectron inopinatum*	VU	1010	F07	
Germain's Peacock-pheasant *Polyplectron germaini* CITES II	VU	1014	F06	
Malaysian Peacock-pheasant *Polyplectron malacense* CITES II	VU	1019	F07	
Bornean Peacock-pheasant *Polyplectron schleiermacheri* CITES II	EN	1026	F07	
Palawan Peacock-pheasant *Polyplectron emphanum* CITES I	VU	1034	F09	
Crested Argus *Rheinardia ocellata* CITES I	VU	1040	F06	
			F07	
Green Peafowl *Pavo muticus* CITES II	VU	1052	F06	
			F07	
Sumba Buttonquail *Turnix everetti*	EN	1088	F08	
Siberian Crane *Grus leucogeranus* [2] CMS I,II; CITES I	CR	1090	W01	
			W02	v
			W03	
			W04	v
			W05	
			W06	
			W07	
			W08	
			W11	v
			W12	
Sarus Crane *Grus antigone* [1] CMS II; CITES II	VU	1118	W11	
			W12	
			W14	
			W15	v
			W16	
			W17	v
			W18	
			W19	v
			W20	v
White-naped Crane *Grus vipio* CMS I,II; CITES I	VU	1151	W02	
			W03	
			W04	

Species	Category	RDB	Region	
White-naped Crane (continued)			W05	
			W06	
			W07	
			W08	
			W10	v
Hooded Crane *Grus monacha* CMS I,II; CITES I	VU	1174	F01	
			W02	
			W03	
			W04	
			W05	
			W06	
			W07	
			W08	
			W09	v
			W10	v
			W14	v
Black-necked Crane *Grus nigricollis* CMS I,II; CITES I	VU	1198	W09	
Red-crowned Crane *Grus japonensis* CMS I,II; CITES I	EN	1226	W02	
			W03	
			W04	v
			W05	
			W06	
			W07	
			W08	v
			W09	v
			W10	v
Swinhoe's Rail *Coturnicops exquisitus*	VU	1254	W02	
			W03	
			W04	?
			W05	
			W06	
			W08	?
			W10	
Okinawa Rail *Gallirallus okinawae*	EN	1260	F02	
Snoring Rail *Aramidopsis plateni*	VU	1265	F08	
Blue-faced Rail *Gymnocrex rosenbergii*	VU	1269	F08	
Talaud Rail *Gymnocrex talaudensis*	EN	1273	F08	
Invisible Rail *Habroptila wallacii*	VU	1275	F08	
Masked Finfoot *Heliopais personata*	VU	1278	F06	
			F07	
Great Bustard *Otis tarda* [3] CMS I*,II; CITES I	VU	1294	G01	
			W04	v
			W06	?
			W07	?
			W08	?
			W10	v
			W11	v
Great Indian Bustard *Ardeotis nigriceps* CITES I	EN	1321	G03	
Bengal Florican *Houbaropsis bengalensis* CITES I	EN	1345	G02	
			W18	?
Lesser Florican *Sypheotides indica* CITES II	EN	1368	G03	
Javanese Lapwing *Vanellus macropterus*	CR	1383	W20	EX?
Sociable Lapwing *Vanellus gregarius* [2] CMS I,II; CITES I	VU	1387	G03	
Ryukyu Woodcock *Scolopax mira*	VU	1395	F02	
Moluccan Woodcock *Scolopax rochussenii*	EN	1400	F08	
Wood Snipe *Gallinago nemoricola* CITES II	VU	1402	F04	
			F05	
Spotted Greenshank *Tringa guttifer* CMS I,II; CITES I	EN	1415	W02	
			W04	
			W06	
			W10	
			W14	v
			W15	
			W17	
			W19	
			W20	
Spoon-billed Sandpiper *Eurynorhynchus pygmeus* CMS I,II; CITES I	VU	1433	W01	
			W02	
			W04	
Spoon-billed Sandpiper (continued)			W06	
			W10	
			W13	
			W14	v
			W15	
			W17	
			W19	v
			W20	
Jerdon's Courser *Rhinoptilus bitorquatus*	CR	1454	G03	
Saunders's Gull *Larus saundersi* CMS I; CITES I	VU	1458	W02	v
			W04	
			W06	
			W08	v
			W10	
Relict Gull *Larus relictus* [2] CMS I; CITES I	VU	1478	W02	v
			W04	v
			W05	
			W06	
			W07	v
			W10	v
Red-legged Kittiwake *Rissa brevirostris* [4]	VU	2433	S01	
Chinese Crested-tern *Sterna bernsteini* CMS I; CITES I	CR	1488	S01	
Indian Skimmer *Rynchops albicollis*	VU	1493	W11	
			W12	
			W14	
			W15	
			W16	
			W17	v
			W18	v
Japanese Murrelet *Synthliboramphus wumizusume* CMS I; CITES II	VU	1508	S01	
Pale-backed Pigeon *Columba eversmanni* [2]	VU	1517	G01	
			G03	
Nilgiri Wood-pigeon *Columba elphinstonii*	VU	1524	F05	
Sri Lanka Wood-pigeon *Columba torringtoni*	VU	1531	F05	
Pale-capped Pigeon *Columba punicea*	VU	1536	F06	
Silvery Wood-pigeon *Columba argentina*	CR	1550	F07	
Slaty Cuckoo-dove *Turacoena modesta*	VU	1553	F08	
Mindoro Bleeding-heart *Gallicolumba platenae*	CR	1556	F09	
Negros Bleeding-heart *Gallicolumba keayi*	CR	1560	F09	
Mindanao Bleeding-heart *Gallicolumba criniger*	EN	1564	F09	
Sulu Bleeding-heart *Gallicolumba menagei*	CR	1569	F09	
Wetar Ground-dove *Gallicolumba hoedtii*	EN	1572	F08	
Tawitawi Brown-dove *Phapitreron cinereiceps*	CR	1576	F09	
Mindanao Brown-dove *Phapitreron brunneiceps*	VU	1579	F09	
Flores Green-pigeon *Treron floris*	VU	1582	F08	
Timor Green-pigeon *Treron psittacea*	EN	1586	F08	
Large Green-pigeon *Treron capellei*	VU	1589	F07	
Red-naped Fruit-dove *Ptilinopus dohertyi*	VU	1597	F08	
Flame-breasted Fruit-dove *Ptilinopus marchei*	VU	1599	F09	
Carunculated Fruit-dove *Ptilinopus granulifrons*	VU	1603	F08	
Negros Fruit-dove *Ptilinopus arcanus*	CR	1605	F09	
Mindoro Imperial-pigeon *Ducula mindorensis* CITES I	VU	1607	F09	
Spotted Imperial-pigeon *Ducula carola*	VU	1610	F09	
Grey Imperial-pigeon *Ducula pickeringii*	VU	1616	F07	
			F08	
			F09	
Timor Imperial-pigeon *Ducula cineracea*	EN	1622	F08	
Red-and-blue Lory *Eos histrio* CITES I	EN	1624	F08	
Chattering Lory *Lorius garrulus* CITES II	EN	1630	F08	
Purple-naped Lory *Lorius domicella* CITES II	VU	1636	F08	
Blue-fronted Lorikeet *Charmosyna toxopei* CITES II	CR	1640	F08	
Yellow-crested Cockatoo *Cacatua sulphurea* CITES II	CR	1643	F08	

Species	Category	RDB	Region	
Salmon-crested Cockatoo *Cacatua moluccensis* CITES I	VU	1662	F08	
White Cockatoo *Cacatua alba* CITES II	VU	1669	F08	
Philippine Cockatoo *Cacatua haematuropygia* CITES I	CR	1676	F09	
Blue-headed Racquet-tail *Prioniturus platenae* CITES II	VU	1689	F09	
Green Racquet-tail *Prioniturus luconensis* CITES II	VU	1693	F09	
Blue-winged Racquet-tail *Prioniturus verticalis* CITES II	EN	1697	F09	
Black-lored Parrot *Tanygnathus gramineus* CITES II	VU	1700	F08	
Sangihe Hanging-parrot *Loriculus catamene* CITES II	EN	1703	F08	
Flores Hanging-parrot *Loriculus flosculus* CITES II	EN	1706	F08	
Red-faced Malkoha *Phaenicophaeus pyrrhocephalus*	VU	1709	F05	
Sumatran Ground-cuckoo *Carpococcyx viridis*	CR	1715	F07	
Short-toed Coucal *Centropus rectunguis*	VU	1718	F07	
Black-hooded Coucal *Centropus steerii*	CR	1723	F09	
Sunda Coucal *Centropus nigrorufus*	VU	1726	W20	
Green-billed Coucal *Centropus chlororhynchus*	VU	1730	F05	
Sulawesi Golden Owl *Tyto inexspectata* CITES II	VU	1735	F08	
Taliabu Masked-owl *Tyto nigrobrunnea* CITES II	EN	1738	F08	
Siau Scops-owl *Otus siaoensis* CITES II	CR	1740	F08	
White-fronted Scops-owl *Otus sagittatus* CITES II	VU	1742	F07	
Javan Scops-owl *Otus angelinae* CITES II	VU	1747	F07	
Flores Scops-owl *Otus alfredi* CITES II	EN	1750	F08	
Giant Scops-owl *Mimizuku gurneyi* CITES I	VU	1753	F09	
Philippine Eagle-owl *Bubo philippensis* CITES II	VU	1757	F09	
Blakiston's Fish-owl *Ketupa blakistoni* CITES II	EN	1761	F01	
Forest Owlet *Heteroglaux blewitti* CITES I	CR	1772	F05	
Cinnabar Hawk-owl *Ninox ios* CITES II	VU	1776	F08	
Sulawesi Eared-nightjar *Eurostopodus diabolicus*	VU	1778	F08	
Sunda Nightjar *Caprimulgus concretus*	VU	1781	F07	
Dark-rumped Swift *Apus acuticauda*	VU	1784	F04	
Blue-banded Kingfisher *Alcedo euryzona*	VU	1788	F07	
Silvery Kingfisher *Alcedo argentata*	VU	1796	F09	
Philippine Dwarf Kingfisher *Ceyx melanurus*	VU	1801	F09	
Rufous-lored Kingfisher *Todiramphus winchelli*	VU	1806	F09	
Sombre Kingfisher *Todiramphus funebris*	VU	1812	F08	
Blue-capped Kingfisher *Actenoides hombroni*	VU	1815	F09	
Purple Dollarbird *Eurystomus azureus*	VU	1819	F08	
Palawan Hornbill *Anthracoceros marchei* CITES II	VU	1822	F09	
Sulu Hornbill *Anthracoceros montani* CITES II	CR	1826	F09	
Mindoro Tarictic *Penelopides mindorensis* CITES II	EN	1830	F09	
Visayan Tarictic *Penelopides panini* CITES II	EN	1833	F09	
Rufous-necked Hornbill *Aceros nipalensis* CITES I	VU	1838	F04	
Visayan Wrinkled Hornbill *Aceros waldeni* CITES II	CR	1854	F09	
Narcondam Hornbill *Aceros narcondami* CITES II	VU	1860	F06	
Sumba Hornbill *Aceros everetti* CITES II	VU	1863	F08	
Plain-pouched Hornbill *Aceros subruficollis* CITES I	VU	1868	F07	
Sulu Woodpecker *Picoides ramsayi*	VU	1876	F09	
Okinawa Woodpecker *Sapheopipo noguchii*	CR	1879	F02	
Visayan Broadbill *Eurylaimus samarensis*	VU	1884	F09	
Mindanao Broadbill *Eurylaimus steerii*	VU	1887	F09	
Schneider's Pitta *Pitta schneideri*	VU	1891	F07	
Gurney's Pitta *Pitta gurneyi* CITES I	CR	1894	F07	
Blue-headed Pitta *Pitta baudii*	VU	1910	F07	
Azure-breasted Pitta *Pitta steerii*	VU	1915	F09	
Whiskered Pitta *Pitta kochi* CITES I	VU	1919	F09	
Graceful Pitta *Pitta venusta*	VU	1923	F07	
Fairy Pitta *Pitta nympha* CITES II	VU	1926	F02 F03 F07	
White-eyed River-martin *Eurychelidon sirintarae* CITES I	CR	1942	W17	?
White-winged Cuckoo-shrike *Coracina ostenta*	VU	1948	F09	
Straw-headed Bulbul *Pycnonotus zeylanicus* CITES II	VU	1951	F07	
Taiwan Bulbul *Pycnonotus taivanus*	VU	1966	F03	
Yellow-throated Bulbul *Pycnonotus xantholaemus*	VU	1969	G03	
Hook-billed Bulbul *Setornis criniger*	VU	1974	F07	
Streak-breasted Bulbul *Ixos siquijorensis*	EN	1979	F09	
Nicobar Bulbul *Hypsipetes nicobariensis*	VU	1982	F06	
Philippine Leafbird *Chloropsis flavipennis*	VU	1985	F09	
Sri Lanka Whistling-thrush *Myophonus blighi*	EN	1989	F05	
Malaysian Whistling-thrush *Myophonus robinsoni*	VU	1993	F07	
Ashy Thrush *Zoothera cinerea*	VU	1995	F09	
Amami Thrush *Zoothera major*	CR	1999	F02	
Grey-sided Thrush *Turdus feae* CMS II; CITES II	VU	2002	F04	
Izu Thrush *Turdus celaenops*	VU	2009	F02	
Rusty-bellied Shortwing *Brachypteryx hyperythra*	VU	2015	F04	
White-bellied Shortwing *Brachypteryx major*	VU	2019	F05	
Rufous-headed Robin *Luscinia ruficeps* CMS II; CITES II	VU	2023	F04 F07	
Black-throated Blue Robin *Luscinia obscura* CITES II	VU	2028	F04	
Black Shama *Copsychus cebuensis*	EN	2032	F09	
Luzon Water-redstart *Rhyacornis bicolor*	VU	2035	F09	
Sumatran Cochoa *Cochoa beccarii*	VU	2039	F07	
Javan Cochoa *Cochoa azurea*	VU	2042	F07	
White-browed Bushchat *Saxicola macrorhyncha*	VU	2045	G03	
White-throated Bushchat *Saxicola insignis* CMS II; CITES II	VU	2051	G01 G02	
Ashy-headed Laughingthrush *Garrulax cinereifrons*	VU	2061	F05	
Snowy-cheeked Laughingthrush *Garrulax sukatschewi*	VU	2065	F04	
White-speckled Laughingthrush *Garrulax bieti*	VU	2069	F04	
Rufous-breasted Laughingthrush *Garrulax cachinnans*	EN	2072	F05	
Golden-winged Laughingthrush *Garrulax ngoclinhensis*	VU	2076	F06	
Chestnut-eared Laughingthrush *Garrulax konkakinhensis*	VU	—	F06	
Collared Laughingthrush *Garrulax yersini*	EN	2078	F06	
Omei Shan Liocichla *Liocichla omeiensis* CITES II	VU	2080	F03	
Black-browed Babbler *Malacocincla perspicillata*	VU	2084	F07	
Marsh Babbler *Pellorneum palustre*	VU	2086	G02	
Bornean Wren-babbler *Ptilocichla leucogrammica*	VU	2090	F07	
Falcated Wren-babbler *Ptilocichla falcata*	VU	2094	F09	
Rusty-throated Wren-babbler *Spelaeornis badeigularis*	VU	2097	F04	
Tawny-breasted Wren-babbler *Spelaeornis longicaudatus*	VU	2099	F04	
Flame-templed Babbler *Dasycrotapha speciosa*	EN	2102	F09	
Negros Striped-babbler *Stachyris nigrorum*	EN	2106	F09	
Snowy-throated Babbler *Stachyris oglei*	VU	2109	F04	
Jerdon's Babbler *Chrysomma altirostre*	VU	2112	G02 W11 W16	 ? ?
Slender-billed Babbler *Turdoides longirostris*	VU	2120	G02	
Black-crowned Barwing *Actinodura sodangorum*	VU	2125	F06	
Gold-fronted Fulvetta *Alcippe variegaticeps*	VU	2128	F03	
Grey-crowned Crocias *Crocias langbianis*	EN	2133	F06	
Black-breasted Parrotbill *Paradoxornis flavirostris*	VU	2136	G02	

Species	Category	RDB	Region
Grey-hooded Parrotbill *Paradoxornis zappeyi*	VU	2142	F04
Rusty-throated Parrotbill *Paradoxornis przewalskii*	VU	2146	F04
Grey-crowned Prinia *Prinia cinereocapilla*	VU	2150	G02
Styan's Grasshopper-warbler *Locustella pleskei* CMS II; CITES II	VU	2154	W02, W04, W06, W10 ?
Streaked Reed-warbler *Acrocephalus sorghophilus* CMS II; CITES II	VU	2160	W06, W10 ?, W19 ?
Manchurian Reed-warbler *Acrocephalus tangorum* CMS II	VU	2165	W03, W05, W06, W10 ?, W17, W18
Izu Leaf-warbler *Phylloscopus ijimae* CMS II; CITES II	VU	2170	F02, F09
Hainan Leaf-warbler *Phylloscopus hainanus*	VU	2176	F06
Marsh Grassbird *Megalurus pryeri* CMS II; CITES II	VU	2179	W03, W04, W05 v, W06, W08
Bristled Grass-warbler *Chaetornis striatus*	VU	2187	G02
Broad-tailed Grassbird *Schoenicola platyura*	VU	2195	F05
Brown-chested Jungle-flycatcher *Rhinomyias brunneata* CMS II; CITES II	VU	2200	F03, F07
White-browed Jungle-flycatcher *Rhinomyias insignis*	VU	2207	F09
White-throated Jungle-flycatcher *Rhinomyias albigularis*	EN	2210	F09
Ashy-breasted Flycatcher *Muscicapa randi*	VU	2214	F09
Kashmir Flycatcher *Ficedula subrubra* CMS II; CITES II	VU	2217	F04, F05
Little Slaty Flycatcher *Ficedula basilanica*	VU	2225	F09
Damar Flycatcher *Ficedula henrici*	VU	2229	F08
Palawan Flycatcher *Ficedula platenae*	VU	2231	F09
Lompobatang Flycatcher *Ficedula bonthaina*	EN	2234	F08
Matinan Flycatcher *Cyornis sanfordi*	EN	2236	F08
Rueck's Blue-flycatcher *Cyornis ruckii* CITES II	CR	2238	F07
Large-billed Blue-flycatcher *Cyornis caerulatus*	VU	2241	F07
Celestial Monarch *Hypothymis coelestis*	VU	2245	F09
Caerulean Paradise-flycatcher *Eutrichomyias rowleyi*	CR	2250	F08
Flores Monarch *Monarcha sacerdotum*	EN	2255	F08
White-tipped Monarch *Monarcha everetti*	EN	2258	F08
Black-chinned Monarch *Monarcha boanensis*	CR	2260	F08
Sangihe Shrike-thrush *Colluricincla sanghirensis*	CR	2262	F08
White-naped Tit *Parus nuchalis*	VU	2265	G03
White-browed Nuthatch *Sitta victoriae*	EN	2271	F04
Giant Nuthatch *Sitta magna*	VU	2273	F04
Beautiful Nuthatch *Sitta formosa*	VU	2279	F04
Cebu Flowerpecker *Dicaeum quadricolor*	CR	2288	F09
Visayan Flowerpecker *Dicaeum haematostictum*	VU	2294	F09
Scarlet-collared Flowerpecker *Dicaeum retrocinctum*	VU	2298	F09
Elegant Sunbird *Aethopyga duyvenbodei*	EN	2302	F08
Sangihe White-eye *Zosterops nehrkorni*	CR	2305	F08
Rufous-throated White-eye *Madanga ruficollis*	EN	2307	F08
Bonin White-eye *Apalopteron familiare*	VU	2309	F02
Dusky Friarbird *Philemon fuscicapillus*	VU	2314	F08
Rufous-backed Bunting *Emberiza jankowskii*	VU	2317	F01
Yellow Bunting *Emberiza sulphurata*	VU	2322	F02, F09
Green Avadavat *Amandava formosa* CITES II	VU	2334	G03
Green-faced Parrotfinch *Erythrura viridifacies*	VU	2340	F09
Java Sparrow *Padda oryzivora* CITES II	VU	2345	F07
Timor Sparrow *Padda fuscata*	VU	2355	F08
Finn's Weaver *Ploceus megarhynchus*	VU	2358	G02
White-faced Starling *Sturnus albofrontatus*	VU	2365	F05
Black-winged Starling *Sturnus melanopterus*	EN	2369	F07
Bali Starling *Leucopsar rothschildi* CITES I	CR	2375	F07
Isabela Oriole *Oriolus isabellae*	EN	2392	F09
Silver Oriole *Oriolus mellianus*	VU	2396	F03, F06
Amami Jay *Garrulus lidthi*	VU	2402	F02
Sichuan Jay *Perisoreus internigrans*	VU	2406	F04
Sri Lanka Magpie *Urocissa ornata*	VU	2410	F05
Banggai Crow *Corvus unicolor*	EN	2415	F08
Flores Crow *Corvus florensis*	EN	2417	F08

KEY

= confined to this region

= also breeds in other region(s)

= non-breeding visitor from another region

= region estimated to support >90% of global breeding population; = 50–90%; = 10–50%; = <10%; percentage unknown

= region estimated to support >90% of global non-breeding population; = 50–90%; = 10–50%; = <10%; = percentage unknown

= region estimated to support >90% of global population on passage; = 50–90%; = 0–50%; = <10%; ? = percentage unknown

v = vagrant

EX? = probably extinct

CITES I = Listed on Appendix I of CITES (see pp.32–33)
CITES II = Listed on Appendix II of CITES
CMS I = Listed on Appendix I of CMS (see p.33)
CMS II = Listed on Appendix II of CMS
CMS I, II = Listed on both Appendices of CMS
CMS I*, II = The 'Middle-European population' of Great Bustard is on Appendix I of CMS, but its other populations are on Appendix II

FOOTNOTES

1 Ranges outside the Asia region into Australasia
2 Ranges outside the Asia region into Central Asia
3 Ranges outside the Asia region into Europe and western Asia
4 Ranges outside the Asia region into the north Pacific Ocean

REFERENCES

This book is a synthesis of the conservation recommendations in BirdLife International (2001). Readers who wish to find the source of the information presented here are advised to check the relevant species accounts in that book, in which more than 7,000 bibliographic references are cited. The information on Important Bird Areas (IBAs) was taken from the national inventories listed below, together with draft material for the directory of *Important Bird Areas in Asia*, which is scheduled for publication early in 2004. Some additional references used in the compilation of this book are listed below, together with the references cited in the introductory sections. Note that many people who reviewed the habitat accounts (who are thanked in the Acknowledgements) provided new information and suggested additional conservation measures.

Introduction

Balmford, A., Bruner, A., Cooper, P., Costanza, R., Farner, S., Green, R. E., Jenkins, M., Jefferiss, P., Jessamy, V., Madden, J., Munro, K., Myers, N., Naeem, S., Paavola, J., Rayment, M., Rosendo, S., Roughgarden, J., Trumper, K. and Turner, K. R. (2002) Economic reasons for conserving wild nature. *Science* 297: 950–953.

BirdLife International (2000) *Threatened birds of the world.* Barcelona and Cambridge, U.K.: Lynx Edicions and BirdLife International.

BirdLife International (2001) *Threatened birds of Asia: the BirdLife International Red Data Book.* Cambridge, U.K.: BirdLife International.

Collar, N. J. (2003) Beyond value: biodiversity and the freedom of the mind. *Global Ecology & Biogeography* 12: 265–269.

Contreras-Hermosilla, A. (2000) *The underlying causes of forest decline.* Bogor, Indonesia: Center for International Forestry Research.

Costanza, R., d'Arge, R., de Groot, R., Farber, S., Grasso, M., Hannon, B., Limburg, K., Naeem, S., O'Neill, R. V., Paruelo, J., Raskin, R. G., Sutton, P. and van den Belt, M. (1997) The value of the world's ecosystem services and natural capital. *Nature* 387: 253–260.

Eames, J. C. and Eames, C. (2001) A new species of laughingthrush (Passeriformes: Garrulacinae) from the Central Highlands of Vietnam. *Bull. Brit. Orn. Club* 121(1): 10–23.

ICBP (1992) *Putting biodiversity on the map: priority areas for global conservation.* Cambridge, U.K.: International Council for Bird Preservation.

IUCN (2001) *IUCN Red List categories and criteria: version 3.1.* Gland, Switzerland and Cambridge, U.K.: IUCN Species Survival Commission.

IUCN (2002) 2002 IUCN Red List of Threatened Species. Available: http/www.redlist.org.

IUCN/SSC = IUCN Species Survival Commission (1994) *IUCN Red List categories, as approved by the 40th meeting of the IUCN Council, Gland, Switzerland, 30 November 1994.* [Gland, Switzerland:] IUCN–The World Conservation Union.

Mittermeier, R. A., Myers, N., Gil, P. R. and Mittermeier, C. G. (1999) *Hotspots: Earth's biologically richest and most endangered ecosystems.* Mexico City: CEMEX and Conservation International.

Olson, D. M. and Dinerstein, E. (1998) The Global 200: a representation approach to conserving the earth's most biologically valuable ecoregions. *Conservation Biology* 12: 502–515.

Sizer, N. (2000) *Perverse habits: the G8 and subsidies that harm forests and economies.* World Resources Institute.

Stattersfield, A. J., Crosby, M. J., Long, A. J. and Wege, D. C. (1998) *Endemic Bird Areas of the world: priorities for biodiversity conservation.* Cambridge, U.K.: BirdLife International.

Wikramanayake, E., Dinerstein, E., Loucks, C. J., Olsen, D. M., Morrison, J., Lamoreux, J., McKnight, M. and Hedao, P. (2002) *Terrestrial Ecoregions of the Indo-Pacific: a conservation assessment.* Washington: Island Press.

Forest, grassland and wetland accounts

[with relevant codes]

Barter, M. (2002) *Shorebirds of the Yellow Sea: Importance, threats and conservation status.* Canberra, Australia: Wetlands International Global Series 9, International Wader Studies 12. [W06]

Critical Ecosystem Partnership Fund (2001a) *Ecosystem Profile: Sumatra, Sundaland.* Available: http://www.cepf.net/xp/cepf/ [F07]

Critical Ecosystem Partnership Fund (2001b) *Ecosystem Profile: the Philippines.* Available: http://www.cepf.net/xp/cepf/ [F09]

Critical Ecosystem Partnership Fund (2002) *Ecosystem Profile: Mountains of Southwest China.* Available: http://www.cepf.net/xp/cepf/ [F03, F04]

Geatz, R. (2002) High energy on the edge of the Himalayas. *Nature Conservancy Magazine* 52(2): 28–37. [F03, F04]

Holmes, D. and Rombang, W. M. (2001) *Daerah Penting bagi Burung: Sumatera.* Bogor, Indonesia: PKA/BirdLife International Indonesia Programme. (in Indonesian) [F07, W20]

Holmes, D., Rombang, W. M. and Octaviani, D. (2001) *Daerah Penting bagi Burung di Kalimantan.* Bogor, Indonesia: PKA/BirdLife International Indonesia Programme. (in Indonesian) [F07, W20]

IUCN/WCMC (1997) *Designing an optimum protected areas system for Sri Lanka's natural forests.* Colombo: IUCN/WCMC/FAO. [F05]

Lambert, F. R. and Collar, N. J. (2002) The future for Sundaic lowland forest birds: long-term effects of commercial logging and fragmentation. *Forktail* 18: 127–146. [F07]

Laurie, A. (2001) Brandt's Vole outbreaks and control in China – an ecological approach. Report presented at Sukhbaatar Aimag Brandt's Vole Seminar, 11 October 2001. [G01]

Mallari, N. A. D., Tabaranza, B. R. and Crosby, M. J. (2001) *Key conservation sites in the Philippines: a Haribon Foundation and BirdLife International directory of Important Bird Areas.* Manila: Bookmark, Inc. [F09, W19]

Newell, J. and Wilson, E. (1996) *The Russian Far East: forests, biodiversity hotspots, and industrial developments.* Tokyo: Friends of the Earth – Japan. [F01, W02, W03]

Rithe, K. (2003) Saving the Forest Owlet. *Sanctuary Asia* February 2003: 30–33. [F05]

Rodgers, W. A., Panwar, H. S. and Mathur, V. B. (2000) *Wildlife Protected Area Network in India: a Review.* Dehra Dun: Wildlife Institute of India. [F05]

Rombang, W. M. and Rudyanto (1999) *Daerah Penting bagi Burung di Jawa dan Bali.* Bogor, Indonesia: PKA/BirdLife International Indonesia Programme. (in Indonesian) [F07, W20]

Seng Kim Hout, Pech Bunnat, Poole, C. M., Tordoff, A. W., Davidson, P. and Delattre, E. (2003) *Directory of Important Bird Areas in Cambodia: key sites for conservation.* Phnom Penh: Department of Forestry and Wildlife, Department of Nature Conservation and Protection, BirdLife International in Indochina and the Wildlife Conservation Society Cambodia Programme. [F06, W18]

Sujatnika, Jepson, P., Soehartono, T., Crosby, M. and Mardiastuti, A. (1995) *Conserving Indonesian biodiversity: the Endemic Bird Area approach.* Bogor: BirdLife International–Indonesia Programme. [F07, F08, W20]

Tordoff, A. W., ed. (2002) *Directory of Important Bird Areas in Vietnam: key sites for conservation.* Hanoi: BirdLife International in Indochina and the Institute of Ecology and Biological Resources. [F06, W10, W18]

Trainor, C. R. (2002) *A preliminary list of Important Bird Areas in East Timor.* Report to BirdLife International Asia Programme. [F08]